Fusion Fiasco

EXPLORATIONS IN NUCLEAR RESEARCH, VOL. 2

The behind-the-scenes story of the 1989-1990 fusion fiasco

Steven B. Krivit

Edited by Michael J. Ravnitzky

Pacific Oaks
Press
SAN RAFAEL, CALIFORNIA

Pacific Oaks Press / *New Energy Times* / www.newenergytimes.com
369-B 3rd St. #556, San Rafael, CA 94901
Library of Congress Control Number: 2015916493

Krivit, Steven B., author.
    Fusion fiasco / Steven B. Krivit ; editors, Michael
Ravnitzky, Cynthia Goldstein, Mat Nieuwenhoven.
    pages cm -- (Explorations in nuclear research ; vol. 2)
    Includes bibliographical references and index.
    LCCN 2015916493
    ISBN 9780976054559 (hbk.)
    ISBN 9780976054528 (pbk.)
    ISBN 9780976054566 (Kindle)
    ISBN 9780976054535 (ePUB)

    1. Low-energy nuclear reactions--Research--History.
    2. Cold fusion--Research--History. 3. Science--Social
    aspects.   I. Ravnitzky, Michael, editor.   II. Goldstein, Cynthia, editor.   III.
Nieuwenhoven, Mat, editor.   IV. Title.   V. Series: Krivit, Steven B.
    Explorations in nuclear research ; v. 2.

    QC794.8.L69K75 2016   539.7'5
    QBI16-600061

Cover design: Lucien G. Frisch (Photograph: © Jahoo | Dreamstime.com)
Interior design template: Book Design Templates Inc.
Typeset in Crimson 11 pt., designed by Sebastian Kosch
Editors: Michael Ravnitzky (Developmental Editor) Cynthia Goldstein (Copy Editor), Mat Nieuwenhoven (Technical Editor)
Index: Laura Shelly

Also by Steven B. Krivit

Hacking the Atom: Explorations in Nuclear Research, Vol. 1 (2016)

Lost History: Explorations in Nuclear Research, Vol. 3 (2016)

Nuclear Energy Encyclopedia: Science, Technology, and Applications,
Steven B. Krivit, Editor-in-Chief; Jay H. Lehr, Series Editor,
Wiley Series on Energy (2011)

*American Chemical Society Symposium Series: Low-Energy Nuclear Reactions
and New Energy Technologies Sourcebook (Vol. 2),*
Jan Marwan, Steven B. Krivit, editors (2009)

*American Chemical Society Symposium Series: Low-Energy Nuclear Reactions
Sourcebook (Vol. 1),*
Jan Marwan, Steven B. Krivit, editors (2008)

The Rebirth of Cold Fusion: Real Science, Real Hope, Real Energy,
by Steven B. Krivit and Nadine Winocur (2004)

www.NewEnergyTimes.com (since 2000)

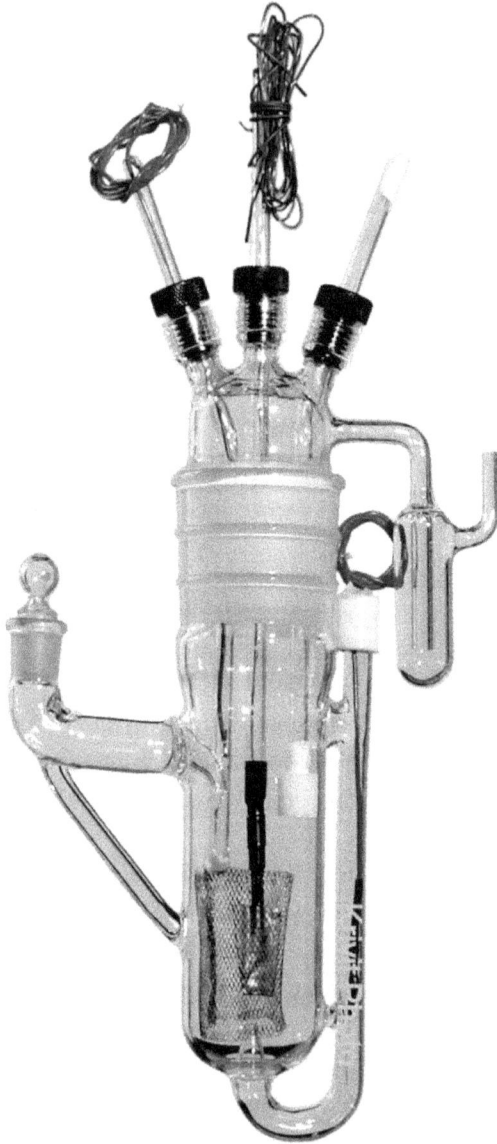

Electrolytic co-deposition cell designed and used by Stanislaw Szpak and Pamela Mosier-Boss at the U.S. Navy Space and Naval Warfare Systems Center (1990) Photo: S.B. Krivit

A new scientific truth does not triumph by convincing its opponents and making them see the light, but rather because its opponents eventually die, and a new generation grows up that is familiar with it.

Max Karl Ernst Ludwig Planck (23 April 1858 – 4 October 1947)

To Jess
Steven B. Krivit

To my sons Nathan and Max Ravnitzky
Michael J. Ravnitzky

Acknowledgments

Acknowledgments for the Series

Several key people have assisted me in the past few years on these books. They are my personal heroes. First is my copy editor, Cynthia Goldstein, who has been part of my team since 2004. Cynthia is my secret weapon who, year after year, helps make my writing intelligible.

Three other people composed the core team that made these books possible. Initially, they responded to my request to critique each chapter as Cynthia and I produced them. All of them exceeded my expectations in each of their unique contributions. None of them is a scientist, but all of them have some scientific background or technical training. I asked them to find all possible flaws and errors, and indeed they brought much to my attention. Of course, if any errors remain, they are my fault alone.

Michael Ravnitzky, in Maryland, accepted my invitation to critically review the draft chapters. After I convinced him that I really did welcome every critique he had to offer, he provided an invaluable outpouring that contributed immensely to the development of the books. Michael was a dream to work with. I am forever in his debt for his insights, wisdom, and his relentless, constructive and diplomatic suggestions, and so many other thoughtful contributions.

Mat Nieuwenhoven, in the Netherlands, has an eye for details like I have not imagined possible. His dedication to this project and helping make it as technically accurate as possible was heroic.

Lucien G. Frisch, in Germany, could see and understand the larger vision of these books from the very beginning, as soon as he read the first completed chapter. Sometimes, he could see the purpose of these books more clearly than I did, as I was on occasion too close to it to see it myself. I thank him not only for his visionary guidance but also for his brilliant artistry that graces the covers of these books.

I wish to extend my heartfelt appreciation and gratitude to the following reference librarians, who have provided me with so much support for all three volumes: my hero Randy Souther, at the Gleeson Library, Geschke Center, University of San Francisco; Lorna Whyte, Diane Delara and Pam Klein at the San Rafael Public Library; and Lorna

Lippes, an independent researcher. I owe much appreciation to Libby Dechman, at the Library of Congress, for her help in establishing the new cataloging subject heading "low-energy nuclear reactions," and to Madelyn Wilson, DOE-OSTI FOIA officer. Thanks to all of you; I have a newfound appreciation for reference librarians and public libraries.

I wish to thank Larry, Dee, Mirabai and Tracy for their assistance in helping me see the path on which I was travelling. Thank you, Flori, Jessica, Christine, Al and Sean, for your encouragement and support. Last but not least, many thanks to Sophie Wilson for crucial conversations that have helped me understand how to tell these stories.

Acknowledgments for This Volume

I owe immense gratitude to several people who have since died. They diligently recorded and preserved this history: Jerry Bishop (*Wall Street Journal*), Hal Fox (*New Energy News, Journal of New Energy*), and Eugene Mallove (*Cold Fusion* magazine and *Infinite Energy* magazine). I wish to thank Christy Frazier, who continues to preserve the legacy of Gene's work. I thank Jed Rothwell for his efforts to access to scientific papers.

I owe much gratitude to Dieter Britz for his thoughtfulness, beginning in 1989, in tracking and meticulously indexing the published scientific papers that contain this history. I am deeply indebted to Ron Marshall for his painstaking review of the final manuscript.

I also wish to acknowledge the many scientists with whom I have spoken and whom I have learned from in the last 16 years. It has been a privilege and an honor to record their stories and discoveries. Thanks to theoretical particle physicist Frank Close for our many stimulating conversations about the Jones-Fleischmann-Pons conflict and the Fleischmann-Pons gamma-ray matter.

I am also indebted to two people for access to extensive archival documents from 1989. The first is Bruce Lewenstein, the creator of the Cornell University Cold Fusion Archive. The second is physicist Richard Garwin, who allowed me access to and permitted me to create a public archive of his extensive files on the 1989 "cold fusion" conflict.

Thanks to author Gary Taubes for his helpful record of aspects of this history. He arrived at a different conclusion, but given the same circumstances, I might have done the same in 1993.

Table of Contents

Edward Teller's Statement (1989)

On March 23, 1989, electrochemists Martin Fleischmann and Stanley Pons, at the University of Utah, claimed they had achieved sustained nuclear fusion in a benchtop lab experiment. Six months later, in October, Edward Teller (1908-2003) presented his thoughts about the idea of "cold fusion" at a scientific workshop sponsored by the National Science Foundation and the Electric Power Research Institute (EPRI).

Teller was a preeminent physicist who made many contributions to science but is best-known for his key role in the Manhattan Project, which developed the first atomic bomb, during World War II. Although most theoretical physicists quickly dismissed the idea of "cold fusion" as inconceivable, Teller was open to the possibility that something about it, not necessarily fusion, was real. Teller envisioned the possibility that, underneath the mistakes and confusion, a new field of science was emerging.

A substantial amount of high-quality research into what was called "cold fusion" existed by October 1989. As Teller strongly suggests below, some of that research showed unambiguous signs of nuclear reactions. Some of the most important results from the research performed in 1989 occurred in national labs in the U.S., Italy, and in India. Most of these results have been known, until now, by only a handful of people.

Teller, who likely had access to almost all of the experimental results, was able to connect the dots. As other physicists did, he pondered the improbability (Gamow factor) of two subatomic particles overcoming the seemingly impossible electrostatic barrier that normally keeps such particles from fusing. But something else, he thought, had to be going on.

Teller was one of the few people in 1989 who was able to remain objective about the new science despite the fact that it was so poorly understood. He was also remarkably prescient and speculated that the answer would not be in strong nuclear reactions but in a neutral particle of small mass (like a neutron) that would catalyze the reaction.

Statement of Dr. Edward Teller:

We are further than ever from a real agreement on cold fusion. What has been seen has a wide divergence in results. I do not remember any case in my lifetime in science when so many experts have differed for such a long time on such relatively simple and inexpensive experiments. We are seeing a great deal of variability in the results — whether due to surface effects or cracks or small changes in some unknown parameter. The experiments differ in many more ways than a simple theorist can explain.

I feel like the visitor looking at the giraffe and concluding, "There ain't no such animal." According to nuclear theory — from the point of view of the Gamow factor — there cannot be such an effect. The Gamow factor is not as simple as it is normally considered. Indeed, one must consider the temperature average over the Gamow factors. But before the hydrogen nuclei really have a chance of interacting with each other, they must be within a fraction of an angstrom, and at that point, the Gamow factor has a value of about 10^{-50}. On that basis alone, what we are seeing must be a series of mistakes.

But this is not the end of the controversy. Some of the good experiments show that something is really wrong with the branching of $d + d \rightarrow t + h$ and $d + d \rightarrow He^3 + n$. While I will not exclude a small variation in the ratio, the actual value reported is 10^8! Proton-producing reactions (the tritium branch) being 10^8 times more likely than neutron-producing reactions. This is simply out of the question if D+D fusion is what is happening.

However, the history of science and experimental physics is full of examples of predictions that things are impossible and yet have happened. I remember what Ernest Lawrence once said about me: "When Teller says it is impossible, he is frequently wrong. When he says it can be done, he is always right."

But what if we are presented with the fact that the results are correct? Then we will have to ask ourselves, What are the minimum changes which we need to make in nuclear physics to explain the facts? If the giraffe exists, how does his heart pump blood into his brain? If the results are correct, then you must assume that nucleons can interact not just when they touch. We need to be able to explain how the nucleons interact at distances as great as 1/10 of an angstrom.

I think it would help if we postulated that the nuclei can interact at 10^4 nuclear radii and that the interaction is not through tunneling but some exchange of "particles" that can extend outside of the nucleus. It will be remarkable but not impossible that "quarks" could exchange or interact at 10^{-9} cm with very low probability. This would be a low probability but still much greater than the Gamow factor. The probability that this could result in cold fusion is possible even if it is unlikely. If there is such an effect, we will then learn something very important. This would be a scientific discovery of the first order, the kind for which we are willing to spend 5×10^9 dollars (Superconducting Super Collider).

I therefore applaud the National Science Foundation and the Electric Power Research Institute for maintaining enough interest and enough support so that a real clarification of the apparent contradictions can be pursued. If that clarification would lead to something on which we can agree and to a reaction probability that is small but much bigger than the Gamow factor would allow, this would be a great discovery. Perhaps a neutral particle of small mass and marginal stability is catalyzing the reaction.

You will have not modified any strong nuclear reactions, but you may have opened up an interesting new field (i.e., the very improbable actions of nuclei on each other). So I am arguing for a continuation of an effort, primarily for the sake of pure science. And, of course, where there is pure science, sometimes, at an unknown point, applications may also follow.

But, according to my hunch, this is a very unclear and low-probability road into a thoroughly new area. The low probability has to be balanced against the great novelty. But to think beyond that and ask what is the practical application, what this very unknown area of nuclear physics may produce, that, I claim, is completely premature. Thank you very much.

— Edward Teller, October 1989 (Teller, 1989)

Introduction

"Science advances one funeral at a time."
Paraphrased from Max Karl Ernst Ludwig Planck

In 100 years of chemistry and physics, most scientists thought nuclear reactions could occur only in high-energy physics experiments and in massive nuclear reactors. But new research shows otherwise: Nuclear reactions can also occur in small, benchtop experiments.

Research shows that, unlike fusion or fission, these low-energy nuclear reactions (LENRs) can release their energy without emitting harmful radiation or greenhouse gases or causing nuclear chain reactions.

Few scientific topics in the last 100 years have created more conflict than this one. Changes in scientific thinking rarely take place without a fight, and this one is occurring now. This book offers readers a ringside seat.

Perhaps the biggest surprise is that the research, labeled incorrectly in 1989 as "cold fusion," never stopped, although it had been pronounced dead again and again. Despite confusing data, highly irreproducible results, and more than a few hot tempers, significant, valid science took place, much of it unrecognized at the time. In some cases, the data were buried for many years.

Archival References

Much of what happened behind the scenes in the 1989 "cold fusion" conflict is disclosed in this book publicly for the first time. Despite the dozen-plus books on the subject, this is the first book to rely extensively on archival material rather than discussions with, or memories from, the key players. Sources include:

> Documents from the Cornell Cold Fusion Archive. The archive contains 40,000 pages of documents.
> Internal documents (5,000 pages) from the summer/fall 1989 Department of Energy "cold fusion" review.

Proceedings of the 1989 National Science Foundation/
Electric Power Research Institute workshop.
Audio recordings from the first "cold fusion" workshop,
which took place in Erice, Italy, on April 12, 1989.
Audio and video recordings of key events during the 1989
American Chemical Society, American Physical Society,
and Electrochemical Society meetings.

Three Books

This is the second book in a three-book series. Each book stands alone, and covers a distinct period of scientific exploration. They are being published in reverse chronological order.

- *Hacking the Atom: Explorations in Nuclear Research, Vol. 1 (1990-2015)*
- *Fusion Fiasco: Explorations in Nuclear Research, Vol. 2 (1989-1990)*
- *Lost History: Explorations in Nuclear Research, Vol. 3 (1912-1927)*

It's Not Fusion

I have found no experimental evidence to support the "cold fusion" idea that deuterium nuclei (a form of hydrogen) fuse at room temperature at high rates. Nor have I found a viable theory that explains how deuterium-deuterium (D+D) fusion might occur in electrochemical cells.

Experimental Nuclear Data

However, I have found an abundance of experimental data, some of it from U.S. Department of Energy (DOE) national laboratories, including Oak Ridge, Lawrence Livermore and Los Alamos, that provides well-measured evidence of previously unrecognized nuclear phenomena.

Viable Theory

One theory appears to explain most of the anomalous phenomena reported in the field. It has nothing to do with D+D fusion. I discuss this theory in Vol. 1, *Hacking the Atom*. It does not involve few-body, strong-interaction fission or fusion. Instead, it involves many-body, collective electroweak interactions that can enable high rates of nuclear transmutation processes, under moderate conditions, in electrochemical cells and other types of systems.

Nuclear Evidence

The most convincing evidence for this new nuclear science is not the measurement of excess heat but the measurement of nuclear products. These include isotopic shifts, elemental transmutations, tritium production and, sometimes, production of tritium and neutrons from the same experiments. For many years, the early proponents of this new science argued for its validity on the basis of excess-heat measurements. This series of books makes no such argument; instead, it reveals the evidence of direct nuclear products.

Volume 2: *Fusion Fiasco*

This volume focuses nearly exclusively on the 1989 "cold fusion" history. This science conflict began when electrochemist Martin Fleischmann, retired from the University of Southampton, England, and his colleague Stanley Pons, chairman of the University of Utah Chemistry Department, announced in a press conference that they had created a sustained fusion reaction in a modified test tube.

When I began writing Vol. 1, *Hacking the Atom*, in 2012, I didn't think another book was needed to tell the old 1989 story. I was certain that I and other authors had covered it thoroughly. I was wrong.

Two things happened. First, I found related chemistry-based transmutation research that took place in the 1910s and 1920s. That material became its own book, *Lost History*. That research is a remarkable precursor to the research that came 60 years later.

Newly Uncovered Facts

The second thing was that, as I was drafting what was to be a brief chapter for the 1989 "cold fusion" history, I began checking facts with some scientists who were involved at the outset. One of them was Richard Garwin, a prominent U.S. physicist, a research fellow at IBM, and a consultant for many high-level science and nuclear projects for the U.S. government.

Garwin shared hundreds of "cold fusion" documents, most of which were internal documents used in the 1989 DOE-sponsored review and which had remained out of sight for more than two decades.

These documents reveal the behind-the-scenes activity and the real story of this crucial event. The documents include reports from researchers at DOE laboratories like Lawrence Berkeley Laboratory who observed nuclear phenomena that they described as "false positive, up to eight times background."

Garwin also sent me audio recordings of the first "cold fusion" workshop in 1989, which took place in Erice, Italy. No detailed accounts of the Erice workshop seem ever to have been published. These recordings reveal another side to this otherwise-bitter debate: enthusiastic, friendly, collaborative relationships between physicists and chemists.

The DOE-sponsored review has been discussed in every historical account. However, another little-known review took place in October 1989, at a Washington, D.C., workshop. Data presented at this workshop does not appear to have been reported in any other book.

Renowned physicist Edward Teller participated in this workshop and, after hearing about isotopic shifts observed by scientists at two independent national laboratories, concluded that nuclear effects were taking place. He also had a hunch about an explanation for the mechanism. The facts concerning these two government-sponsored reviews have been buried for two-and-a-half decades.

Jerrold Footlick, an author and former editor at *Newsweek,* sent me audio tapes of his interviews with former staff members at the University of Utah. They reveal the special interests behind the infamous University of Utah fusion press conference.

Letters sent by Fleischmann to his good friend electrochemist John Bockris reveal Fleischmann's actual motive for attempting electrolytic fusion. (Hint: it was not to find a novel source of energy.)

After I had many extensive conversations and a meeting at Oxford University with theoretical particle physicist Frank Close, he provided me with several documents that shed new light on his accusations that Fleischmann and Pons had manipulated a gamma-ray graph.

Close also helped me sort out two other sensitive matters in this history: Pons' accusation that 1) Steven Jones had pirated his and Fleischmann's ideas and that 2) his graduate student, Marvin Hawkins, had stolen Pons and Fleischmann's lab books.

The book also includes a section on magnetic and inertial confinement thermonuclear fusion research, whose goal is the development of large-scale fusion reactors. The researchers hope to use the same process as the sun to create vast amounts of energy on Earth. These hoped-for power plants are promoted as the ultimate power source.

Scientists and administrators working on these projects regularly claim improvements and breakthroughs. This apparent progress is used to justify funding of fusion programs that have probably consumed at least $100 billion in public expenditures worldwide in the past 50 years.

In fact, researchers today are nowhere near the energetic breakeven point required to create commercially viable fusion reactors. Through a peculiar form of dual accounting for energy balances, the researchers have played games with the words commonly understood by the general public and U.S. legislators. In doing so, they have created a false impression of progress when reporting their experimental results.

Science Wins

If any scientist who was involved in the 1989 "cold fusion" conflict escaped without bruises or embarrassments, I don't know about it. Nevertheless, there are clear winners: science and the scientific process.

The infallibility of nature, coupled with mankind's ability to use analytical tools and critical thinking, reveals the power of science to illuminate nature's truths and to advance humanity.

A Note on Sources

Sources are identified by author and date parenthetically in the text and most can be obtained from these resources:

Cornell Cold Fusion Archive (CCFA) — Located at Cornell University, this physical archive contains 10,000 pages of original source documents, dozens of audio and video recordings, and physical objects.

New Energy Times Richard Garwin Cold Fusion Archive — Portions of this digital archive are available on the *New Energy Times* Web site. The full archive is available at the Internet Archive Web site. It contains 5,000 pages of original source documents.

New Energy Times 1989 Archives — Several sets of digital archive materials with links to other electronic documents and digital recordings on the Internet Archive Web site.

Volume 1: *Hacking the Atom*

Volume 1, *Hacking the Atom,* is the story of how the science initially and erroneously called "cold fusion" continued to progress slowly but incrementally after its near-death in 1989.

The most significant early advances were heavy- and light-water electrolysis experiments, performed at Hokkaido University in Japan and at the University of Illinois, respectively. The data revealed that a variety of nuclear transmutations — an increase or decrease in the number of protons in an atomic nucleus that change one element to another — were occurring in the low-energy nuclear reaction experiments, providing crucial insights into the new science.

Beginning in 1999, a new method of gas-loading LENR experiments performed at Mitsubishi Heavy Industries in Japan revealed even more convincing evidence of nuclear transmutations in LENRs.

In 2004, the U.S. DOE, responding to a request from five "cold fusion" researchers, sponsored a second review of the field. The review did not change the position of the U.S. government, but it did reawaken worldwide interest in the topic.

In 2005, a preprint of a promising theory was released on arXiv by theoreticians Allan Widom and Lewis G. Larsen, and in 2006, it was published in the *European Physical Journal C — Particles and Fields.*

In succeeding years, many scientists who had observed unexplained nuclear phenomena defended their belief in the D+D "cold fusion" idea, even when all evidence was to the contrary. *Hacking the Atom* explains these events.

Volume 3: *Lost History*

Lost History, tells the story of research that took place 100 years ago, a story that is surprisingly similar to events in the modern era reported in Vols. 1 and 2. It has been obscured and omitted from history books for nearly a century.

In the 1910s and 1920s, this research was known both in scientific circles and by the general public. It was reported in popular newspapers and magazines, such as the *New York Times* and *Scientific American.*

Papers were published in the top scientific journals of the day, including *Physical Review, Science* and *Nature.* Prominent scientists in the U.S., Europe and Japan and even Nobel Prize recipients participated in this research. In the 1930s, however, it was all dismissed as error, primarily because the theory was not understood and the experimental results were difficult to repeat.

This historical era is best understood with the benefit of the conceptual insights from the modern era as explained in Vols. 1 and 2. The sources used in Vol. 3 are primarily published scientific papers from that era, and for this reason, Vol. 3 is geared toward a more technical and academic audience.

Preparation for a Paradigm Shift

Focus on U.S. Activity

This research began in the United States. However, less than 24 hours after "cold fusion" was announced, researchers around the globe, particularly in France, Italy, Russia, India, China and Japan, began work on their own experiments and theories. This history, as well as the current research, was and remains an international activity; however, reporting the full international scope in this series is impractical.

No Practical Devices Yet

LENRs may someday lead to practical energy or heating devices. Research shows that LENRs can reach local surface temperatures of 4,000-6,000 K and boil metals (palladium, nickel and tungsten) in small numbers of randomly scattered microscopic LENR-active hot-spot sites on the surfaces of laboratory devices. To date, routine production of excess heat in laboratory apparatus at levels greater than 1 Watt has been more difficult.

Although some people seem to understand the basic science of LENRs, much engineering research and development is needed for the science to evolve into practical device design and reproducible fabrication. Today, there are no commercially practical LENR reactor devices, even though some people and organizations in and associated with the field have episodically made such claims since 1989. I have investigated many such claims of commercially viable devices in recent years and found them to be unsubstantiated.

Nevertheless, the body of scientific data suggests that someone or some companies will eventually commercialize the technology for thermal power generation applications.

Welcome to the Journey

I have independently investigated and reported on this subject for 16 years. I invite scientists and non-scientists alike to join me on this journey of scientific exploration and discovery. It is my pleasure to share this adventure with you now.

Steven B. Krivit
San Rafael, California
Sept. 1, 2016

CHAPTER 1

A Science Controversy Like No Other

"Cold Fusion" in History

On a Thursday afternoon in the spring of 1989, two bold electrochemists announced in a press conference at the University of Utah that they had demonstrated the remarkable feat of nuclear fusion in a simple apparatus that resembled a test tube. The claim by Martin Fleischmann and Stanley Pons was heralded by the worldwide news media as a potential energy source that was clean and produced no greenhouse gases, hard radiation, or nuclear waste. Deuterium, a form of hydrogen abundantly present in all of the earth's oceans, promised a virtually unlimited supply of fuel.

Some people at the time described the claim as second in importance only to the discovery of fire. Others described it as the most exciting event in nuclear physics since the discovery of fission.

The quest to tame nuclear fusion was not new. Controlled nuclear fusion has been the dream of scientists since the 1950s. Despite billions of dollars spent on thermonuclear fusion experiments, none has resulted in net power production. (Chapter 3) Yet Fleischmann and Pons said they created fusion in a glass tube submerged in a Rubbermaid plastic bucket, for $100,000.

Most scientists, especially those who knew anything about nuclear physics, thought the claims were ludicrous. Many experts thought Fleischmann and Pons were simply frauds. They had good reason for their doubts: The evidence offered by Fleischmann and Pons looked nothing like fusion. The mechanism for how it allegedly worked was unclear. Moreover, if the two chemists were right, their claim meant a rewrite of science textbooks.

Nevertheless, the news media initially promoted the claim as real and as the solution for the world's energy troubles, acid rain, global warming and wars for Middle East oil.

Americans had not forgotten the disruption of waiting in long lines for gasoline during the 1973 oil crisis. Ironically, the day after the Utah "fusion" announcement, an Exxon oil tanker, the Valdez, hit a reef off Alaska and created the worst oil spill the country had ever seen.

A week after the Exxon Valdez *oil spill, as oil-covered birds washed up on the Alaska shore and volunteers did their best to save the wildlife, cartoonist Dick Locher captured the moment.*

The news was on the cover of *Time* magazine, *Newsweek* and *Business Week*, as well as every TV channel, radio station and local newspaper. However, most scientists could not repeat the experiment. Making matters worse, scientists who had doubts couldn't comment intelligently on the "fusion" claim because Fleischmann and Pons' preliminary note had not yet published and available details were sketchy.

Within two months, the consensus among physicists was that the idea was dead. Within four months, a federally appointed panel of experts in the U.S. agreed. Within a year, the whole spectacle disappeared from the public spotlight, succumbing to accusations of fraud, delusion and incompetence.

But it didn't die. The research went underground. Fleischmann and Pons were invited by the Toyoda family and were given a lab in the south of France in which to continue their work privately.

A few researchers around the world had early success in their replication attempts. These researchers worked out of the public spotlight, sometimes during off-hours in their labs and with whatever materials they could gather without going through official channels. In experiments built with their own hands, with data taken by their own instruments, and with results observed with their own eyes, they saw the glimmers of a new science.

The Lost History

Few people in 1989 knew that there had been a precursor to these events nearly a century earlier. Between 1912 and 1927, scientists performed chemistry experiments and observed the transmutation of elements using low-energy methods. The data was hard to understand, the experiments were difficult to repeat, and the claims were suspiciously similar to the unscientific claims of medieval alchemists. As a result, the research was assumed to be wrong and was forgotten. Most modern science textbooks and histories of science do not even mention it. When I was doing research for this book, I had known of only four scientists from the 1920s who had done such research. I expected their work would compose only a single chapter in this book.

I was astonished to learn that there was much more. The lost research spanned two decades, from the beginning of the first discoveries in atomic science. It involved a dozen scientists, some of them Nobel laureates. The papers had been published in the most prestigious journals. The news was well-known to the general public, with stories appearing in a variety of publications ranging from the *New*

York Times to small-town newspapers. There were even reports of transmuting inexpensive base metals into gold and the discovery of a whitish, as-yet-unidentified metal with a mass similar to gold's.

That story composes the third book in this series, *Lost History: Explorations in Nuclear Research, Vol. 3.* The scientific papers from a hundred years ago are wonderfully descriptive. The research and instruments were simpler back then. However, the papers are more challenging to read because the scientists lacked the understanding we have and the terminology we use now. Nevertheless, the evidence indicates that at least some of these scientists succeeded to transmute elements in their low-energy experiments. They reported the production of rare gases and even an as-yet-unidentified gas with a mass of 3 times that of ordinary hydrogen.

Historians have wrongly credited world-famous Sir Ernest Rutherford for the first man-made nuclear transmutation, when all the original scientific papers make clear that the actual credit belongs to one of his students.

Wendt and Irion

In 1922, American scientists Gerald L. Wendt and Clarence E. Irion synthesized helium using the exploding electrical conductor method. Despite doubts and criticism, no one unambiguously identified any error in their 21 successful experiments. Nuclear evidence from exploding conductor experiments was confirmed 80 years later by researchers at the Kurchatov Institute in Russia. (Urutskoev, 2002)

Paneth and Peters

The rare scientist in 1989 who might have been aware of any older low-energy transmutation research seemed to know only about the claims of German chemists Fritz Paneth and Kurt Peters. In 1926, they claimed to have transmuted palladium and hydrogen into helium. When, in 1989, curious observers looked deeper into the history, they learned that Paneth and Peters seemed to have retracted their claims in

1927. Indeed, they did abandon their claims, but the story does not end there.

As *Lost History* shows, Paneth and Peters analyzed possible errors and wrong assumptions, and were able to explain most of their experimental runs. After making their best efforts to explain their "mistaken" conclusions, they assumed that a rational explanation for their remaining apparent positive results would eventually present itself:

> For the rest of the positive tests, even today, we cannot give an explanation. But since the majority of our experiments have explained themselves in a "natural" way, we think it probable that it will also happen for our outstanding (unexplained up to now) experiments. (Paneth, Peters, and Günther, 1927)

The anomalous results were never explained.

Asleep for 60 Years

After Paneth and Peters, low-energy nuclear transmutation research was dormant for decades. Soon, thermonuclear fusion and nuclear fission were discovered. Scientists learned how to artificially accelerate particles and to transmute elements. These high-energy physics-based transmutations were repeatable, controllable, and understandable.

Soon after the discovery of the neutron came the concept of the nuclear chain reaction. After that came the atomic (nuclear fission) bomb, then the hydrogen (thermonuclear fusion) bomb, and the peaceful use of atomic energy for nuclear power, first in military submarines and then in land-based generating plants. High-energy physics earned a place in science; low-energy-based nuclear transmutation (as discussed in *Lost History*) didn't and was largely forgotten. Nuclear physicists in the U.S. gained additional prestige and influence as a result of their involvement in the Manhattan project and their development of the atomic bomb.

With one exception in 1951, (Sternglass, 1957) the idea of (what is now called) low-energy nuclear reactions (LENRs) remained dormant until Fleischmann and Pons' 1989 announcement. This 60-year gap is not as surprising as it seems. The early low-energy experiments were difficult, inconsistent and in conflict with accepted theory.

Startling News in 1989

Participants and observers of the 1989 "cold fusion" conflict, even experienced nuclear scientists, had no context for the startling news that chemistry experiments could produce nuclear reactions. The idea seemed to come out of thin air and appeared to contradict well-established physical law; it suggested a new scientific paradigm. The ensuing conflict, due in part to a disquieting series of events, a rush to claim credit, and the prevalence of new communication technologies — fax and e-mail — was unprecedented in modern science.

The reaction to the news, particularly among learned men and women of science, was not unlike the suggestion that the Earth revolved around the sun when it was believed otherwise. The 1989 "cold fusion" conflict was an ugly, painful period in the history of science for nearly all of its participants, which include scientists, journalists and program managers.

Synopsis

The field's two progenitors, electrochemists B. Stanley Pons (b. 1943), at the time the chairman of the Chemistry Department at the University of Utah, and Martin Fleischmann (1927-2012), professor emeritus from the University of Southampton, U.K., and a Fellow of the Royal Society, were dismissed as cranks by the scientific community six weeks after announcing their claim. The stigma of "pathological science" remained attached to them. They were not able to return to academic research and continue the work that had so captured their interest and passion.

Within days of the March 23, 1989, announcement of "cold fusion," a number of science authorities predicted that the entire idea of "cold fusion" was moments away from its death. Naysayers questioned why something so apparently wrong and unscientific could persist for so long. This book answers that question.

Science Controversies

Sixty days into the "cold fusion" conflict, Isaac Asimov, a prolific, well-respected American author, science essayist, and professor of biochemistry at Boston University, wrote a letter to the *Los Angeles Times* that offered a good comparison of the "cold fusion" controversy with other science controversies:

> The current controversy over cold fusion is a very exciting example of science in progress. Some investigations confirm it; other investigations don't. There is excitement on one side, denunciation on the other. Is that the way science works? Loud squabbling? Angry accusations and rebuttals? Sometimes yes.

Asimov mentioned the erroneous interpretation of canals on the surface of Mars, nonexistent "N-rays" that were proposed as a discovery similar to X-rays, and the idea of "polywater" proposed by Soviet physicist, Boris V. Derjagin. He had claimed that "polywater" was a new form of water, much denser than ordinary water, and that it had a much higher boiling point, 500° C instead of 100° C. Within a few years, the entire "polywater" idea had been discredited.

There was so much excitement with "cold fusion" that three major scientific societies squeezed in impromptu sessions to discuss "cold fusion" at their respective conferences within weeks of the announcement. Two of them had record attendance.

Clayton Callis, the president of the American Chemical Society, introduced the topic in a special symposium on April 12, 1989, to 7,000 eager chemists and 150 perplexed reporters in the Dallas Convention Center arena. "This scientific meeting," Callis said, "has to be a precedent-setting event for the American Chemical Society, both in attendance and in general interest."

Journalist Patt Morrison, writing for the *Los Angeles Times* on May 9, 1989, attended the "cold fusion" session at the Electrochemical Society meeting a few weeks later. "The quest for the cold fire of fusion has moved to Los Angeles," Morrison wrote, "to a meeting of the Electrochemical Society, which in all of its 87 years has surely seen nothing like this."

Just how big a story was "cold fusion"? According to one physicist at the time, it was the first time that the three major U.S. weekly news magazines had the same story on their covers since the 1963 assassination of President John F. Kennedy.

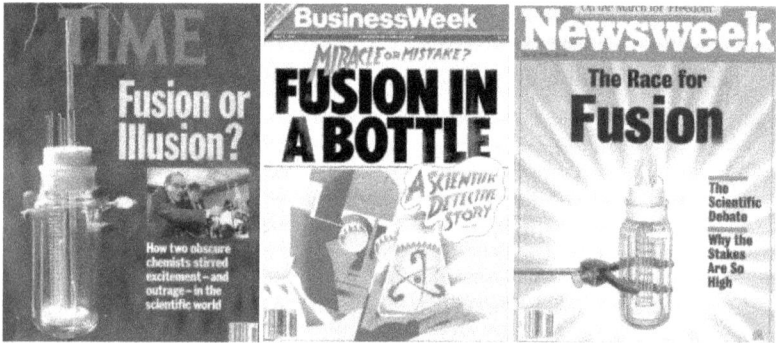

The "cold fusion" conflict made the covers of three U.S. news magazines in the first week of May 1989.

Distinction Between "Cold Fusion" and LENRs

There is a crucial distinction between the idea of "cold fusion" and LENR research. "Cold fusion" is the hypothetical idea that deuterons or protons can somehow overcome high Coulomb barriers and engage in charged-particle fusion reactions at or near room temperature at high rates. This idea directly contravenes current scientific understanding. There has been no experimental evidence consistent with fusion.

LENRs identify these phenomena without ascribing them to fusion. LENRs are based on electroweak interactions and neutron-capture processes. The Coulomb barrier does not apply to neutral particles, such as neutrons, and no laws of physics are violated.

The first book in this series, *Hacking the Atom: Explorations in Nuclear Research, Vol. 1,* goes much deeper into the discussions of these processes. When I refer to the history of this field, I will, however, call the topic "cold fusion," because that was the term used at the time.

Fossil-Fuel Alternative?

The primary pursuit of LENRs is the determination of the possibility of a new source of energy. Laboratory experiments show that LENRs have the potential to produce useful energy but without typical harmful effects of conventional nuclear energy.

Fossil fuels — oil, natural gas and coal —provide the lion's share of world primary energy consumption. Fossil fuels are a nonrenewable energy source. A few facts help illustrate the current predicament: In the United States, crude oil production peaked around 1975. In the United Kingdom, coal production peaked around 1910. Yet we have built, and are living in, a civilization based on the erroneous assumption of unlimited availability of fossil fuels. The potential negative consequences for these two near-term opposing factors are staggering.

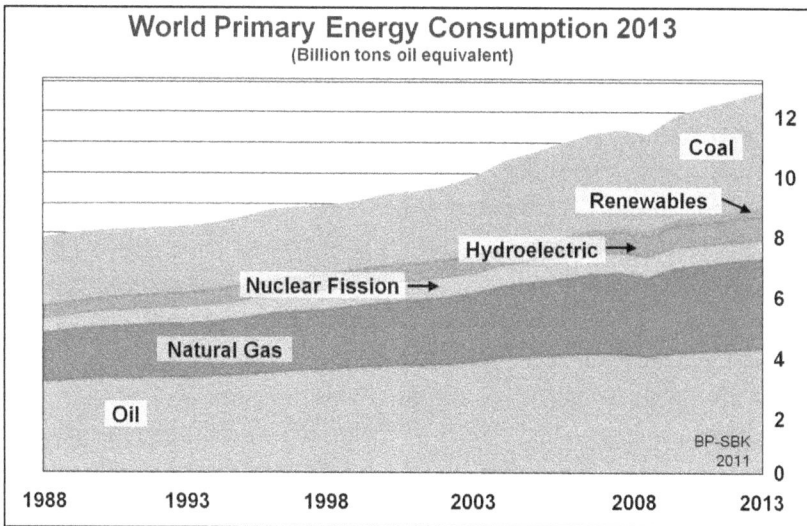

World Primary Energy Consumption 2013
(Billion tons oil equivalent)

June 2014 BP graph of world primary energy consumption.
(Labels added by S. Krivit)

In addition to reporting energy consumption, in its Statistical Review of World Energy, June 2014, BP also presented data on the remaining global fossil fuel resources. There are 53 years left of oil, 55 years of natural gas, and 113 years of coal.

Renewables such as wind and solar likely will become more cost-competitive, and if breakthroughs in the cost, performance and lifespan of batteries occur, solar and wind may provide service when the wind doesn't blow and the sun doesn't shine.

Some people suggest that nuclear fission is the best source of baseload electrical energy. With breeder reactors using uranium-thorium fuels, coupled with nuclear materials reprocessing, fission reactors can power the world for hundreds of years. Other people suggest that fission is a risky and problematic technology — an unsafe and unnecessary source of energy.

Scientists have tried to harness controlled thermonuclear fusion as a source of energy since 1951. Unfortunately, after tens of billions of dollars and nearly seven decades of research, no thermonuclear fusion experiment in the world has ever produced a single watt of power in excess of the total power input.

So what are the remaining options? The bad news is that there are none. There are no known, proven alternatives for wide-scale electrical production or alternatives for liquid-fuel-based transportation, despite occasional claims from some technology promoters. On April 12, 1989, physicist Harold Furth, the director of the Princeton University Plasma Physics Laboratory, spoke before 7,000 chemists who were hearing about "cold fusion" for the first time, at the Dallas, Texas, convention center. Furth had dedicated much of his professional career to the harnessing of thermonuclear fusion. A futurist, he saw the connections among mankind's past, present and future relationships with energy.

At the time Furth gave this talk, the populations of China and India had not yet developed their insatiable hunger for energy and interconnectedness. The Internet as we know it today did not exist. Tim Berners-Lee created the first web browser that same year. The digital mobile phone network did not exist. Here's what Furth said at the end of his talk. I do not have copies of the slides he displayed:

Finally, I should say something about why we're interested in fusion at all. This graph, which was made at Lawrence Livermore National Laboratory is quite useful. It takes a fairly optimistic view of world energy consumption. It assumes that the world population will level off at a mere 10 billion and that these people will be content to live at two-thirds of the present U.S. standard of living, [as measured by] energy consumption. That means that there's going to be quite a rise in annual energy consumption.

The second curve shows you how we're doing with energy that's easily available by taking fossil fuels out of the ground, or hydroelectric, or burning up the readily burnable uranium [for fission plants]. You see we're doing awfully well in consuming that, and it's about to keel over and go the other way. So there's a gap that will develop. That gap can be bridged, to some extent, by things like solar power, which is very effective for peak heat loads in nice weather. But in rainy weather, and at night, you would like some baseload power source.

Fission is a good choice, but it has made itself unpopular, and in some people's minds — many people's minds — fusion would be an even better choice because it would provide a long-term solution that promises both to be economical and to have no highly adverse environmental impact. Now, how soon do we need it? The divergence between power needed and power available will take place, according to this graph, around 2040. Knowing how quickly the utilities industry responds to new ideas, that means you better know exactly what you want to do 30 years earlier, namely around 2010. So, clearly, there isn't that much time to be lost in identifying a really promising energy source for meeting this long-range problem.

One thing that distracts me about this graph is that, in a mere 300 years, we will have blown the entire bank deposit of the fossil-fuel energy bank that was laid down over 400 million years so that humanity could advance to a high level of civilization. We and our immediate descendents have the extraordinary privilege of blowing this entire bank account in one-millionth of the time it took to accumulate.

So I visualize our descendents in the year, let's say, 2350, looking back and thinking about us [laughter from the audience] and wondering what we had in mind — wondering whether there would be anything redeeming to be said about us. I think one of the few redeeming things that could be said was that we devoted some very small fraction of this bank to developing a new energy source that could keep civilization going after we've blown all this stuff. If by chance we could succeed in developing fusion as that energy source, then people may think we weren't altogether bad. I very much hope it works out that way. Thank you. (Furth, 1989)

Furth, like Asimov, saw the potential significance of the Fleischmann-Pons claim. Asimov saw its relevance to the history of science. Furth sought its relevance to global energy concerns and the intractable dependency on fossil fuels by the human species.

Poster Child for Bad Science

The term "cold fusion" represents the epitome of bad science. The "scientific fiasco of the century," as one author called the topic, is used to teach ethics in science, specifically how science should not be performed and reported. In the world of science, "cold fusion" implies fraud, delusion and nonsense.

Behind the image of bad science and human drama, a new scientific phenomenon has emerged and it has nothing to do with fusion.

Why was this new science so difficult to see and understand? How did the concept of "cold fusion" get its reputation as pathological science? Was it deserved? How did the new science come to light? And why did it take more than two decades to emerge? These are questions I will answer in this book.

CHAPTER 2

Unfriendly Competition or Scientific Piracy?

University Administrator Accuses Jones

At 1 in the afternoon on March 23, 1989, the University of Utah held a press conference at which Martin Fleischmann and Stanley Pons claimed they had discovered a new approach to "nuclear fusion." Their idea of fusion was wrong, but they did discover a new phenomenon that released sizable and otherwise-unexplained levels of energy, in the form of heat. This chapter reviews the events that took place in the seven months leading to that press conference. These events provide insight into the manner in which Fleischmann and Pons announced their work.

Although Fleischmann and Pons naturally wanted credit for their claim, a University of Utah press conference was not the way they wanted to share the news. The press conference was precipitated by a developing conflict with Steven Earl Jones, a physicist at nearby Brigham Young University. Jones had been working on a different concept: muon-catalyzed fusion. Competitive pressure from Jones, actual and perceived, compelled the University of Utah, and Fleischmann and Pons, to announce their work 18 months before they were ready to do so.

Administrators and the two electrochemists at the University of Utah thought that Jones was attempting to pirate their idea. As it turned out, Jones did use some of Fleischmann and Pons' ideas. Jones tried to claim the unannounced research findings as his own after he learned about it from Fleischmann and Pons. In March 1989, people at the University of Utah accused Jones of piracy, and by April, the story was in the local papers. Jones vehemently denied the accusations.

This chapter presents the complete story for the first time. The events reported here are supported by additional references in "Timeline of the Early Conflict Between Steven Jones and Martin Fleischmann/Stanley Pons," available at the *New Energy Times* Web site.)

Jones Needs New Work

As Jones tells this history, he began his electrolytic fusion work several years before Fleischmann and Pons announced their claim in 1989. This is true; he had. Jones had worked with his colleague Johann Rafelski, a theoretical physicist at the University of Arizona, since the summer of 1983. But by 1989, Jones had gone as far as he could with muon-catalyzed fusion. It was clear among scientists that muon-catalyzed fusion offered no promise as a practical energy source. The JASON group, a highly respected group of physicists that advises the federal government under contract, conducted a review of muon-catalyzed fusion and on Aug. 1, 1988, advised terminating funding in that area. Jones needed a new line of research.

Since 1982, Jones had maintained a relationship with Ryszard Gajewski, project director of the U.S. Department of Energy's Advanced Energy Projects Division. During that time, Gajewski had provided $1.9 million to fund Jones' research. He knew that Jones was seeking a new research topic.

University Rivalry

In May 1988, after spending $100,000 of their own money for their research, Fleischmann and Pons sent a funding proposal to the Office of Naval Research. Pons then sent the proposal to Jerry Smith, a program manager in DOE's condensed matter physics, in the materials sciences division. Smith suggested that a more appropriate place for the proposal was the Department of Energy's Office of Advanced Energy Projects.

The proposal ended up at the DOE on Gajewski's desk. Frank Close gave me more details in a 2009 e-mail.

"While I was at a dinner party at Rafelski's house in 1991 or 1992," Close wrote, "Rafelski produced the document, which contained a DOE cover page, with receipt date. The cover sheet contained a date stamp which immediately astonished me. It was dated August 23, 1988."

According to documents obtained from DOE through a Freedom of Information Act request, Aug. 23 was actually the date Pons signed the proposal. The next day, Jones resumed his interest in electrolytic fusion.

On Aug. 24, according to Jones' lab books, Jones called a meeting of his staff to discuss using electrolysis in their fusion research. Gajewski thought that Jones and Rafelski would be good reviewers because of their experience in muon-catalyzed fusion and, within a few weeks, he sent the Fleischmann-Pons proposal to them to review.

I interviewed Gajewski by phone on April 29, 2009. He told me that his normal procedure would have been to telephone a potential reviewer before sending out a proposal, but two decades later, he could not remember any details of any phone calls. According to Jones' lab books, Jones restarted fusion research when he received the copy of the Fleischmann-Pons proposal on Sept. 20, 1988.

Jones and Rafelski were now, along with other colleagues at Brigham Young University, working on electrolytic fusion. By December 1988, Jones had adopted three concepts that originated with other scientists.

The origin of the specific composition for his electrolyte came from the geologic-piezonuclear fusion idea of Paul Palmer, a Brigham Young University physicist in the Department of Physics and Astronomy. The impetus to use electrolysis came from the Fleischmann-Pons proposal. The idea to use palladium cathodes also originated with the Fleischmann-Pons proposal.

On Dec. 9, 1988, before they had collected any significant data, Jones and Rafelski discussed filing a patent with Palmer, but independent of Fleischmann and Pons, for "stimulating nuclear fusion by means of flow of hydrogen isotopes in metal lattice." (Taubes, 48) The following day, Jones wrote a draft proposal and sent it to Gajewski, suggesting that he had found a shortcut to nuclear fusion.

"In conclusion," Jones wrote, "we have demonstrated for the first time that nuclear fusion occurs when hydrogen and deuterium are electrolytically loaded into a metallic foil. This remarkable process obviates the need for elaborate machinery to generate and contain either plasmas or muons to induce fusion. We are now exploring means to enhance the fusion yield of this new process." (Taubes, 48, 49)

On Dec. 16, 1988, Gajewski called Pons, told him about Jones' work, and suggested that Pons work together with Jones. According to Taubes, until Gajewski's phone call, Pons had been reluctant to tell even his close friends about his "cold fusion" research. Even when Pons told a close colleague about his fusion research in September 1987, Pons swore him to secrecy. Now, Gajewski was calling Pons and suggesting that he collaborate with one of his reviewers. As distasteful as that must have sounded to Pons, he called Jones. They apparently had an amicable conversation. (Taubes, 50)

In mid-December, Edward F. "Joe" Redish, a theoretical physicist at the University of Maryland, invited Jones to speak at the May 1989 meeting of the American Physical Society about muon-catalyzed fusion. (Close, 68, 358) Redish did not invite Jones to speak about his new fusion work. When I interviewed Redish in 2014, he told me that he was unaware of Jones' newer work when he sent the invitation to Jones in 1989. Jones got his first significant data — a burst of neutrons — in mid-January 1989, with run No. 6. (Close, 68)

Jones submitted an abstract on Feb. 2, 1989, for the APS meeting. In the abstract, Jones first mentioned muon-catalyzed fusion, then introduced his new cold fusion work. Jones' abstract, which omitted Palmer, Rafelski or any other co-authors, claimed a shortcut to nuclear fusion without the need for high-temperature plasmas. Here is an excerpt from his abstract:

> We have shown that nuclear fusion between hydrogen isotopes can be induced by binding the nuclei closely together for a sufficiently long time, without the need for high-temperature plasmas. ... We have also accumulated considerable evidence for a new form of cold nuclear fusion,

which occurs when hydrogen isotopes are loaded into materials, notably crystalline solids (without muons). Implications of these findings on geophysics and fusion research will be considered. (*APS Abstracts*)

Jones had raised the suspicions of Fleischmann and Pons months earlier when he, as an anonymous reviewer for their proposal, responded not just with a critique but also probed specific technical questions. In 1991, Eugene Mallove, the editor of *Infinite Energy* magazine, interviewed Fleischmann.

"We had some very positive refereed comments," Fleischmann said. "One referee wrote some comments to us. I was standing with Stan in the kitchen, and I said to him, 'This referee is Steven Jones from BYU! If we answer his question No. 7, we'll tell him why we think fusion takes place in Jupiter. If we answer question No. 1, we'll tell him how to set it up in the lab. What do we do?' In [our proposal], there are also all sorts of other things. I said [to Stan], 'I'm not very happy about this whole situation.'"

The concerns expressed by Fleischmann highlight potential ethical issues if, in fact, Jones' questions were for the purpose of advancing his own research. As a reviewer, Jones was in a position to delay Fleischmann and Pons' research and enhance his own research. Jones had not disclosed to Fleischmann and Pons his potential conflict of interest. Sometime in February, Gajewski contacted Jones and directed him to, among other things, get help from an electrochemist. Jones did not seek the help of Fleischmann or Pons, as Gajewski might have hoped. Instead, Jones sought out Douglas Bennion at BYU.

Questions of Piracy

Norm Brown, head of the Technology Transfer Office at the University of Utah, called Lee Phillips, BYU's technology transfer official, on Feb. 3. Phillips told science writer Gary Taubes that Brown said he needed to talk to Phillips "about the possibility of one of [BYU's] professors pirating one of [the University of Utah's] professors' stuff."

(Taubes, 59) Brown then called Gajewski about the concern. (Taubes, 60)

At BYU, Taubes wrote, "Jones met with a patent review committee headed by Lee Phillips, regarding a 'Device to Produce Controlled Nuclear Fusion.' Jones and Palmer presented this reactor-to-be, according to the minutes of the meeting, as though there was no prior work in the field, other than that by a 'Russian author.'" Ordinarily, the inventor is responsible for disclosing relevant prior art, such as patents or published papers.

As expected, the patent committee encouraged Jones to file before anyone else did. (Taubes, 60) Sometime around Feb. 10, Gajewski placed a hold on the forthcoming funding for Pons and Fleischmann because of Gajewski's concern about the brewing conflict. (Taubes, 60)

Meeting in the BYU Lab

On Feb. 23, Fleischmann and Pons accepted Jones' Dec. 16, 1988, invitation to meet and discuss their respective research. They met in a BYU lab with Jones, Palmer, J. Bart Czirr, a radiation detection expert, and Daniel Decker, chairman of the BYU Physics Department. During lunch, Jones told Fleischmann and Pons that he was ready to publish his data. Fleischmann argued against Jones going public because, Fleischmann said, the field would become flooded with too many scientists. It was a weak argument, but eventually Fleischmann was more forthcoming.

"Finally," Taubes wrote, "Fleischmann appealed directly to Jones' professed sense of fairness. He said that he and Pons had worked on it for years, as well, but they needed 18 more months. If Jones went public before then, he and BYU would receive all the credit.

Jones argued that he had been invited to give a talk at the American Physical Society meeting, and he wanted to submit a paper beforehand. Gajewski at the Department of Energy was telling him that, if he wanted DOE funding on cold fusion, publishing a paper would help tremendously. He had no choice, [he said,] but to publish." (Taubes, 62, 64) But Jones had not, in fact, been invited to APS to talk about his new fusion work, as I confirmed with the chairman of the APS session. He had been invited to talk about muon-catalyzed fusion.

At the end of the Feb. 23 meeting at BYU, according to Taubes, Jones suggested that Pons and Fleischmann return to BYU on Monday morning with a working cell. Jones suggested that he would test it with their neutron detector, and, if he observed neutrons, they would write up the result jointly. (Taubes, 64, 65) Fleischmann reluctantly agreed. That Monday meeting never happened.

A page from Palmer's lab book gives his perspective on the Feb. 23 meeting: "Visit by Stanley Pons and Martin Fleischmann. U of U. This was a fun day. Pons very quiet. Fleischmann an old-time con-artist — maybe. At least he is so good that neither Bart, Steve nor I could tell whether or not we were conned. But we knew we'd been conned, but we didn't know how." (Taubes, 65) Taubes summarized the Feb. 23 meeting:

> The [Feb. 23] outing at BYU strengthened Pons and Fleischmann's convictions on two points. It confirmed their suspicion that Jones had pirated their theory for inducing fusion. They reached this conclusion because Jones, a physicist, was indeed doing electrochemistry, not to mention doing it with palladium electrodes, just as they were. (It probably [contributed to Fleischmann and Pons' suspicions] that Jones kept referring to his two or three years of concerted effort, [yet] Pons and Fleischmann saw [only] an amateurish experimental setup that couldn't have represented more than a few months of work.) Why would a physicist think of doing anything electrochemical? (Taubes, 68)

Presidential Intervention

The following week, on Friday, March 3, 1989, Chase Peterson, president of the University of Utah, stepped in and attempted to arrange a summit conference with Jeffrey Holland, the president of BYU, to see whether the two universities could work things out amicably. Peterson did not immediately reach Holland and instead spoke with Jae Ballif, the BYU provost and a popular physics professor. (Taubes, 74)

The summit took place in Provo at BYU on Monday, March 6. Representing the University of Utah were Peterson, Fleischmann, Pons, and Joe Taylor, the outgoing dean of the College of Science. Attending from BYU were Jones, Palmer, Czirr, Holland, Ballif and Lamond Tullis, associate academic vice president. Peterson led the meeting.

"Peterson said he believed that a satisfactory agreement between [BYU] and the [University of Utah] could be reached," Taubes wrote. "He sketched out his views on how the scientific credit could be shared, hoping that his two chemists and the BYU physicist could publish simultaneous articles announcing the discovery. He hoped that they could work out the issue of patents and future research." (Taubes, 77)

But Jones had his own agenda. Jones told Peterson that he had been invited to talk about his electrolytic fusion work at the May 1989 American Physical Society meeting. Peterson knew this would constitute a public disclosure and could legally impair or jeopardize the University of Utah's patent claim.

Peterson knew that the American Chemical Society spring meeting would take place in mid-April, and this could be a chance for Fleischmann and Pons to announce their claim. But Peterson also knew that the abstracts for the APS meeting would publish in early April. If Jones' abstract went public first, the University of Utah could lose international patent rights. Peterson asked Jones to wait to publish his data. Peterson said that Fleischmann and Pons needed another 18 months to complete their work properly. The publication date of the APS abstracts was thus a key factor that determined the date of the University of Utah press conference. (Taubes, 79)

But Jones wouldn't budge. He told Peterson that he was going public in May, with or without Fleischmann and Pons. As an alternative, Peterson suggested that the two groups publish simultaneous papers in a journal. (Taubes, 80) Notes from Palmer's lab book again give the perspective of the BYU contingent.

"Bart [Czirr] said Pons and Fleischmann are going crying to the president for him to come," Palmer wrote, "to bear down on us and stop us before we get beyond our discovery of the heat engine that ... may power industry. They don't want us to get the Nobel Prize, when it is in their grasp." (Taubes, 76)

Back in Salt Lake City, the University of Utah representatives were convinced that Jones was a scoundrel and that he was trying to steal their glory and intellectual property rights.

This lack of cooperation, and later actions by Jones, set off a premature rush to publicity and publication that damaged the credibility of the scientists and, to a lesser degree, the University of Utah.

Fleischmann-Pons Publication Path

On March 10, apparently by coincidence, Ron Fawcett, a professor of electrochemistry at University of California, Davis, and the American editor of the *Journal of Electroanalytical Chemistry*, called Pons.

According to what Fawcett told reporters, he had called Pons to talk about one of Pons' students who had come to U.C. Davis, for a job interview. Pons took the opportunity to tell Fawcett about his fusion research and about his need for rapid publication of his and Fleischmann's preliminary note. Fawcett welcomed the manuscript.

Pons shipped the manuscript overnight to Fawcett. The next day, Pons faxed it (or possibly just the cover page) to Roger Parsons, the U.K. editor. The journal officially received it on March 13. The manuscript was accepted sometime before March 23. (Fogle, 1991, 27) The University of Utah filed its first cold fusion patent the same day. It filed a second cold fusion patent by March 21. (Taubes, 95)

Fleischmann was now under a great deal of pressure to make sense of the results, and the data were confusing. Fleischmann had a good relationship with colleagues at the U.K. Harwell laboratory, and he had hoped that they might provide one of the first independent replications. One of his former students, electrochemist David Williams, worked at the lab, and Fleischmann sent him a fax with his and Pons' preliminary heat data. "This information," Fleischmann wrote, "is still very incomplete so you can imagine how annoyed we are to be rushed into premature publication." (Taubes, 88)

By March 15, Jones had become defensive, as shown in his letter to Jae Ballif, the BYU provost. "It is true that, after more than two years of [my] funded work on cold nuclear fusion," Jones wrote, "in particular, on fusion during electrolytic infusion of isotopic hydrogen into metals, I was asked by the Department of Energy funding agent to review a proposal by Pons and Fleischmann. ... I do not believe that I have incorporated any of their original ideas into my research."

But Jones had not personally worked on that earlier electrolytic fusion research. And the researchers in his group who did, Palmer and Rod Price, completed their experiments in September 1986. Jones' group restarted electrolytic fusion experiments only in September 1988, after hearing of the Fleischmann-Pons proposal. There had been a two-year gap between experiments. Jones claimed the delay was because they were building a new neutron detector. The detector was finished, according to Jones, coincidentally, just as they learned about the Fleischmann-Pons research.

Planning the Press Conference

Peterson called a meeting in his office on March 16. Joining him were Peter Dehlinger, an attorney from Palo Alto, California, and two attorneys whom Pons had requested, C. Gary Triggs and Gary Sawyer, a North Carolina patent expert. Also attending were Jim Brophy, the vice president for research, Norm Brown, the head of the Office of Technology Transfer, Pons and Fleischmann. They knew what they were up against. (Taubes, 93)

"The abstract that Jones had submitted to the American Physical Society," Taubes wrote, "even with two meager sentences on cold fusion, might constitute a public disclosure once it appeared in print. It was scheduled [to print] the first week of April. That became the deadline for whatever had to be done, which gave the [University of Utah] three weeks [to stake its claim]. The option of having Pons unveil [his and Fleischmann's work] at the American Chemical Society meeting in the second week of April was no longer viable. Jones' APS abstract would appear a week before the [ACS] meeting." (Taubes, 96)

"Initially," Taubes wrote, "it was Peterson who suggested the public announcement, but the three lawyers apparently embraced its wisdom. [Peter] Dehlinger, [the attorney who filed the first patent applications for the University,] later supposed that the decision might have gone the same way even if everyone but Peterson had been [opposed to] a press conference. But such was not the case. 'The fact is,' Dehlinger said, 'the three lawyers were arguing that there is no second place in this kind of business. Either you're there first, or no one remembers you.' Dehlinger's recollection of the meeting also had Fleischmann 'almost in tears' as the consensus finally emerged that they would call a press conference." (Taubes, 96-7)

Feeling their backs against the wall and suspicious that Jones was trying to claim patent and intellectual priority on the fruits of their labor, Fleischmann, Pons, the University of Utah administrators, and their attorneys secretly made plans to go public with their claim as soon as possible. They brought Pam Fogle, the University of Utah's news director, to the meeting, and together they scheduled a press conference, despite her objections, for the afternoon of March 23.

The audio recording of an interview given by Fogle to author Jerrold Footlick reveals that the administrators did not tell her about the underlying Jones conflict. Instead, the administrators allowed her to believe that the urgency for the press conference was only because news of Fleischmann and Pons' research findings had begun "leaking out."

University of Utah Press Conference

Fogle moved into high gear on March 17. She began by having her science writer interview Fleischmann and Pons. On March 20, Pons and Fleischmann submitted a revised version of their manuscript to the *Journal of Electroanalytical Chemistry*, which received it on March 22. An initial version of Fogle's press release was ready by the morning of March 20. (Close, 101)

According to Taubes, on Tuesday, March 21, Pons called Jones to confirm that he was still planning to publish his paper. Jones said yes, and in those pre-email days, they agreed to rendezvous at the Federal

Express office at the Salt Lake City airport at 2 p.m. on March 24 to send off the papers. (Taubes, 100)

On Wednesday, March 22, Fogle began calling reporters (but not sending out the press release) and telling them about the press conference. The press release was scheduled to go out at midnight. She had wanted to give reporters the bare minimum of time to get to the press conference in order to minimize the chances of the story leaking out before the press conference. Her plan failed. One of the reporters she spoke with was Jerry Bishop, a seasoned reporter with the *Wall Street Journal.* Bishop immediately began doing his own research. He learned that Jones had some history with the topic, and he reached Palmer on the phone. Bishop asked Palmer what he knew about the University of Utah's announcement. This is most likely when and how Jones learned — certainly to his surprise and dismay — what the University of Utah was up to. (Taubes, 101)

Adding to Jones' disappointment that Wednesday, a false rumor circulated that his work had confirmed that of Fleischmann and Pons. Someone (unidentified) at the DOE learned about the forthcoming University of Utah press conference and that, according to the rumor, a reviewer of the Fleischmann-Pons DOE proposal had confirmed the Fleischmann-Pons result. The person at DOE called Jones, who must certainly have been outraged by that point. (Close, 102) Jones began compiling his history of fusion research that day. (Taubes, 148)

After March 23, the people at BYU who had been involved in the cold fusion negotiations believed, as Lamond Tullis, associate academic vice president, put it, that they had been "had." Tullis said they were filled with "righteous indignation" and "stunned incredulity."

Jae Ballif, the provost, said that he felt "devastated that, for whatever reason, agreements between honorable people were not kept." Only Steve Jones professed to be relieved: Now he no longer had to wonder what Pons and Fleischmann were up to. He knew, and it was done. What moved Jones to righteous indignation was the continuation of the "rumors" that he had engaged in scientific larceny. (Taubes, 147-8)

At 1 p.m. on Thursday, March 23, lights for the television cameras switched on, and Jim Brophy, the University of Utah vice president for research, took to the podium, followed soon after by Chase Peterson,

Pons and Fleischmann. None of them mentioned anything about the underlying conflict with Jones. Their mission was clear: establish priority for Fleischmann and Pons' intellectual primacy and the University of Utah's intellectual property. A few of Peterson's words at the podium were more significant than most people in the room realized.

Peterson spoke about "questions as to where all the credit lies and where the ownership lies." He said that, "if it turns out to lie with the University of Utah, as we think it will, then we would do all in our power to have this exploited, by ourselves and others, for the benefit of cheap energy with little cost to the world's ecology."

Although my copy of the videotape from the press conference does not show it, Taubes has information about another key statement made there. Taubes wrote that "a reporter asked whether Pons and Fleischmann were aware of any similar work going on elsewhere." Taubes refers to a recording of the conference. "Let's see. I'll answer it," Brophy said. "We're not aware of any such experiments going on. There are none reported in the literature." (Taubes, 104-5)

Because Brophy, among other people, believed that the Palmer-Jones experiment was similar to that of Fleischmann and Pons, Brophy's statement was a lie. Within hours of the press conference, the Jones group had completed and faxed a copy of its manuscript to *Nature*. Unlike Jones' APS abstract, Jones' manuscript now included seven co-authors. (Taubes, 105) No one knew at the time that the Palmer-Jones work, although it used electrolysis, palladium, and deuterium, was significantly different from the Fleischmann-Pons experiment and, more important, so were the results.

Fleischmann and Pons used a simple combination of heavy water and salt in their electrolyte. It was based on Fleischmann's long-standing interest in the behavior of deuterium in palladium. Palmer and Jones used a complicated "witch's brew" of a dozen metal salts; it was based on Palmer's speculation that fusion takes place within the Earth. Instead of using nearly 100% D_2O, as Fleischmann and Pons did, Jones used only 10% D_2O in a solution of H_2O.

Fleischmann and Pons said they observed large amounts of excess heat and a few excess neutrons. The Palmer-Jones group measured a few excess neutrons but no excess heat. Eventually, it became clear that the Palmer-Jones experiment offered no promise as a source of energy. But that distinction didn't reveal itself immediately.

BYU Press Conference

The following week, on March 29, 1989, Paul Richards, the head of BYU's press office, invited local news media to BYU to learn about the Jones group's work. Richards showed the news media a copy of Jones and Rafelski's 1987 *Scientific American* article titled "Cold Nuclear Fusion." Thus, they appeared as qualified authorities on "cold fusion." (Taubes, 150) The reporters did not know or recognize that the 1987 Jones-Rafelski work described muon-catalyzed fusion rather than the electrolytic fusion — as conceived by Palmer and promoted and claimed by Jones.

The reporters were unaware that the Palmer-Jones experiment used a different electrolyte from the Fleischmann-Pons experiment. The reporters did not know that this difference, among other crucial differences, was intrinsic to the production of excess heat in the Fleischmann-Pons experiment.

In short, these technical differences meant there was no reasonable expectation that the Palmer-Jones experiment should have performed like the Fleischmann-Pons experiment and produced any excess heat. But few reporters could have known or understood this level of detail at that time.

To add to the confusion, when Jones spoke to reporters and members of Congress, he always compared his results with those of Fleischmann and Pons. In doing so, he implied that his group's experiment was essentially the same as that of Fleischmann and Pons. But it was not.

Jones told the news media that his work showed poor potential as an energy source. Jones had reversed himself. No longer was his work a shortcut to fusion energy or a remarkable process that obviated the need for conventional fusion reactors. By portraying his experiment as a hopeless route to fusion energy and by saying that his experiment was equivalent to that of Fleischmann and Pons, Jones was taking Fleischmann and Pons down with him. If Jones wasn't going to get credit for a big fusion discovery, neither were Fleischmann and Pons.

"Jones reiterated to the reporters," Taubes wrote, "that the BYU results were much less dramatic than the Utah results. Did they promise energy salvation? 'Not by a long shot,' Jones said, rolling his eyes." (Taubes, 150) The reporters had no idea of the crucial differences between the design and the results of the two groups' experiments.

Innocent Jones

On April 2, the conflict between Fleischmann and Pons and Jones escalated. A local Salt Lake City paper, the *Deseret News,* printed a story in which professors from the University of Utah suggested that Jones had pirated ideas from Fleischmann and Pons.

On April 18, Jones got the BYU public communications office to release his two-page version of his fusion research history in an attempt to counter the accusations of piracy. In his history, Jones omitted all references to his and Rafelski's thoughts about trying — after they had received Pons and Fleischmann's proposal — to restart their own electrolysis work, to patent "cold fusion" first. He mentioned nothing in his history about the fact that he had claimed to the Department of Energy that he had been the first to demonstrate electrolytic fusion.

Jones did not mention that his group had a nearly two-year lapse in its electrolytic fusion work. The hiatus ended exactly when he and Rafelski received the Fleischmann-Pons proposal. Jones also wrote about the origins of his work with fusion but did not mention Palmer. Within a few years, Jones expanded and revised his version of his history.

Jones Backpedals

Jones continued to ride the wave triggered by the Fleischmann-Pons announcement. A week later, on April 26, Fleischmann and Pons were in Washington, D.C., testifying before Congress about the prospects of "cold fusion" as a new energy source.

Jones had been asked to testify there, too. Jones was an expert, he told Congress, on the Fleischmann-Pons type of research. That was not true; however, nobody but Jones knew it, and neither Fleischmann nor Pons called Jones out. When Jones testified, he picked up a small plant on the table and, using a poetic metaphor, said that the plant would never grow into a tree. In the same way, Jones said, "cold fusion" would never be a practical source of energy.

The following week, on May 1, Jones finally got his chance to present his scientific paper at the American Physical Society meeting in Baltimore, Maryland. Jones emphasized that his group had worked on "cold fusion" since 1986, and he showed notarized pages of his lab books in an attempt to convince those in attendance that he had a legitimate history in the work. Jones also made direct comparisons of the results of his experiments to the results of Fleischmann and Pons' experiments, perpetuating the falsehood that the two were the same — and therefore, the two sets of results should have been similar. This falsehood was unknowingly perpetuated by the news service for the American Institute of Physics, the parent organization of the American Physical Society.

"Steven Jones of Brigham Young University," the AIP News Service wrote, "reported that new measurements of reactions in an electrolytic cell were consistent with his earlier observations (*Nature*, Vol. 338, 737, April 27, 1989) of neutrons from cold fusion reactions, albeit at a rate many orders of magnitude less than for the University of Utah."

As discouraging as this sounded for Fleischmann and Pons, the news may have been a relief for thermonuclear fusion researchers who were worried about losing their federal funding to any scientist with a beaker and a couple of electrodes. Jones went out of his way to tell everyone at the Baltimore APS meeting how trivial his heat effects were. He was warmly received by his fellow physicists.

Jones assured his colleagues that room-temperature fusion was no threat to plasma fusion research. Emmett Black, a researcher with General Electric Research, was at the meeting and later wrote a report summarizing the meeting. "Jones opened the session with a discussion of his own work," Black wrote, "which he emphatically said would NOT lead to a new power source but does present interesting new physics."

Jones came across convincingly. He appeared as the cautionary, conservative, likable scientist whom people could trust and believe. On camera, he rarely appeared without his trademark gentle smile, soft voice, and occasional jovial chuckle.

He also leveraged the growing anger among physicists against Fleischmann and Pons. He compared his tiny-heat claim to the large-heat claim of Fleischmann and Pons. He audaciously told the physicists and the media that there was only one chance in 2 million that his measurements were wrong and that, therefore, in his expert opinion, there was only one chance in 2 million that Fleischmann and Pons were right.

Jones' Version of History

Sometime around May 1989, Jones wrote a second version of his fusion research history. No co-authors are listed on the document.

In 2003, I met Jones in Cambridge, Massachusetts, at the 10th International Conference on Cold Fusion. I knew nothing of this history at the time. I also knew very little nuclear physics then. While we were sitting at a table during a break, he gave me an impromptu lesson on the structure of the atom and the fundamental concepts of nuclear fusion. A few months later, he wrote a third version of his cold fusion history and sent me a copy.

Around that time, a few "cold fusion" researchers were preparing to petition the Department of Energy to take a second look at "cold fusion" research. The "cold fusion" scientists invited Jones to participate. I suspect that this Department of Energy review was Jones' impetus to update his history. In his third version of his history, Jones listed three people as authors, with the following header:

> By BYU Professors Jae Ballif, William Evenson, and Steven Jones (This history was originally written April-May 1989 by Professors Jae Ballif, William Evenson, Steven Jones, with revisions in March 2004.)

In 2014, I sent an e-mail to Jones and asked him why versions 1 and 2 of his history did not list any co-authors. A couple days later, Jones e-mailed me back. "The authors were myself, Jae Ballif and William Evenson (all BYU physicists). Our names should have been listed; perhaps these were early drafts," Jones wrote.

I telephoned Ballif on Dec. 20, 2013, and asked him whether he remembered writing an account of the history of the BYU cold fusion history. He told me, "I wrote a lot of memos, but I didn't write any history. I never did." I telephoned Evenson the same day. He denied contributing to Jones' history of fusion research. I sent version 1 and version 2 of Jones' history to both of them for verification. Here are their responses.

"I do not believe I have ever seen either of these documents," Evenson wrote. "I speculate that the longer one was prepared by Steven Jones, possibly reviewed by Jae Ballif, and perhaps others. Ballif wrote back to me and said that he wasn't involved."

"I received and read the two articles you sent on the subject of cold nuclear fusion," Ballif wrote. "I was not [an] author or co-author of either of these articles."

Because my chronology relies heavily on Taubes', I checked with Jones about the accuracy of the Taubes book. I asked Jones whether there were any significant factual errors or omissions about him in Taubes' book.

"Yes," Jones wrote, "there were problems in Taubes' book, so three professors undertook to write a brief history of cold fusion at BYU, in order to set the record straight: professors Jae Ballif, William Evenson, and myself."

Beyond the factual inconsistency about the authorship, Jones' explanation is chronologically impossible. Jones wrote his history in mid-1989. Taubes completed his first draft two years later, in mid-1991.

I exchanged several more e-mails with Jones and Taubes. In the end, Jones failed to identify any errors in Taubes' book.

CHAPTER 3

Not Even a Watt

Thermonuclear Fusion 50 Years Later

The news of "cold fusion" arrived in the shadow of decades of unfulfilled promises in thermonuclear fusion research. Since the 1970s, thermonuclear fusion researchers and advocates had been saying that practical fusion reactors were just two decades away.

Ethan Siegel, a professor of physics and astronomy at Lewis & Clark College, who has a Ph.D. in astrophysics, wrote about the progress of fusion on the Forbes.com blog on Aug. 27, 2015: "The reality is we've moved ever closer to ... the breakeven [power] point in nuclear fusion — where we get out as much [power] as we put in." Yet there is still no practical fusion reactor, and no experimental reactor has produced a single watt in excess of the total power required to operate the reactor.

As I was checking basic facts about the claimed steady technical progress in fusion research — which I had assumed were correct — I discovered an astonishing discrepancy between what was publicly reported and the actual progress in net power produced by fusion.

From Heat to Electricity

On Dec. 18, 1957, in Shippingport, Pennsylvania, the first commercial nuclear fission power plant produced electricity. Today, about 30 countries operate about 450 nuclear fission reactors to generate a large portion of their electrical energy.

In very simple terms, fission reactions emit energetic neutrons which, when captured, create heat. That heat is used to boil water and produce steam to drive turbines, which then create electricity. The production capacity of fission reactors can be measured in both produced heat and produced electricity. The rate of each is given as a measure of power, in watts. A commercial nuclear fusion reactor, in theory, would also emit neutrons. These neutrons would also produce heat that could then be extracted to produce electricity.

When scientists speak about nuclear reactions measured in watts, unless the power from that nuclear reaction has been harnessed to a turbine, that power refers only to heat, not to electricity. There have been no electrical power-generating nuclear fusion reactors. As a result, no electrical power has been produced by nuclear fusion. Only energetic neutrons and heat have been produced by nuclear fusion reactors and, at most, for only a few seconds at a time.

Fusion proponents say that they face considerable technical challenges, which they failed to anticipate, as well as unsteady levels of government funding. They also defend the field on the basis that the work has added to basic science research. At other times, they also have said that they understand the science and that the challenges before them are only engineering details.

Two Incorrect Representations

Magnitude-of-Power Representation

The public understanding and support of fusion research is predicated on two incorrect concepts. The first is that any heat, above and beyond the total system input power, has been produced. It hasn't, not even for a split second. Experiments have always consumed more electrical energy than energy, in heat, they have produced. Among laypeople familiar with thermonuclear fusion research, only a few are aware that no fusion reactor has ever produced a single watt of power, after subtracting the total power going into the reactor. No fusion reactor has ever demonstrated a power gain — only a power loss.

Some fusion proponents say that reactors would have shown a gain if they had been run with more potent fuel, but this has not been tested.

Net-Power Representation

The second incorrect concept is the amount of electrical power that fusion experiments consume during tests. The most expensive scientific experiment on Earth, the International Thermonuclear Experimental Reactor (ITER), a fusion machine estimated to cost $21 billion, is under construction in Cadarache, France. Financial and technical support for the project comes from the European Union, China, India, Japan, South Korea, Russia and the United States. ITER is intended only as a science experiment; it will not produce any electrical power.

Two Fusion Methods

The term "hot fusion" was not used before early 1989 because the corollary term "cold fusion" was not a common term. Therefore, no distinction was needed. Thermonuclear fusion research was and is conducted using two approaches: magnetic confinement fusion and inertial confinement fusion.

The daunting challenge with both approaches is to confine hydrogen isotopes closely enough, densely enough, and long enough that the strong nuclear force can overcome the electromagnetic force. The electromagnetic force normally keeps atomic nuclei separate — and the physical world from collapsing in on itself. However, if nuclei can be squeezed together, under specific conditions, the strong nuclear force overcomes the electromagnetic force and pulls the nuclei together. This is thermonuclear fusion, and it is the process that scientists believe occurs in the sun and other stars.

Since 1950, scientists have dreamed of harnessing energy from thermonuclear fusion on Earth. Fusion energy, in theory, seems ideal: free of combustion products, producing minimal waste, and using a substance found in ocean water as a virtually unlimited source of fuel. Its enthusiasts envision it as the Holy Grail of energy.

But creating fusion on Earth requires building devices that can contain temperatures in the millions of degrees to make ions collide and fuse. The problem is that no material on Earth can withstand such sustained temperatures. A lot of serious effort has gone into studying ways to solve this problem.

Historically, neither the magnetic fusion research program nor the inertial confinement fusion program has been presented to the general public for the purposes of science research. The primary mission of the U.S. inertial confinement fusion program is, in fact, to facilitate accurate simulation of selected physical weapons effects without resorting to underground testing of thermonuclear devices. The U.S. program is funded through the Department of Energy's weapons program and operated by the National Nuclear Security Administration.

Magnetically Confined Fusion

Magnetically confined fusion experiments generally use devices based on a design developed in the Soviet Union, called a "tokamak," an acronym for the Russian name for a toroidal chamber containing magnetic field coils. Seen from the inside, it looks like a large doughnut-shaped machine and can be as large as a three-story house.

The Princeton Plasma Physics Laboratory was the flagship facility in the U.S. trying to demonstrate sustained nuclear fusion. After researchers went as far as they could with the Princeton Tokamak Fusion Test Reactor (TFTR), it was shut down in 1997. In the United Kingdom, the Joint European Torus (JET) tokamak fared slightly better, and it achieved its peak performance that same year.

Different Meanings for the Same Words

People working in the thermonuclear fusion field often display graphs and depict their research progress with the terms "fusion power" and "megawatts." Clarification is required.

To a layperson, the term "fusion power" means net positive power that comes out of a fusion reactor — something they may hope to use someday to power their homes. Fusion researchers, however, have assigned a different meaning to the phrase "fusion power." This double meaning has caused widespread confusion among the lay public.

When people see figures in megawatts, they assume that this means electrical output, but it actually describes measurement of the heat output. However, fusion researchers rarely, if ever, label their graphs with "heat out"; instead, they denote the value as "fusion power."

Another point of confusion is that, when the public reads or hears that megawatts of "fusion power" have been produced from reactors, it assumes that power values are net figures. This is not the case. When thermonuclear fusion researchers use the term "fusion power," it has nothing to do with actual net power produced by a fusion reactor. This will be explained in detail later in this chapter.

Net-Power Representation to U.S. Congress, 1993

On May 5, 1993, the U.S. House of Representatives Committee on Science, Space, and Technology Subcommittee on Energy held a hearing to review the current state of fusion development, understand proposals for more funding, and look at alternative fusion processes.

N. Anne Davies, Department of Energy associate director for fusion energy, gave the representatives the Department of Energy's budget request for fusion energy for fiscal year 1994.

"A year ago," Davies said, "when we were here, we reported on results from the experiments in JET. In that experiment, two megawatts of fusion power was produced in a short pulse length of about two seconds. Dr. Rebut, who's going to appear before you later today, was the director of JET at that time, and he is now the director of ITER. At Princeton, we expect to begin the deuterium-tritium experiments in September, with the production of 10 megawatts or more of fusion power by next year."

Next, Paul-Henri Rebut, the director of the ITER Design Activities, testified. "Since the mid-'70s," Rebut said, "a 1,000-fold increase has been achieved in the overall performance of experimental fusion devices.

"On November 9, 1991, a deuterium-tritium fuel mixture, including only 10 percent of tritium in JET, produced over a megawatt of fusion power for more than two seconds. These achievements are the result of the determined pursuit of strong and focused programs with involvement of industry."

The first point that requires clarification is that the JET reactor did not produce electrical power; it only produced heat from the emitted neutrons. However, Davies and Rebut omitted a key fact: Many more megawatts of electricity went into the entire system to produce the 1-2 megawatts of fusion heat. JET, in fact, did not produce any overall power; it consumed power. In their testimony, neither Davies nor Rebut mentioned the multi-megawatts of power the device consumed.

Net-Power Representation to U.S. Congress, 2009

In the 2009 House hearing on energy, Edmund J. Synakowski, associate director for Fusion Energy Sciences in the Office of Science of the U.S. Department of Energy, testified in both an oral and a written statement.

"JET soon announced to the world the generation of a few million watts of fusion power," Synakowski said, "enough to power thousands of homes. The race was on. TFTR at Princeton began its experimental campaign with the deuterium-tritium fuel mix and completed it with experiments in 1994 that generated over 10 million watts of fusion power. The JET experiment ultimately created a record 16 million watts of fusion power in 1997, a result enabled by the larger size of the device as compared to TFTR."

As did Davies' and Rebut's, Synakowski's testimony gave the clear but erroneous impression that JET had produced net power in the millions of watts. Synakowski presented a picture of an exciting and practical source of energy. Then he added a qualifying statement to keep his comment — for anybody who understood his lingo — marginally technically honest.

> More power was used to heat and control the plasma in each of these cases than was used to create the fusion reactions themselves. The figure of merit ... Q, relates to the fusion power created to the power used to heat the plasma. The JET experiment yielded a Q of about 0.6.

Fusion experts knew what "Q" meant, in the context he provided.

Other people might not have been aware that "Q" had two meanings, which I will explain in a moment. For nonexperts, including members of Congress, the terms "figure of merit" and "Q" would be meaningless.

All they were likely to have understood from Synakowski is that the fusion reactors had produced millions of watts of net power, enough, as Synakowski said, to light thousands of homes, even for just a few seconds. They may not have realized that the reactors had produced no net power but instead consumed millions of watts.

Net-Power Representation to European Union

Members of the European Union research commission may have believed a similar erroneous concept. An archived webpage from the official European Union energy research section said that JET had accomplished its mission and that "the scientific and technical basis has now been laid for demonstrating net fusion energy production."

Farther down the Web page, the text says, "JET's fusion output was 1.7 megawatts. Beaten in 1994 by the American TFTR installation (10.7 MW), JET took the lead again three years later by attaining 16.1 MW (i.e., 65% of the power injected), using a new operational technique."

At this point, a non-expert might get confused. The headline gave the impression that JET had demonstrated "net fusion energy." The text gave three values of megawatts of produced fusion power. But what did that phrase "power injected" mean? Was it an important detail?

Magnitude-of-Power Representation

The fact is that JET did not produce a net output of 1.7 or 16.1 MW. TFTR did not produce a net output of 10.7 MW. Neither produced any net output based on total system input power. That's the first error.

But this fact was not deeply buried; I knew it, and so did a lot of people who had a modest understanding of thermonuclear fusion research. It is, of course, well-known to people working in the field.

Before investigating this matter, I was certain that the 65% number was correct. I thought that, at its best, JET lost only 35% of the total power that went into the system. That number seemed to be understood widely among people who had modest knowledge of the subject.

But in that JET experiment, the ratio of heating power out to total system power in was not 65%, not even close. That's the second error.

Net-Power Representation

Here is how the second error came to light. One of my technical editors asked whether I had information about the progress that had been made in increasing the net fusion power over the decades.

I sent an e-mail to Stephen O. Dean, the director of Fusion Power Associates, a nonprofit research and educational foundation, and asked whether he had such information. He didn't. Slowly, the picture came into focus. I soon learned how important the insiders' phrase "power injected" or "heating power" or "applied fusion power" was.

"The applied fusion power," Dean wrote, "is not a relevant measure of progress since these have all been experiments not designed for net [power]. The input referred to is just the input to the plasma and does not include the power to operate the equipment."

I was confused. I thought that the numbers — for example, the 65% cited for JET — reflected total net power. I asked him whether he knew the best total net power for those devices. He didn't. I asked him whether this meant that JET's and TFTR's peaks were based on the input heating power rather than the total input electrical power. Yes, it did, he wrote. Now I was concerned.

As I soon learned, in addition to the power required to heat the plasma, power is consumed in tokamaks by a variety of processes. The greatest among these is the power required to create and maintain the magnetic field that suspends the plasma within the toroidal chamber.

Two Methods of Accounting

At first, I didn't believe that fusion researchers normally accounted for only a fraction of the total input power when they stated net power values. I called a plasma fusion physicist who worked for General

Atomics and asked him to explain this. He corroborated what Dean had told me. It was true.

Yes, people in the magnetic fusion research industry, since the 1970s, have always used applied heating power rather than total system input power when reporting their progress. I asked the fusion physicist whether he knew how much greater the actual total system input power was than the heating input power. He guessed that, typically, total input power was about 10 times as much. If this was correct, then fusion results had been exaggerated by an order of magnitude for decades.

I sent an inquiry to Nick Holloway, the media manager for the Communications Group of the Culham Centre for Fusion Energy, which operates the Joint European Torus. I told him that I understood JET had generated 16 MW fusion power with 24 MW applied heating power input. I asked him whether he could tell me about how much total input electrical power was required to make that much power.

"We don't have the electrical power input figure for this pulse to hand unfortunately," Holloway wrote. "Below is some information from my colleague Chris D. Warrick on JET's typical electrical power levels, so it will be of this order. But if you do need the exact input figure we can find out." Here is Warrick's e-mail:

> The general answer is that a JET pulse typically requires ~700 MW of electrical power to run. The vast majority of this goes into feeding the copper magnetic coils and the rest into subsystems and energizing the heating systems. In future machines, the copper coils will be replaced with superconducting coils – which will ensure the total input power is dramatically reduced. I don't have on hand the specific numbers for this particular pulse.

Order of Magnitude Difference

Holloway and Warrick had confirmed it: The total input power was an order of magnitude larger than applied heating power, as was the value which has been universally used to represent the state of the art in thermonuclear fusion research.

The total system input power used for JET's world-record fusion experiment was about 700 MW. Thus, a more accurate summary of the most successful thermonuclear fusion experiment is this:

With a total input power of ~700 MW, JET produced 16 MW of fusion power, resulting in a net consumption of ~684 MW of power, for a duration of 100 milliseconds. In other words, the JET tokamak consumed ~98% of the total power given to it. The "fusion power" it produced, in heat, was ~2% of the total power input.

As most of the public would understand the term "fusion power," JET produced none. (This calculation assumes, for the sake of example, that the number of ~700 MW is a precise value to three significant digits, which it most likely is not.)

The truth about the overall efficiency of the reactors has been so well-hidden that even Charles Seife, the author of a pessimistic book on fusion, missed it. He, too, was unaware that the researchers were reporting their power input based on applied thermal power input rather than the total electrical power input. Seife thought that the best JET experiment had lost between 10% and 40% of the input power.

> JET got 6 watts out for every 10 it put in. It was a record, and a remarkable achievement, but a net loss of 40 percent of [power] is not the hallmark of a great power plant. Scientists would claim — after twiddling with the definition of the [power] put into the system — that the loss was as little as 10%. This might be so, but it still wasn't breakeven; JET was losing energy, not making it.

Seife had no idea that JET actually lost about 98% of the input [power], rather than 10% to 40% of the input [power]. The shorthand typically used to describe energy production in the fusion community has created a mistaken view of its success among most observers.

ITER's Net-Power Mystery

On Oct. 22, 2015, Lev Grossman wrote about ITER in *Time* magazine. Like Seife, he seemed to have no idea that people in fusion research had used two methods of accounting to report their results:

> The goal for [tokamak] machines is to pass the break-even point, where the reactor puts out more energy than it takes to run it. The big tokamaks came close in the 1990s, but nobody has quite done it yet.

Two percent is not close to 100%. Based on the way fusion researchers have permitted, if not fostered, this misunderstanding, it is not Grossman's fault. This led Grossman, like many science journalists, to misunderstand the projections for ITER:

> The gain (the ratio of [power] out to [power] in) of a commercial fusion plant would have to be in the 15-to-20 range; right now, ITER's target gain is 10. To date, no fusion reactor has reached a ratio of 1, the break-even point. Then there's the question of how exactly to extract that energy from the reactor in the form of heat, so that it can plug into the existing infrastructure.

ITER will never plug into the existing infrastructure, with the exception of drawing power. The target gain of 10 is only the expected ratio of fusion power out to heating power in; it has nothing to do with total net power. ITER will never approach the output necessary for a commercial fusion plant; the target gain for ITER is a flat zero.

The Japan Atomic Energy Agency, one of the participants in ITER, asked the question, "Will ITER make more energy than it consumes?" on its Web site. Here is the answer the site provided: "ITER is about equivalent to a zero (net) power reactor, when the plasma is burning."

According to the Japan Atomic Energy Agency Web site, ITER will produce 500 MW of heat in pulses that last up to 400 seconds.

According to the ITER Web site, the power supply for ITER is planned to have a capacity of up to 620 MW for peak periods of 30 seconds.

The total installed power for ITER will require much more electricity. According to a technical document, "Power Converters for ITER," written by Ivone Benfatto, working with the European Fusion Development Agreement in Garching, Germany, the total installed power will be about 1.8 GVA. It will draw that power from a dozen hydroelectric and nuclear fission power plants in the nearby Rhône Valley.

Proponents hope that the first working fusion reactor to generate electrical power, called DEMO (DEMOnstration Power Plant), will come sometime after ITER. Proponents have not specified a date when such a reactor might finally prove that nuclear fusion can produce net energy on Earth.

According to an official European Commission Website, DEMO "will be designed to produce up to 500 megawatts of electricity, which will require a thermal output of around 1,500 megawatts." The reason for the requirement of 1,500 MW thermal output is that conversion of heat to electricity is generally only 30% efficient.

This means that fusion scientists envision going from ITER, which is projected to produce zero net power for 400 seconds, to DEMO, which will continuously produce 1,500 MW thermal net power and become the "first commercial fusion power to reach the grid."

Private Fusion Investments

Grossman's *Time* magazine article discussed a handful of other private investments in fusion research.

Lockheed Martin, for example, announced in October 2014 that it would be delivering a working prototype of a fusion reactor within five years. On Dec. 23, 2014, I questioned Geneva Greene and Heather Kelso, in Lockheed Martin media relations, about their progress.

I asked specific questions about power input and output. They had no answers. "We have not released our quantitative data and do not have publicly releasable data to provide at this time," Kelso wrote.

Inertially Confined Fusion

The other main fusion approach has been to use powerful lasers to bombard, compress and collapse tiny pellets of hydrogen-isotope fuels. In the U.S., the largest and most powerful laser system ever constructed was designed and built for this purpose. It is located at the Lawrence Livermore National Laboratory in California and is called the National Ignition Facility. It's as long as a football field and is three stories high. The National Ignition Facility also has never produced any net power.

There has been confusion about the progress of inertial confinement fusion, as well. Philip Ball, writing for *Nature* on Feb. 12, 2014, reported on claimed progress at the National Ignition Facility, in his article "Laser Fusion Experiment Extracts Net Energy From Fuel."

"Using the world's most powerful assembly of lasers," Ball wrote, "a team of researchers say they have, for the first time, extracted more energy from controlled nuclear fusion than was absorbed by the fuel to trigger it — crossing an important symbolic threshold on the long path toward exploiting this virtually boundless source of energy."

The headline and Ball's lead paragraph give the impression that the researchers extracted more power, as well as energy out than was put into the system. But this is not true. The second paragraph said that NIF was "still a way off from the much harder and long-sought goal of 'ignition,' the break-even point beyond which a fusion reactor can generate more energy than is put in. Many other steps in the current experiments dissipate energy before it even reaches the nuclear fuel." By speaking indirectly about "a fusion reactor" and suggesting that energy losses were the result of "dissipation," Ball obscured the fact that the NIF experiment did not "extract net energy."

Here is a simple overview: Energy from the grid (a) powers the lasers (b), which then inject energy into the fuel target to create fusion reactions (c). Ball omitted component (a) of the equation. Buried at the very bottom of the article, where few readers would notice, is the most important information: Ball quoted a Lawrence Livermore researcher who revealed the ratio of total power in to total power out. "Our total gain — fusion [power] out divided by laser [power] in — is only about

1%," the researcher said. The total system gain would be much lower, taking into account the total system input power.

In 2016, the U.S. Department of Energy wrote that "The question is *if* the NIF will be able to reach ignition in its current configuration and not *when* it will occur." (emphasis in original) (NNSA, 2016)

U.S. Fusion Funding

From 1951 to 2014, U.S. taxpayers spent $36 billion for magnetic and inertial fusion research, according to Fusion Power Associates. Globally, an estimate for fusion energy research is at least $100 billion. This number excludes any funding for classified fusion research.

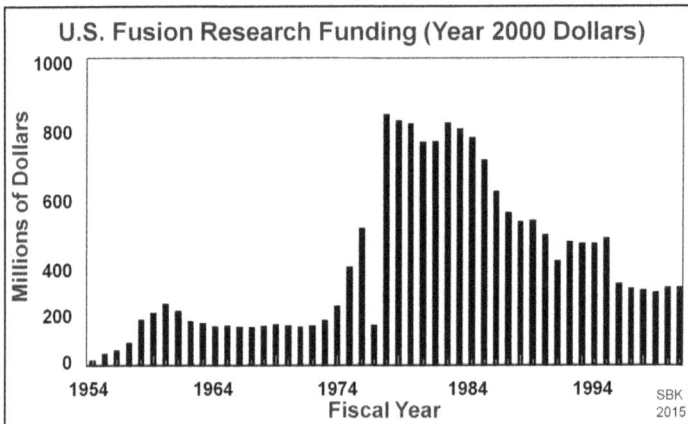

Fusion funding (1954-2001) by Richard E. Rowberg (Congressional Research)

Fusion funding got a strong boost after the 1973 OPEC oil crisis. The yearly funding level peaked in 1978 at $1.1 billion (adjusted for inflation), stayed there for a few years, then dropped sharply and, by 1989, was at $0.7 billion.

When news of "cold fusion" broke in 1989, people in the thermonuclear fusion business could not have been happy that two chemists with a glass tube reported not only that they had achieved a sustained fusion reaction but also that they produced more total power (in heat) than their total electrical power input. And that they'd done it out of their own pockets with a mere $100,000.

CHAPTER 4

Send Lawyers, Heavy Water and Money

Behind the Scenes in Utah

In mid-March 1989, very few people knew about the brewing trouble between the University of Utah and Brigham Young University. Among those was the University of Utah contingent, which included electrochemists Stanley Pons, the chairman of the Chemistry Department, his colleague Martin Fleischmann, a visiting research professor, Chase Peterson, the president, James Brophy, the vice president for research, Norm Brown, the director of Technology Transfer, attorneys representing the university, and two private attorneys representing Pons.

Steven Earl Jones, a physicist at Brigham Young University, and a few administrators there had an inkling that trouble was brewing.

The University of Utah news director, Pam Fogle, knew some of what was happening. In 1995, she was interviewed by Jerrold Footlick, the author of the 1997 book *Truth and Consequences: How Colleges and Universities Meet Public Crises.*

According to Fogle's interview with Footlick, the only reason Fogle knew about the press conference was that rumors had started to leak out about the "fusion discovery" and university administrators thought that the story could no longer be contained. This matches what the university administrators said during the press conference.

In 1989, Fogle was interviewed by Tom Gieryn, the author of the

1999 book *Cultural Boundaries of Science: Credibility on the Line*. According to the interview Fogle gave to Gieryn, Fogle had even heard the fusion rumors at her church.

But these two interviews reveal that the more-informed group withheld from her the primary reason for the university's forthcoming press conference: the perceived threat from Jones' impending publication, the desire to protect the university's intellectual property, and the use of the press conference to establish priority.

Fleischmann's Reluctance

Fleischmann reluctantly went along with the administrators' and attorneys' idea to hold a press conference. I talked with him about this in 2003. Gary Taubes, the author of the 1993 book *Bad Science: The Short Life and Weird Times of Cold Fusion*, spoke with one of the attorneys who had been in the meetings and said that Fleischmann was close to tears when he realized the inevitable direction of the meeting's outcome. All three of our accounts match.

When I met Fleischmann for the first time, on Aug. 24, 2003, I asked him about the press conference. He wasn't willing to divulge much and gave me a one-sentence reply. "I really didn't want to do it this way," Fleischmann said. "I did not want to do this project this way."

One of Fleischmann's friends and closest colleagues, John O'Mara Bockris, gave me copies of some letters Fleischmann had sent to him. Fleischmann's Feb. 7, 1991, letter explains in great detail his and Pons' perspective going in to the press conference.

The Situation Before March 1989 and Its Relevance to Secrecy

It may well be that all our concern about national security is a load of baloney, but this is what we believed (and still believe), and at the time, we thought it would be totally irresponsible of us if we had not kept the project secret. At various times, I have tried to find out whether there might be some substance to our line of thinking, and I have always been met with a smiling and tight-lipped response.

As you know, we ran out of money and made an application to the DOE in [August] 1988. (Incidentally, we knew quite a lot more about what we were doing than we put in that application). Our approach to the DOE really precipitated our problems. But even if news about what we were doing had not leaked out that way, it is doubtful whether we could have kept the lid on until September 1990. One member of Stan's group could not keep his mouth shut (I know who it is), and I was even stopped on the campus by one person saying, "I hear you are working on fusion."

Patents and Their Bearing on Secrecy — The March 1989 Fiasco

Much has been said about the Steven Jones saga. Stan and I felt that [his group] really had nothing of substance and [that we] should have waited 18 months, that is, to September 1990, but the Jones group felt that they had to publish their work — perhaps they were constrained by funding? Jones, at that time, considered that they had observed neutrons but no heat although, quite clearly, they had done no calorimetry at BYU. It was Stan's and my view that this forced us to tell the University of Utah about the work, and the patent issue was taken out of our hands.

It seems quite clear to me that the university had to seek patent protection. Whatever the consequences of taking this line may have been for the project, the university could not have stood aside and just let the work go. Can you imagine the hullabaloo there would have been if we had published our work without seeking patent protection? However, what was so unfortunate was that the patent issue took on its own dimensions and became the driving force of events leading to the press conference and all that razzmatazz.

You know that Stan and I were opposed to such a high-profile approach. To be quite frank, we were utterly exhausted, and I think our judgment was clouded. It is true that I tried to get hold of Sir George Porter to try to have a high-level discussion about the issues, but I failed to do so.

[Porter was the president of the prestigious Royal Society of London. Fleischmann was hoping that Porter could leverage Porter's influence and convince the University of Utah administrators to cancel the press conference because of its potential security implications.]

However, having said all that, we have to admit that in the end we went along with the approach, and we must take the blame. I went home as soon as possible, as I had had enough of the high-profile activity. [This appears inaccurate. According to University of Utah news director Pam Fogle, the press conference had to be scheduled by March 23 because Fleischmann already had plans to go back home to England for Easter.]

I would add, though, that if we had had no press conference, the effects would have been just the same but there would have been a time delay. There is one more point about the patents which I would like to make here. You will know that Stan and I will not benefit financially from this work. Everybody in Utah knows this, but this does not stop them from casting aspersions on our motives. From the strictly selfish point of view, it is therefore of no concern to us whether the patents succeed or fail, and I really do not know why we are so concerned about them. I suppose our concern is due to a misplaced sense of loyalty.

Explosive Intention

Another part of Fleischmann's letter to Bockris will be revelatory to all but a dozen people in the world. Fleischmann and Pons have been celebrated by their fans, some of them quite passionately, for the benevolent quest of finding a new source of clean, unlimited energy. This is not what they initially had in mind.

"Our working hypothesis," Fleischmann wrote, "was that fusion might take place in clusters of deuterium metal. Deuterium metal was already predicted to be a potential nuclear explosive in the 1970s."

Fleischmann further clarified the true initial intention of his and Pons' hypothesis, somewhat reluctantly, when I interviewed Fleischmann and a colleague of his, electrochemist Michael McKubre, in 2003. Here's an excerpt of the conversation:

MCKUBRE: Do you think there is a commercial object at the end of this tunnel?

FLEISCHMANN: That's the difficulty of this field. It was clear very early on that there was the possibility of developing an energy source. I did not expect that, incidentally. I expected the applications to lie severely in, shall we say, "Give unto Caesar that which is Caesar's."

MCKUBRE: The energy release would be rather more rapid?

FLEISCHMANN: The implications of the energy release would constrain the subject to lie in the area of national security. But what turned out was that there might be a possibility of having a reliable clean energy source. And it was clear that this was feasible right from the early work.

Only after Fleischmann and Pons realized that the heat from their reactions released slowly, rather than with the velocity required for an explosive, did they consider the prospects of an energy source.

Additional Funds Needed

In November 1996, Christopher P. Tinsley of *Infinite Energy* magazine interviewed Fleischmann, and he provided additional details about his and Pons' situation in 1988 before the March 1989 University of Utah fusion press conference:

Stan and I funded the first phase of the work ourselves. It was secret. We reckoned we would get our first [scientific] answers for about $100,000, which was as much as we could afford to spend. In the summer of 1988, we reckoned that we would need $600,000 to complete the first phase by about September 1990. We planned to review the question of publication in September 1990. We had, at that time, and continued to have all the way through, tremendous hang-ups about whether this work should be published at all. In fact, in 1988, we went through several discussions about whether the work should be classified for reasons of national security. ... We also had to inform the American Department of Energy and I had to inform the Harwell laboratory in the U.K. about this work. So I said, "Let's kill many birds with one stone. Let's write a research application rather than a patent," which we submitted to the DOE. Initially, it didn't go to the DOE, but it ended up there in August 1988.

And that, of course, brought us into this conflict situation with another scientist [Steven Jones] who had been interested in the subject previously. He had not done the experiment in a way in which he could possibly have succeeded, mainly because he had used 10% D_2O in H_2O. Of course, he would have had hardly any deuterium in the lattice — and so he started to work on this topic again.

There is nothing wrong with his restarting his work, incidentally. People object to that, but I don't object to that at all. However, I think that he should have disclosed his intention to restart his work when he refereed our proposal. What was hard for Stan and me was that Jones wanted to disclose his results in March 1989. Stan and I were still working in secret at that time, but, because of this development, we had to inform the University of Utah because we thought that the university might need to take patent protection.

They said yes, so then the patent became the driving force. And it was the patent consideration which produced the press conference, the "prior claim." I was not in favor of that at all, but it was that which produced the press conference.

News Director Speaks

In her interview with Thomas Gieryn, Pam Fogle explained what it was like, from her perspective, in the weeks leading up to the press conference:

It was in the winter, either December or January. I was talking with the vice president of research [Jim Brophy], and he indicated to me that, in the near future, a large announcement would be coming up about fusion.

He asked me if I knew what fusion was. I had covered physics so I knew what fusion was, but I couldn't figure out anybody in the Physics Department who would be working on fusion. Then he said, "Well, it's not coming from Physics. It's in the Chemistry Department." So we talked about it only in the vaguest of terms at that time.

I knew that my science writer would need time to develop a story because I figured it would be complicated. That was Barbara Shelley, and she and I both shared responsibility for dealing with the media on the fusion story. ... For several weeks, we tried to pursue the story, to get a draft of the journal manuscript, to conduct interviews with researchers, and in each case, we were told, "You really will have to wait until it comes time to make the announcement."

Once the date of Thursday, March 23 had been decided for the news conference, March 17, the Friday before, was the first time that Barbara was allowed to interview Pons and Fleischmann.

By this time, I had already heard rumors around campus. I began hearing stories about people hearing things at church that there was a big announcement coming from the university. Bits and pieces of information were coming out.

It's our job keep the pulse of what's happening on campus, and now I'm getting worried if my writer will have adequate time to develop the story. So Barbara did extensive background on traditional fusion, having no idea what she was going to face when she got into the interview. ...

She ended up working the whole weekend to develop the story. And I remember her comment to me, when I went in on Sunday to check and see how things were going. She was having a little trouble developing the story. She said to me, "We need a new definition of fusion."

Pamela Fogle, 1989 University of Utah News Director

It was an astute observation for a science writer who knew almost nothing about the topic. It wasn't that a new definition of fusion was needed; it was clear to the writer that Fleischmann and Pons' results didn't look anything like fusion.

Science Writers Versus Attorneys

After the participants in the March 16 meeting agreed (some reluctantly) to move forward with a public announcement and press conference, they brought in Fogle to join them. Fogle summarized some of the main problems in her interview with Gieryn:

> We had considerable disagreement with the attorneys on the content. There were some pieces of information in the draft about the experiment that would have been very helpful in terms of blunting a portion of the controversy that occurred. A few things come to mind.
>
> We had a description of how long it took the experiment to gear up. That would have resolved a controversy because people expected to set up their electrochemical cells, add the electricity, and the thing would start producing heat overnight.
>
> Well, in fact, it can take anywhere from a week to several months, depending on the size of the palladium rods. In the draft press release, there was a better description of the apparatus and the procedure, but those too got watered down.
>
> The attorneys also wanted to eliminate all of the personal details about the two scientists and how they came up with their ideas while taking a hike up Millcreek Canyon. At that point, I put my foot down. I said, "I'm sorry, this is the heart and soul of the news release, and it humanizes them. There is no legal reason that I can see why these details should be excluded." So that remained in the final version of the press release.

Patent Attorneys in Charge

Fogle went into greater detail about her difficulties when she was interviewed by Footlick. Here is the relevant excerpt:

FOGLE: Six days before the news conference, we knew what story we had, and we knew that we were going to have a news conference. So we had to write the news release, figure out which reporters to notify, figure out how to notify them, and do that within that six-day time frame. These are all the details that you ordinarily follow when you're making a major scientific announcement, all of which ordinarily lend the sort of credibility to the work that you're doing.

One of the first questions I asked at the March 16 meeting was, "Has this been submitted to a scientific journal?" And the answer was, "Yes, the *Journal of Electroanalytical Chemistry.*" My next question was, "When was it scheduled for publication?" It was scheduled for May, they told me. I asked them, "Could we wait until then?" The answer was no.

FOOTLICK: Because things are leaking out?

FOGLE: Yes. So I backed up and said, OK, usually if you're going to make an announcement like this, the journal will pull the article if you do it without approval or if you don't time it with the publication. Fleishmann called the U.S.-based editor and got that editor's approval to hold the news conference. The editor eventually moved up the publication date from May 1 to April 10.

This was a refereed journal. When we drafted the news release, we had the paper's title, the name of the journal, and the publication date. These are part of what got excised by attorneys — there were three batches of attorneys: East Coast, West Coast and to the south — who reviewed this release, which is something I've never had happened before, either.

FOOTLICK: That's crazy.

FOGLE: I've never have lawyers review a news release. They took all of that information out. In fact, they took it out entirely. The one concession that I got, knowing it had to be in there, was that — and it ended up relegated to a cryptic line on the second page — they had written a paper and it been accepted and it was to appear some time. I don't know whether we said May or what was said. But it was a concession that today I would fight over, significantly fight over. There were details of the experiment that were in the paper that were perfectly appropriate to be in the news release but were excised.

FOOTLICK: Now, the reason that you didn't do an advance release with an embargo was that somebody was afraid it would be broken?

FOGLE: Right.

FOOTLICK: Now, who was afraid of that? The lawyers?

FOGLE: Oh, no. Everyone. Brophy, the president. Everybody said, "This is too big a story." All reporters would break it. They all said it would be broken. And I knew it would be broken, too, but my point was that the reporters should have a copy of the release.

By not having a copy of the release in the hands of the science writers, they don't have anything to go on. They have their Rolodex of people that they talk to, but they had no way to know what was the content of the release so that they could bounce it off their experts. And even then, that wouldn't have helped a whole lot because this was out of chemistry and most of their experts would have been in physics, anyway.

I made a couple of phone calls to some key science writers around the country, people that I have worked with before on other stories. I had an approved paragraph that I was allowed to use, encouraged to use, and encouraged not to expand on.

My instructions were to invite them to come to the news conference. The first call I made, as I recall, was to Jerry Bishop of the *Wall Street Journal*. When he started asking me questions, I realized that, despite my instructions from the attorneys, I needed to answer his questions, and so I did to the best of my ability.

FOOTLICK: What things were beyond your control?

FOGLE: The lawyers were beyond my control.

FOOTLICK: Yes. OK, let me stop there. You said your news release was reviewed by three teams of lawyers. How did the lawyers get in so early? Or should I ask Peterson that?

FOGLE: You might want to ask Peterson.

FOOTLICK: It seems to me astonishing that you have coveys of lawyers this early.

FOGLE: I think that the administration and the lawyers were

working on this as early as January of 1989. This announcement wasn't a surprise to the lawyers. I think a lot of ground had already been covered.

FOOTLICK: But you didn't know?

FOGLE: No. I didn't know it until six days before the news conference. Brophy had told me in January that there was going to be a major story. He would not tell me who.

FOOTLICK: He wouldn't trust you to tell you that?

FOGLE: Right. In January, he would not tell me who or what department. I went back to him the next day and said, "You know, Jim, I can't do anything to prepare for this story until I have a little bit of information. I've got a science writer who's going to need to research the area. I need to know what area."

He told me at that point, toward the end of January, that there was major work going on in the Chemistry Department on room-temperature fusion. I knew about hot fusion. I did not know about any other kind. So I sent the science writer to the library, and I said, "Learn everything you can about fusion, hot or otherwise."

FOOTLICK: There wasn't much about room-temperature fusion anyway to look at.

FOGLE: Not much, but there was more than enough about hot, and you read about the hot fusion, and you can draw out some questions.

FOOTLICK: So this is a couple of months in advance when you're doing this?

FOGLE: Yes, but I had no idea beyond that what was happening. Then six days prior to the news conference, we found out all the details. The lawyers were involved early January.

FOOTLICK: I want to get back to the things that Fleischmann and Pons wanted to get done. They said they wanted 18 months more before making the announcement. Why did they submit the article if they wanted 18 months more?

FOGLE: I'm not sure that there was a whole lot in their original journal article. I'm not a chemist, but I've got a copy of the article. I read it, and of course I didn't understand most of it, but it didn't seem to me that the article was all that illuminating.

FOOTLICK: So they did have more work they wanted to do, and this was premature?

FOGLE: I know that at least Pons thought it was premature. There was the leaking-out business, but there was something that happened that I never have resolved to my satisfaction. [Fogle did not know the depth of the conflict with Steven Jones.]

FOGLE: One of the first calls to reporters I made was to Ed Yeates, who is the science reporter at KSL-TV, a CBS affiliate, Channel 5. I had worked with Ed since 1979 and knew him really well.

We talked a lot about the story coming up and what content was going to be available. We had made arrangements the morning of the news conference for Pons' laboratory to be open for still photography, for the camera crews to come up and take whatever shots they wanted to take, video, whatever they want to prior to the news conference because we knew that, after the news conference, it would be just impossible to get into the room.

So Ed and I were working on all these arrangements and talking it out. I'd already called the other two stations, and Ed also has his Rolodex, and he was calling a lot of people. One of the people that Ed relied on heavily was a researcher at Los Alamos National Laboratory.

So I'm still at my desk at about 5 or 5:30 or something on Wednesday evening, and I get a call from Ed, who says that the story is on the international wires. As far as he was concerned, the embargo was broken, and he was going to run an advance on the 6 o'clock evening news. So Ed and I agreed that, since the embargo was broken, it was fair game, he could do whatever he wanted to. I flipped on the TV news in my office, and here is a computer-generated image of the experiment, even though he had not been up to the laboratory yet. He had not even seen it yet!

I'm sitting there, and I'm going, "My lord! That's what that experiment looks like!" Ed told me that he had gotten a description of the experiment from a researcher at Los Alamos.

So now I knew that it wasn't just Utah or the local community who knew about this. There were other scientists in the United States that Stan had communicated with who knew about the experiment.

In fact, the news had traveled internationally. A few days before the press conference, Fleischmann leaked the story to the *Financial Times* of London.

"Our claim of achieving cold fusion is no more preposterous than, say, Noriega claiming his candidate won..."

Cartoon courtesy Bruce Beattie

CHAPTER 5

The News Heard Round the World

University Announces "Fusion" Discovery

C live Cookson, a journalist with the *Financial Times* of London, was the first to break the March 1989 "cold fusion" story. Looking back at the history on March 24, 2009, Cookson, who had long since lost enthusiasm for the subject, wrote that his story on "cold fusion" "would have been the greatest scoop of my journalistic career — if it had been true." He explained how he got the story:

> The weekend before Easter 1989 [March 18-19], I was visiting my parents in the English countryside. I happened to answer the phone before dinner. The caller was Martin Fleischmann, a colleague of my father's at Southampton University, where both were chemistry professors.
>
> But I hardly recognized Martin's voice — such was the tone of suppressed excitement. He told me at once that he was in America, and he did not want to talk to my father but to me — in confidence. And, once I'd assured him that I'd divulge nothing until he agreed, out poured his extraordinary tale of what became known as "cold fusion."
>
> He wanted my advice, as a science journalist, about how to reveal to the world what he was convinced would be one of the most important research breakthroughs of the century

— an unlimited new source of clean energy from nuclear fusion on a laboratory scale.

Martin was both excited and upset, because the University of Utah, where he was working on cold fusion with Stan Pons, wanted to hold a press conference in five days' time, to announce their breakthrough. Martin, on the other hand, felt the press conference would be premature, because they hadn't yet published their work in a peer-reviewed journal. I advised him to resist the university's pressure to hold a press conference, if he possibly could — while being aware that such sensational news might leak out.

On Tuesday, March 20, Martin called me again. He'd lost the battle. There would be a press conference two days later. [Fleischmann was still fighting the decision made on Thursday March 16.]

For various reasons, I could not fly out to Salt Lake City for the occasion, as Martin wanted. So he offered to give me the information in advance, under embargo, so that the FT could publish it in Friday's paper like everyone else. Only then did it occur to me that there wouldn't be an FT on Friday. Unlike American papers — or indeed most British papers — we never publish on Good Friday.

I quickly realized that I could turn this situation to my journalistic advantage, if I persuaded Martin to let me publish in Thursday's paper. And he agreed, accepting my argument that cool, calm coverage in the FT would help to set the tone for the press conference later that day. Martin faxed over some technical information, including a diagram of the Utah test-tube fusion apparatus. ...The scoop duly appeared in the FT.

Sometime around 4 on Wednesday afternoon on March 22, 1989, Mountain Time, Cookson's stories appeared on the international newswires. Ed Yeates, a broadcast journalist with KSL-TV in Salt Lake City, had already received a telephone call from the University of Utah news director Pam Fogle.

Fogle had embargoed the story for March 23, at 1 p.m. But now that the story had broken, Yeates informed Fogle, he was going forward, and running his story on the 6 p.m. news. The next morning, Jerry Bishop, whom Fogle had also spoken with on March 22, ran his story in the *Wall Street Journal.* These were the first three stories on "cold fusion."

Bishop was the first journalist to use the term "cold fusion." He knew that Jones and Rafelski had used the term "cold fusion" to describe their work, and, without realizing the differences between the two approaches, Bishop assumed that the term applied to both.

Thereafter, all news media, then the public and the scientific community, called the Pons-Fleischmann work "cold fusion." It is important to note that neither Pons and Fleischmann nor the University of Utah used the term "cold fusion" in their press release, at the press conference or in their published preliminary note. It is therefore inaccurate to call the event the "Cold Fusion Press Conference," as most historians have done.

The University of Utah Fusion Press Conference

At the University of Utah fusion press conference, Martin Fleischmann and B. Stanley Pons announced that they had accomplished, for the first time ever, sustained, controlled nuclear fusion on Earth. They also claimed a net positive power output for the first time from a fusion experiment. Pam Fogle, in a draft of an article she wrote in 1991, described the beginning of the press conference:

> On Thursday, March 23, 1989, at 1 o'clock in the afternoon just before the Easter holiday, Pons and Fleischmann announced to about 20 national and local reporters and about 100 intensely interested students, faculty and friends of the university, that they had produced nuclear fusion at room temperature. They were a little nervous, yet anxious to discuss the work that had consumed their time for the last five years.

B. Stanley Pons, speaking at the March 23, 1989, University of Utah fusion press conference. Image: KSL-TV

Author Gary Taubes spoke to many people who attended the press conference though he himself did not attend.

> By 1 p.m., the lobby of the Henry B. Eyring Chemistry Building was overflowing. Visitors were packed half a dozen deep by the rear doors. Seven television cameras stood on tripods along the back wall; klieg lights offered a hint of the surreal.
>
> The 46-year-old chemist [Pons] was wearing a conservative dark blue suit and a polka-dot tie and appeared pale. His hair was cut in a peculiar style that the *Boston Globe* would later liken to a Buster Brown and the *Los Angeles Times* to a Julius Caesar. There was something about him — perhaps the hair, or his nervous manner, or his thick eyeglasses — that made him seem like the archetypal scientific egghead. In his hands, Pons cradled a model of his cold fusion reactor, which looked like ... nothing more than a sophisticated test tube about the size of a highball glass.

Martin Fleischmann, speaking at the March 23, 1989, University of Utah fusion press conference. Image: KSL-TV

Pons cleared his throat and looked down at his notes; he spoke quickly and so quietly that his soft Southern drawl could barely be heard.

Fleischmann, 61, was visiting from the University of Southampton, where he had built a name for himself as one of the most distinguished electrochemists in the world. His thinning brown hair was combed over the crown of his head, and he wore dark-rimmed eyeglasses. He stooped slightly and spoke with a clipped and very distinctive Slavic accent that betrayed his Czechoslovakian birth and English upbringing. Fleischmann seemed tired and ill at ease that afternoon. Of the five men at the table, he may have been the only one who fully grasped the magnitude of their actions. Just that morning, he had warned one of the graduate students that their lives would never be the same after they went public. (Taubes, xvii)

Visionary University President

University of Utah President Chase Peterson introduced the research findings with poise and dignity. If he was anxious from the tumultuous weeks leading up to this moment, it was not apparent; he had arrived calm and composed.

Peterson did not promote the claim as a guaranteed solution to the world's energy problems, but he suggested that the research might someday lead there: "The full story of the research that professor Pons and professor Fleischmann will announce today will not be known for months or years, as others confirm and challenge and enlarge their ideas and their data."

What the Scientists Said

Pons was reportedly nervous speaking in front of a classroom. A press conference was sure to be torturous. Fogle wrote of Pons' reluctance in one of her accounts of the history. "Few people knew," Fogle wrote," that Pons had first insisted on a small, private room for the announcement. ... Less than one hour before the news conference, he relented, and we moved the entire setup to the building's lobby. This gave us much-needed extra room, but it also made Pons and Fleischmann look like media hounds."

Pons looked tired, he fumbled, and he was anything but eloquent. Clearly, there were other places he would rather have been at that moment. At the podium, he glanced at some hastily scribbled notes before uttering the words that set off the storm. "Basically, we've established a sustained nuclear fusion reaction by means which are considerably simpler than conventional techniques," Pons said.

Nothing else he said mattered after this. Fleischmann and Pons had known by that time, as they wrote in their preliminary note, that the "bulk of the energy release" was not from fusion. When their preliminary note eventually published, it made no difference. The words "sustained nuclear fusion" marked Pons and Fleischmann for life.

At the press conference, Fleischmann and Pons briefly explained that they knew of no chemical process that could explain the amount of heat they had measured. They had measured some neutrons and tritium, but neither of these normal fusion products was a dominant signature of their reaction. They didn't have evidence for fusion. They knew it, and so did everyone else who was familiar with nuclear fusion.

What the Scientists Didn't Say

Neutrons, had they been from fusion, would have been so abundant that they would have killed the two chemists. Nevertheless, scientists who tried to repeat the experiment went on a mostly fruitless search for neutrons. Fleischmann and Pons failed to say at the press conference that the rate of neutrons they observed was several orders of magnitude less than what would be expected from fusion reactions.

Making things worse, particularly in the absence of a published paper or a preprint, they told reporters, who became the *de facto* communicators to the scientific community, a gross misstatement. A reporter asked whether a special configuration for the electrode was required to make it work.

"At this stage," Fleischmann said, "no special constraint on the design parameters." Someone else in the audience asked how high the current needed to be. "We have been up to about half an ampere per square centimeter so far," Fleischmann said, "but there are no special technical restrictions on the design."

No communication could have been further from the truth or more disastrous. Fleischmann and Pons had developed the specific cell design, materials preparation, and operating parameters over five years.

What the Administrators Said

Halfway into the press conference, Jim Brophy, the vice president for research at the University of Utah, mentioned Fleischmann and Pons' preliminary note.

"It is being submitted to a journal," Brophy said. "We have chosen to have a press conference this afternoon, frankly because the results are so exciting, to set the record straight, so to speak. The scientific journal paper will have much more detail than you are hearing this afternoon."

The first part of Brophy's comment was wrong and counterproductive. The manuscript was not "being submitted"; it had been submitted, and it had been accepted for publication. Brophy had contributed to the misunderstanding that Fleischmann and Pons had announced their claim in a press conference before their preliminary note was accepted in a peer-reviewed journal.

The second part of Brophy's comment was a lie. The press conference was not taking place because the results were "so exciting." It was taking place to secure the university's intellectual property rights in response to the perceived threat of Steven Jones.

Selling Fusion

Later in the press conference, Peterson moved on to the idea of technology. He was transparent about his and the university's motives:

> We would also like the discovery to benefit of the economy of Utah. "That's not always easy to guarantee because ideas aren't contained by borders, but perhaps ownership and patents are, if indeed there is ownership and there are patents to come out of this. So, if that's the case, we would do all in our power to encourage the further development and the further research of this, in the university, in the universities of Utah, and in commercial operations that would grow up around this.

The University of Utah press release invited interested commercial parties to contact the university's director of technology transfer. Although Fleischmann and Pons caused themselves great harm by claiming "fusion," the promotion of "cold fusion" as technology was the university's, not theirs.

What the Media Reported

Within hours, television news media worldwide delivered images of test tubes superimposed on images of atomic bombs and the fiery surface of the sun. Here's some of what the public heard:

Ted Koppel/Jim Slade (ABC-TV)
"It is a process described as ridiculously simple, which could eventually provide the world with a cheap, easy and clean source of energy."

Tom Bearden (*MacNeil/Lehrer News Hour*)
"A simple tabletop experiment, using materials found in any college chemistry lab, an unlikely and unexpected setting for what could be the most important discovery since fire."

Paul Recer (Associated Press)
"The control of nuclear fusion, the energy secret of the sun and stars, has been the golden fleece for nuclear physicists for more than 30 years. Thousands of scientists in every industrialized nation of the world have spent decades trying to achieve what many believe is the most nearly perfect source of energy possible."

Ian H. Hutchinson (fusion researcher at MIT, quoted by the Associated Press)
"Suppose you were designing jet airplanes, and then you suddenly heard on CBS News that somebody had invented an anti-gravity machine. That's the way we feel."

Science Celebrities

Frank Close, theoretical particle physicist and author of the book *Too Hot to Handle: The Race for Cold Fusion*, wrote that Fleischmann and Pons were treated like celebrities. "Rarely has serious talk of Nobel prizes, fame and fortune followed so fast after a claimed scientific breakthrough," Close wrote. "Before March 23, few people outside of the world of electrochemistry knew Fleischmann and Pons. By the evening, they had become instant celebrities."

Pons' phone rang nonstop. As soon as his secretary hung up one call, the phone would ring again. Perhaps one of the most interesting calls Pons received was from Edward Teller, a pioneer in thermonuclear fusion research and a former director and co-founder of the Lawrence Livermore National Laboratory.

Pons sent a copy of the preprint to Teller, and within 24 hours Teller was in the newspapers and on television, supporting the claim. A day after the press conference, he spoke favorably about the claim to the *Los Angeles Times*. "Initially, my opinion was that it could never happen," Teller said. "I'm extremely happy now, because I see a very good chance that I was completely wrong. The experiment sounds extremely promising."

Teller told the *Times* the day after the press conference that scientists at Livermore had met with him and were mapping out plans to repeat the Fleischmann-Pons experiment.

Maugh and Dye of the *Los Angeles Times* reported that "graduate students took a news story about the discovery and pasted it onto the front page of the *National Enquirer*, along with a story headlined 'Baby Born With a Wooden Leg.' Copies were pasted all over the chemistry building."

Fleischmann was on his way back to England that Friday, but Teller tracked him down before he left the U.S. Fleischmann had boarded a plane in Salt Lake City bound for San Francisco (normally, a two-hour flight), where he was to make a connecting flight to London. When Fleischmann told me about this in 2003, he said that, while they were in the air, the plane seemed to be taking an unusually long time to reach San Francisco. He ended up missing his connection.

He gave more details when Close interviewed him in 1990. Fleischmann rebooked his flight for Saturday and settled into his hotel room, thinking that he could have a moment's peace because nobody knew where he was. As he entered his room, the phone rang. He was bewildered, "Who on earth knows I am here?" It was none other than Teller, who had found him and was eager to speak with him about "cold fusion." (Close, 112)

It Didn't Look Like Fusion

The experimental results that Fleischmann and Pons obtained, while highly suggestive of an unknown nuclear process, did not resemble the products of nuclear fusion. The fact that they were still alive was testament to this. They knew this when they submitted their manuscript to the *Journal of Electroanalytical Chemistry* on March 11.

"The most surprising feature of our results, however," Fleischmann and Pons wrote, "is that reactions observed in thermonuclear fusion are only a small part of the overall reaction scheme and that the bulk of the energy release is due to a hitherto-unknown nuclear process or processes."

Two days later, on March 13, a perplexed Fleischmann sent a fax to his colleague David Williams, at Harwell. "The intriguing thing," Fleischmann wrote, "is that the rate of tritium generation is much less than corresponds to the heat production. Neutron production is still lower. What on earth is going on? Are we seeing some strange neutron + lithium-7 reaction system?"

D+D Fusion Branching Ratios

According to the well-understood theory of deuterium-deuterium nuclear fusion, the reaction paths occur through one of three possible branches. The first branch produces helium-3 and a neutron. The second branch produces tritium and a proton. The third branch produces helium-4 and a gamma ray. In D+D fusion reactions, a neutron is produced, on average, almost 50% of the time, tritium is produced almost 50% of the time, and helium-4 is produced less than 1% of the time. Since the discovery of D+D fusion, these ratios have not changed.

Deuterium+Deuterium Fusion

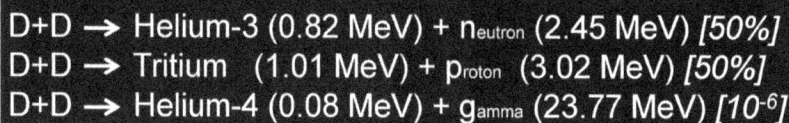

$$D+D \rightarrow \text{Helium-3 (0.82 MeV)} + n_{neutron} \text{ (2.45 MeV) } [50\%]$$
$$D+D \rightarrow \text{Tritium (1.01 MeV)} + p_{proton} \text{ (3.02 MeV) } [50\%]$$
$$D+D \rightarrow \text{Helium-4 (0.08 MeV)} + g_{amma} \text{ (23.77 MeV) } [10^{-6}]$$

Shamed for Press Conference

A poorly researched National Science Foundation-sponsored project at the University of California, Berkeley, examines the typical judgment that Fleischmann and Pons were unethical and had breached scientific conduct for announcing their research in a press conference. (Caldwell and Lindberg, 2010)

There is nothing wrong or unethical about announcing a science claim in a press conference or in a press release, after a manuscript is accepted in a peer-reviewed journal. For example, on Tuesday, Jan. 15, 1957, Richard Garwin, a physicist with IBM Corporation, who was also an associate professor at Columbia University, along with his colleague, physics professor Leon Lederman, announced a discovery about weak interactions among decaying particles at a Columbia University press conference.

Like Fleischmann and Pons, Garwin and Lederman had not published their manuscript at the time of the press conference. However, Garwin and Lederman's manuscript had been neither peer-reviewed nor accepted. This fact is evident because the journal received the manuscript the day of the press conference.

Based on a survey of newspaper archives, there was no outrage against Garwin and Lederman and no accusations of unethical or unscientific behavior. (Garwin et al., 1957) Certainly, the Garwin-Lederman press conference was not unique in the world of science. However, significant factors distinguished the Garwin-Lederman press conference from the Fleischmann-Pons press conference.

Garwin and Lederman were physicists, working in the field of physics. Fleischmann and Pons were chemists, encroaching on the field of physics.

Garwin-Lederman's discovery was about an esoteric property of particle physics, which they claimed had little practical value at that time. Fleischmann-Pons' claim was for fusion, a concept well-known among scientists and the general public. Moreover, its potential global impact was immense. Garwin and Lederman's idea was not novel; they were simply confirming a plausible theory that had been proposed by other scientists.

Fleischmann and Pons, on the other hand, were proposing a new hypothesis as well as claiming experimental support for it. Their idea appeared to contradict fundamental principles in physics.

Last but not least, the Fleischmann-Pons claim — for people who were willing to look past the theoretical discrepancy — presented an imminent threat to continuity of funding for the multibillion-dollar international fusion research alliance with industry, academia, and public funding sources.

President Peterson's Defense

Many years after the press conference, University of Utah president Chase Peterson tried to get me to rewrite that part of the history. In 2003, when I met Fleischmann, he reiterated to me that he was "not at all in favor of the high-publicity route adopted by the University of Utah." He wrote to me that the university made it clear to him that he "had to appear supportive of their position." After I published Fleischmann's comments in my 2004 book, Peterson sent me a letter saying that the blame for the press conference belonged to Pons and Fleischmann. Peterson speculated about what might have happened:

> The university, and I myself, supported Stan and Martin's decision to announce. It would have been impossible for the university to force Martin to call the press conference or to somehow force him to withhold his data from his colleagues. Within a university, and certainly within the University of Utah, the standard of academic freedom would make it impossible, insulting, and legally unsustainable for any dean or president or administrator of any stripe to tell or somehow 'order' any member of the faculty to publish or report anything, research or otherwise, that the faculty person chose not to publish or report.

Peterson's account of what might have happened is not consistent with what did happen. There are too many accounts that contradicted him, not the least of which is the statement by Fogle to Jerry Footlick that she probably would have "been looking for another job" had she been more adamant during the planning meeting.

There is also the account of Dehlinger, reported by Taubes, who "later supposed that the decision to hold the press conference might have gone the same way even if everyone but Peterson" had been opposed to it.

Certainly, Fleischmann and Pons agreed and went along, however reluctantly, with the press conference. This does not mean that they both did so without coercion or pressure. Peterson was clearly a man of vision with a brilliant understanding and appreciation of science. Despite his strong support for Fleischmann and Pons, however, his drive to promote "cold fusion" got the better of him.

In 1990, Peterson transferred half a million dollars from the university to the university's National Cold Fusion Institute and said the money came from an anonymous donor. Hugo Rossi, the dean of science, stopped short of calling it fraud and called it an "apparent deception."

The funds transfer was the last of several actions by Peterson that outraged the faculty. On June 4, 1990, the *Deseret News* wrote that the "academic senate called into question Peterson's ability to lead the university."

"In an overwhelming vote," the *News* wrote, "the senate passed a resolution asking the state Board of Regents and the University Institutional Council whether it was in the best interest of the university for Peterson to continue at the helm." A week later, Petersen announced that he would retire during the following academic year.

CHAPTER 6

The Case of the Missing Lab Books

Pons Suspects Graduate Student

The University of Utah got the attention it wanted from its March 23, 1989, press conference. In the following days, the university news office logged 1,500 phone calls. Within a few weeks, the Technology Transfer Office signed nondisclosure agreements with 40 companies that wanted a private look at the five as-yet-unpublished patent applications.

Disappearing Slides

Not everybody used legal methods to approach the University of Utah for additional information. In the hours following the Thursday afternoon press conference, someone stole Stanley Pons' slides, according to a March 29 news story by JoAnn Jacobsen-Wells in the *Deseret News.* Jacobsen-Wells quoted University of Utah spokeswoman Barbara A. Shelley, who said that the transparencies "had to have been stolen." Shelley said, "They were not mislaid."

"Hundreds of people around the world," Jacobson-Wells wrote, "are hungry for any tidbits of information about the simple lab apparatus that could make scientific history." The slides were never reported found.

Disappearing Lab Books

By Sunday night, according to Taubes, Pons realized that his lab books had disappeared too. Their lab assistant became the prime suspect.

Back in October 1988, Pons had asked 27-year-old graduate student Marvin Hawkins to take on the laborious task of operating the electrolytic cells and taking data. It was a thankless job, particularly because he was confined to an isolated, windowless laboratory several stories below ground. But this worked out well for Pons, who was doing his best to try to keep the project secret. Hawkins had earned Pons' trust, and Pons came to depend on him not only to run the experiments but also to help purchase equipment on Pons' behalf so Pons could, as much as possible, stay below the radar.

There are two versions of the story. Both include Hawkins removing the lab books from campus and driving them down to Provo to give to his brother, to put in his brother's safe deposit box. According to Hawkins, Pons requested that Hawkins remove the lab books to keep them safe. According to Pons, Hawkins removed the lab books without his knowledge and permission.

Written Out of History

Even though Hawkins had worked on the experiments for months, on Tuesday, March 21, two days before the "fusion" press conference, he was shocked to learn from Pons that his name would not appear as an author of the Fleischmann-Pons preliminary note. Pons also told Hawkins then that he would not, contrary to what he expected, be participating in the press release or news conference. Here's Taubes' account:

> After doing at least a good part of the lab work on cold fusion, Hawkins had reason to expect he would be listed as an author on the seminal paper, which would have meant a share of the Nobel Prize and any royalties. Now his contribution would not even be acknowledged.
>
> "Stan said that the university had decided that I had to play a lower-key role in the announcement," Hawkins explained. "The lower-key situation was that I wasn't going to be involved in the press release and that they decided not

to include my name on the paper."

Initially, he shrugged it off. "And then I was walking across the campus," Hawkins said, "and it just dawned on me like a brick that they had written me out."

When I interviewed Hawkins in 2014 and asked him how Pons explained the exclusion of his name from the preliminary note to him, Hawkins was vague, but he pointed in the direction of the university. "I said, 'Hey! What's going on?' And they said, 'Well, the university this and the university that,'" Hawkins said.

There are two other sources of information that shed light on this. The first is a message posted to an electronic newsgroup later that summer by Keith Calvert Ivey, a Yale University student. Ivey had attended a lecture on "cold fusion," and he heard something from a former University of Utah physical chemistry professor (he did not remember the name) who was now at Yale but had just returned from a visit to Utah. Here is Ivey's account:

> The chemistry professor said that one of the main reasons that things were such a mess was the behavior of the University of Utah administration (one of the reasons he left there). He said that the university was trying to keep as many details secret as possible because of patents. He also said that the administration had kept Fleischmann and Pons from putting the name of their student on the paper as an author so as to decrease the number of people involved in possible patent claims.

A memoir from Cheves Walling, a distinguished professor of chemistry at the University of Utah, supports this. Walling had written about a conversation he had with Hawkins in 1989.

"I suggested to him," Walling wrote, "that this might have been a policy matter instigated by the patent lawyers, and asked Hawkins if he considered himself one of the inventors of 'cold fusion.' He said, not at all, he was only interested in the scientific credit, and indicated that he would be willing to sign a paper to that effect." Considering the aggressiveness of the university, Walling's explanation is the most plausible.

Historical accounts have put the blame for the omitted author name exclusively on Fleischmann and Pons. But the quotes from Ivey and Walling shed light on the mystery and are not discussed in those accounts. Other historical accounts have not offered any plausible explanation for why Fleischmann and Pons would have omitted Hawkins' name.

Hawkins was disgruntled, as he told Taubes. "When somebody says, 'Oops, we screwed up. We inadvertently omitted one of our co-authors' names, and we apologize,' that's absolute bull," Hawkins said.

Safeguarding or Stealing?

When I interviewed Hawkins, several inconsistencies in his story became apparent. He told me that he put the lab books in his brother's safe deposit box "a couple of days before the press conference." But Hawkins told Taubes that Fleischmann and Pons asked him after the press conference to safeguard the lab books.

At an unknown point, Hawkins made copies of the lab books. Taubes' account of when Hawkins did this is confusing. On the one hand, Taubes' narrative suggests that Hawkins made copies and left them in the lab after the Thursday press conference and before he put the originals in the safe deposit box. This would have been between Thursday and Saturday. On the other hand, Taubes' narrative suggests that Hawkins left the copies in the lab the following Monday. (Taubes, 1993, 131) Either way, it conflicts with what Hawkins told me.

Perhaps more significant is Hawkins' dubious suggestion that Pons would entrust Hawkins with Pons' most precious possession, after Pons had told Hawkins that his name would not be on the preliminary note, nor would he participate in the press release or news conference.

Hawkins was up in the mountains skiing at Park City on Sunday. When Pons realized on Sunday that the lab books were gone, Pons decided to bring in Mark Anderson, a postdoctoral researcher, to take over the lab work from Hawkins.

Pons spoke with Hawkins on Monday morning and demanded that he return the lab books immediately. But Hawkins said they were in his brother's safe deposit box, and he couldn't get there that day.

Threatened With Arrest

When Pons arrived at his laboratory Tuesday morning, he found only the photocopies and a note from Hawkins, according to Taubes. Pons spoke with Hawkins on the phone again and demanded that he come to the lab right away.

Hawkins told Taubes that, when he and his wife got to Pons' office, Pons said that there was a warrant out for his arrest and that he had better get the books back to him immediately. Hawkins told Taubes that he then consulted an attorney. "The lawyer," Taubes wrote, "told Hawkins to keep the lab books as long as Pons had been given the photocopies."

The fact that Hawkins relayed to Taubes the apparent advice of an attorney to maintain possession of lab books that did not belong to him is suspicious.

Books at Church Office?

Taubes wrote that Pons, after a week of frustration trying to retrieve the books, got a call from someone at the Mormon church headquarters who said the lab books were in the church office and asked Pons to come get them. Taubes got this information from Hugo Rossi, the dean of the College of Science. It is certainly one of the strangest facets of this story, but unfortunately I've been unable to locate additional references that either confirm or deny this fact. Alternatively, I have seen no other accounts that describe how or when Pons inevitably got the lab books back. Fleischmann mentioned the lab book event in a letter to his friend John Bockris on Feb. 7, 1991.

> Let me put you straight about our dear friend Marvin Hawkins. It was he who removed the results to store them in

a safe in southern Utah, so much so that Stan and I originally lost all the prime data. It was only in response to the more dire threats that Marvin produced the notebooks, which we agree were locked away.

Many years later, when Fleischmann was near the end of his life and Pons had long removed himself from the media spotlight, Hawkins was asked to appear in the documentary film about "cold fusion." Oddly, in addition to affirming the group's work, Hawkins also professed deep loyalty to Fleischmann and Pons. "What we did was correct," Hawkins said. "I will defend them at every turn."

Reprinted with permission. © *Bill Griffith*

CHAPTER 7

The Man Behind the Curtain

Garwin's Influential Role

On March 24, Fleischmann and Pons submitted to the journal *Nature* a manuscript nearly identical to the one they had already submitted to the *Journal of Electroanalytical Chemistry*. The version for *Nature* was never published. They abandoned that effort in mid-April after reviewers demanded more revisions than Fleischmann and Pons were willing to make. One of those reviewers was Richard Garwin.

After completing his Ph.D. in physics at the University of Chicago in 1949, Garwin joined the physics faculty there. A few years later, he went to Los Alamos National Laboratory to work on Edward Teller and Stanislaw Ulam's concept for a hydrogen (fusion) bomb.

"In the early 1950s," Teller said, "when I had the first crude design of the hydrogen bomb, Dick Garwin came to Los Alamos and asked me how he could help. Actually, the design I had in mind was not that of a real bomb but of a model for an experiment. I asked Garwin to change this crude design into something approximating a blueprint." (Teller, 1986) At 24, Garwin improved the design, code-named Ivy Mike, and it was constructed and detonated in 1952 from Enewetak, a small atoll in the Pacific Ocean.

After his achievement with the bomb design, he became an advocate for arms control and nuclear disarmament. He has been active in politics and has often been called on to advise the federal government and many presidents. He spent most of his career working at IBM's Thomas J.

Watson Research Laboratory in Yorktown Heights, New York, and was the director of the lab from 1968 to 1969.

Garwin is one of the most well-respected experimental physicists in the United States. He has had a long affiliation with the secretive program known as JASON, funded by the Department of Defense to investigate and advise the federal government on cutting-edge science and technology matters.

When the "cold fusion" news broke, Garwin quickly became one of the most influential people who worked primarily behind the scenes to dispute and deny the experimental data. Teller, in contrast, publicly supported the research.

IBM Attempts to Replicate

On March 23, 1989, when Garwin heard about the "cold fusion" claim, he immediately made five pages of calculations, trying to figure out if, and how, room-temperature fusion might be possible theoretically. He was on the phone the same day with one of his colleagues at the Yorktown Heights laboratory, physicist James (Jim) F. Ziegler, the manager of Radiation Science at the lab. Ziegler told him that he had all the required materials at the IBM lab; Ziegler began "cold fusion" experiments.

On March 27, Ziegler posted a message to an internal IBM electronic newsgroup. "We have tried to duplicate this experiment without success," Ziegler wrote. "We would appreciate hearing from anyone who learns of more details about the experiment and how to conduct it."

Ziegler, like many people, was missing details of the experimental protocol. But even if he had a copy of the Fleischmann-Pons preliminary note, he would have failed because the two chemists omitted three crucial details, which I will discuss in Chapter 10.

Another significant event took place on March 27. Garwin wrote a letter to his colleague Tsung-Dao Lee at Columbia University. Lee and his colleague Chen-Ning Yang had made a discovery in particle physics theory that later earned them a Nobel Prize. Garwin and his colleague Leon Lederman were among the people who repeated the initial experiment and thus confirmed Lee and Yang's theory. Lederman later received the Nobel Prize for another achievement.

The letter to Lee shows that Garwin knew with certainty that the main effect observed by Fleischmann and Pons was the excess heat. He knew that the emitted rate of neutrons reported by Fleischmann and Pons was orders of magnitude lower than what would have been expected from nuclear fusion. Here's the relevant paragraph from Garwin's letter to Lee:

> What is clear is the incompatibility between the reported 4 watts excess heat (over 100 hours or more), claimed by the University of Utah group and the few (if any) neutrons they found. A watt of 2.5 MeV neutrons is about 2.5 x 10**12 per second! At 60 Curies of neutrons (with a relative biological effectiveness of 10), the investigators would be dead from an hour's exposure at one meter distance.

As evidenced by an April 14, 1989, letter to his friend Hans Bethe, a 1967 Nobel Prize winner in physics, Garwin knew that Fleischmann and Pons had claimed only 4,000 neutrons per second rather than 2.5 x 10**12 (2,500,000,000,000.) Bethe had won the prize for his contributions to the theory of energy production and nucleosynthesis in stars.

Had the Fleischmann-Pons experiment actually been the result of a fusion process, they, or their graduate student who performed most of the hands-on research, would have been killed. The lack of neutrons was so obvious that a grim joke called the "dead graduate student problem" quickly surfaced. The "problem" was simply that there was no report of a dead graduate student in the Pons' laboratory. To physicists, this meant that there were no strong neutron emissions. Scientists who could not imagine the possibility of a new nuclear process therefore thought that no dead body means no neutrons. No neutrons mean no validity to the Fleischmann-Pons excess-heat claim.

Although Garwin knew that the reported rate of neutron emissions from the Fleischmann-Pons experiment was far less than the reported generation of excess heat, if it was fusion, he gave specific directions to IBM staff members who were trying to confirm the Fleischmann-Pons experiment. "PLEASE DO NOT TRY TO OBSERVE FUSION BY ITS HEAT EFFECTS," Garwin wrote on March 30. "Neutron detection is a billion times more sensitive."

No advice about the Fleischmann-Pons experiment could have been more misguided. But someone with an e-mail address of WThorne at a Boston IBM location, who we discovered was Bill Thorne, president of Inventu Corp., was paying more attention to the news and told the newsgroup more useful information:

> In the most recent appended message, a comment was made by one of the Utah experimenters (or an associate) regarding the non-lethal quantity of neutrons observed. It was simply that the process that is occurring in their experiment is one that is a new type of fusion (previously unknown or not predicted).
>
> Maybe fusion and fission aren't the only possibilities? How many variants are possible in each? It may be a hoax or just some kind of mistake, but if not, it won't be the first discovery of something that the scientific establishment considered impossible!

On March 31, 1989, someone sent a message to the IBM newsgroup and passed on information, given by Pons in a lecture that day, that the palladium cathodes had to be "electrolyzed for weeks to prepare" in order to work. Ziegler's expectation to see positive results after running an experiment for only three days was grossly unrealistic.

The same day, someone also posted on the IBM newsgroup the announcement from the Pons lecture. It included a sentence from the Fleischmann-Pons JEAC preliminary note: "We conclude that the conventional deuterium fusion reactions are only a small part of the overall reaction scheme and that other nuclear processes must be involved." Anyone on the IBM newsgroup knew not to look for the expected products of fusion.

Nature Manuscript

Three editors at *Nature* handled the March 24, 1989, Fleischmann-Pons manuscript submission. The first was Editor-in-Chief John Maddox. The second was Washington, D.C.-based editor David Lindley and the third was U.K.-based editor Laura Garwin, Richard's daughter.

When she first saw the manuscript, Laura Garwin was immediately concerned that Fleischmann and Pons hadn't reported data and results from light-water experiments. She assumed, incorrectly, that Fleischmann and Pons had done no control experiments. As Fleischmann and Pons wrote in their preliminary note, they had performed "blank experiments using platinum cathodes." They explained later that running the experiment with a platinum cathode was the best "blank."

This control provided the validation for their experimental claim of excess heat. It did not, however, provide the validation for their theoretical claim of deuterium-deuterium fusion. As it turned out, the phenomenon was not deuterium-deuterium fusion, and experiments with hydrogen as well as deuterium could produce excess heat.

From Laura Garwin's perspective, the lack of apparent controls seemed amateurish, as she later told author Gary Taubes.

"I was extremely surprised that the JEAC [version] got published," Laura Garwin said. "It didn't have the elementary control experiment. The obvious thing you see when you look at that paper is, Why didn't they do it with H_2O? Any high school student could've refereed it, because of that obvious [missing] ingredient of the scientific method."

The second problem with the JEAC and the *Nature* manuscripts is that Fleischmann and Pons included a graph of a gamma-ray spectrum that was obviously wrong to any nuclear physicist. I will discuss the gamma-ray matter in Chapter 17. During the next five weeks, these two vulnerabilities — control experiments and the gamma-ray spectrum — became the primary targets for Fleischmann and Pons' critics.

I do not know when Richard Garwin received the Fleischmann-Pons *Nature* manuscript to review, but I know that he sent his comments to Lindley on April 2. As is customary with anonymous peer-review,

Garwin could not tell anybody, including his peers who were attempting experiments, that he was a reviewer and that he had a copy of the Fleischmann-Pons *Nature* manuscript.

On March 30, 1989, Pons gave out preprints of his and Fleischmann's revised JEAC preliminary note, marked with bold lettering, "Confidential," to five of his colleagues. Within a few days, the preprint had been circulated around the world. E-mail was in its infancy, but fax machines had become widely popular. The preprint was faxed and re-faxed, and then again to so many people that, in some cases, the original text became so faded that the only word that remained visible was "Confidential."

Garwin received the JEAC preprint on April 3, and, as he wrote in his letter to Lindley, the papers were nearly identical. "There is nothing in the *Nature* manuscript," Garwin wrote, "that is not in this one — mostly in the same words."

In his comments, Garwin speculated about a mechanism that could explain away Fleischmann and Pons' excess-heat results as an ordinary chemical reaction, though he wrote out his speculation as if it were a proven certainty.

However, in physics, an area in which he was well-qualified, he noticed something wrong: It concerned the gamma-ray peak given in the manuscript. The error was serious, and Garwin and other critics used this error to support their position that all the Fleischmann-Pons claims, including the heat measurements, were wrong. Some critics said the gamma-ray problem was not simply an error but outright fraud.

Invitation From Italy

Within days of the March 23, 1989, Fleischmann-Pons announcement, a friend and colleague of Garwin's, Antonio Zichichi, the founder and director of the Ettore Majorana Centre for Scientific Culture in Erice, Sicily, Italy, and a major figure in Italian and European physics, began planning a one-day conference on "cold fusion."

Zichichi is a Sicilian nuclear physicist who is well-known to the Italian public, and he is a great popularizer of science. Among his prominent positions, he was the president of one of Italy's premier nuclear research laboratories, Istituto Nazionale di Fisica Nucleare.

Despite his background in conventional nuclear physics, Zichichi was open-minded, even enthusiastic about "cold fusion." He asked Garwin to help him organize the conference and bring in speakers. Garwin at first accepted but, on April 5, tried to convince Zichichi to cancel the conference, telling him that none of Garwin's contacts had seen any positive results.

"Dear Nino, Garwin wrote, "regarding your proposed Fusion Forum in Erice, I report that it does not look good. ... Nobody has seen a neutron or gamma-ray and certainly no heating, so they are reluctant or downright unwilling to come to Erice now."

Zichichi was not persuaded to cancel the conference. He responded, "Dear Dick, The meeting in Erice CANNOT be cancelled. You must come because the date has been chosen for you to be there. All other points you mentioned are DETAILS which cannot cancel the meeting. Ciao, Nino."

Koonin's Attempt to Increase Electron Mass

Note: From time to time, I present optional short sections with technical information for readers who are curious about the deeper scientific details. These sections will appear with shaded background like this.

Two days later, on April 7, Garwin sent half a dozen invitations to U.S. scientists — travel expenses courtesy of Zichichi — to attend the Erice workshop. Among the invitees was Steven Koonin, a theoretical nuclear physicist at Caltech. That very day, Koonin and his colleague Michael Nauenberg submitted a manuscript to *Nature* in which they tried to provide a theoretical explanation for both the Jones and the Fleischmann-Pons experiments.

In the process of re-examining theoretically calculated rates for deuterium-deuterium fusion, Koonin and Nauenberg were surprised to find that the reaction rate was 10 orders of magnitude higher than the previous estimate. Even with this correction, however, Koonin and Nauenberg's new estimated reaction rate was substantially lower than what would be necessary to explain the implicit reaction rate that would be commensurate with the heat reported by Fleischmann and Pons, if it was from deuterium-deuterium fusion. In order to further increase the deuterium-deuterium fusion reaction rates through electron screening in the bulk lattice, they performed calculations based on ad hoc assumptions for various values of increased electron mass.

But they couldn't figure out how Fleischmann and Pons got so much heat. "We also find," Koonin and Nauenberg wrote, "that hypothetical enhancements of the electron mass by factors of 5-10 are required to bring cold fusion rates into the range of values claimed experimentally. However, we know of no plausible mechanism for achieving such enhancements."

Although there was general agreement that the idea of increased electron masses could substantially increase deuterium-deuterium fusion rates, at that time no one was able to propose detailed physics that would increase the masses of electrons inside the bulk metallic lattice.

On April 11, Garwin sent Bethe a copy of the Koonin and Nauenberg "cold fusion" theory paper, along with Garwin's own calculations of fusion rates at room temperature. More than colleagues, he and Bethe, and their wives, were social acquaintances. On April 17, Garwin sent Bethe another letter with a draft of his forthcoming *Nature* editorial.

"The astounding claim of Fleischmann," Garwin wrote, "however, is that they obtain the steady evolution of excess heat of some 10 watts per cubic centimeter for 100 hours or more — 4 MJ/cc or 600 eV per atom."

Garwin didn't believe it. Garwin (and Bethe) certainly knew fusion. They were the experts. Fusion was their turf. But Zichichi's plans for the meeting in Erice were in full swing; Garwin had little choice but to go along with them. Before we get to the Erice workshop, we have to backtrack a bit to March 28 and March 31. Fleischmann spoke at two other meetings before he got to Erice.

CHAPTER 8

A Pleasant Break in Europe

Fleischmann Gets Respect on Tour

After Fleischmann left Utah on March 24, Pons was left to contend with a deluge of attention, most of it unwanted. His phone was ringing nonstop. The slides he displayed during the press conference had been stolen. Three days later, he learned that his lab books were missing. Pons called his graduate student, Marvin Hawkins, who said he had taken them for "safekeeping."

Fleischmann, meanwhile, was enjoying relatively pleasant scientific camaraderie in Europe. Contrary to what previous writers have assumed, Fleischmann did not "flee" from Utah in reaction to the press conference. Pam Fogle, the University of Utah news director, explained the correct sequence of events. The press conference had been scheduled around Fleischmann's calendar because he had plans to return home for the Easter holiday.

On March 28, Fleischmann gave a lecture at the U.K. Harwell Atomic Energy Research Establishment laboratory, where he still worked as a part-time consultant. Harwell was the primary laboratory for nuclear energy research in the U.K.

On March 31, Fleischmann gave a lecture at CERN, the European Organization for Nuclear Research, one of the world's largest and most prestigious particle physics laboratories and now the home of the Large Hadron Collider.

On April 12, Fleischmann spoke at the first conference on "cold fusion," a one-day event held at the Ettore Majorana Centre for

Scientific Culture in Erice, Sicily, Italy, organized by Antonio Zichichi and Richard Garwin.

Harwell Scientists: "It's Wrong"

Fleischmann's first lecture was at the Harwell atomic energy lab. Frank Close described the meeting in his 1991 book *Too Hot to Handle: The Race for Cold Fusion.* Twenty people attended what Fleischmann thought would be a private meeting in which he explained his and Pons' claim. The attendees included the ranking heads of various divisions at Harwell. Close did not attend the meeting, but Ron Bullough, the chief scientist of the lab, did and later shared documents and details with Close.

The meeting had little impact on the brewing controversy except for one critical moment: Fleischmann displayed a graph of what he believed to be the energy peak of a gamma-ray spectrum emitted from his and Pons' experiment. Immediately, some scientists in the room said matter-of-factly, "It's wrong." For physicists, it was obvious that the gamma peak was wrong, and they explained this to Fleischmann.

This gamma peak, depicted at an energy of 2.5 MeV, should have been located at 2.2 MeV, based on neutron capture reactions by protons in the water. When he investigated the matter later, Close deduced that Fleischmann called up Pons right away and told him they needed to correct the graph. Chapter 17 will discuss the details of the gamma peak and why it was significant.

Standing Room Only at CERN

Three days later, on March 31, Fleischmann presented his lecture to a packed crowd in the main auditorium at CERN.

The gamma-ray peak was now at 2.2 MeV. The attitude toward "cold fusion" was still friendly and optimistic. Allen Lewis, a researcher at Los Alamos National Laboratory, sent a humorous message to an electronic newsgroup implying that "cold fusion" might pose a threat to thermonuclear fusion research.

"For sale: slightly used Tokomak," Lewis wrote. "Made in 1976, Korinsky octopole confinement design. Less than 1,000 hours' experimental use. Power requirements: greater than 20,000 Gauss flux. $3 million, or best offer. If you call, ask for Al."

Fleischmann lecturing in main auditorium at CERN. Image: CERN Courier.

Fleischmann's talk at CERN was well-received, according to Jon Caves, a computer programmer who worked at CERN:

> This was a strictly scientific seminar. No questions were allowed on the nonscientific aspects of the talk. In fact, the camera crews of various TV stations were asked to leave before the talk began. But they were given a chance to interview Fleischmann after the seminar.
> The efficiency of their cell is "miserable" [quoting Fleischmann], and their best result was 111% of breakeven, that is, 10% more power out than power in, but they predict that, with a properly designed cell, their efficiency could be

over 1,200%, that is, 10 times out what they put in.

I personally could see nothing wrong with his explanation of the phenomena. There is no known chemical reaction which can produce the amount of energy involved. It has to be nuclear fusion. Whether or not this is going to have any practical use is still to be seen.

As Fleischmann said, a lot of work now has to go into understanding why and how this is happening. There were some very worried theoretical physicists leaving the hall after the talk, and there were mumbles about rewriting the theory of quantum mechanics.

Douglas Morrison, Expert in Wrong Results

If Garwin was the hub for "cold fusion" communications among U.S. physicists, Douglas Morrison, a high-energy particle physicist at CERN, was soon to become the hub for Europe.

Morrison was born in 1929 in Glasgow, Scotland. Undoubtedly brilliant, at 15 he obtained the Scottish "Certificate of Fitness," which qualified him for university entry, though he was too young to enter a university. Instead, he took on a research job in a dye and color factory. His expertise in industrial research on rockets and explosives earned him an exemption from military service. Morrison worked at CERN from 1956 until he died in 2001. He was described by his colleagues as a "distinguished and conscientious physicist and a popular figure at CERN." (Schmid and Kellner, 2014)

Before "cold fusion," Morrison had written an electronic newsletter that discussed topics he thought were of interest to other physicists. It was today's equivalent of a blog. His newsletters weren't secret, but they were primarily intended for physics "insiders." When "cold fusion" came along, it immediately became his central focus. When the "cold fusion" news broke, he had a worldwide distribution network in place. According to Morrison, he was "possibly the first person to observe [thermonuclear] fusion in Europe." Unless otherwise noted below, all quotes from Morrison are from his newsletters.

In the first few months of the "cold fusion" story, Morrison had the only regular, organized publication about "cold fusion" news. His network also worked to keep him well-informed because his readers tipped him off to news, particularly in Europe.

Not only was Morrison a broadcaster of sorts, he was also a uniquely qualified scientific analyst. These factors made him one of the most influential voices worldwide on the topic of "cold fusion."

"I had given several serious lectures on 'Wrong Results in Physics,'" Morrison wrote, "and found that they exhibited certain characteristics so that they could be recognized before they had been proved wrong. After the press reports of the Utah announcement, I wondered if this was a case. But after I had heard Fleischmann's talk at CERN, I was inclined to believe that his results were correct."

Morrison's first reaction was to wonder whether "cold fusion" was just another wrong result in physics, but he initially contained his doubts because he was impressed by Fleischmann's talk at CERN. Morrison's suspended judgment didn't last long. Within five weeks, Morrison's attitude shifted. On May 1, 1989, Morrison publicly declared that the work of Fleischmann and Pons was wrong; he called it "pathological science."

From that point, he used his newsletters as a bully pulpit to attack "cold fusion" research. He was quick-witted and eager to engage in debate with "cold fusion" researchers. He not only disseminated useful and valid critique but also propagated misleading and inaccurate criticisms. Soon, he polarized and denigrated the "cold fusion" researchers by labeling them "true believers." Morrison was very clever, and non-experts would have had great difficulty distinguishing facts from falsehoods in his newsletters.

In 1993, Fleischmann and Pons published a paper in the peer-reviewed journal *Physics Letters A.* (Fleischmann and Pons, 1993) Initially, Morrison responded with critical comments on an Internet newsgroup. Sometime later, he participated in the customary etiquette of scientific critique and submitted his comments to the journal, and the journal offered Fleischmann and Pons the chance to reply.

Morrison used hypothetical scenarios to explain away Fleischmann and Pons' claims, but they called him out. "We believe," Fleischmann and Pons wrote, "that the onus is on Douglas Morrison to devise models which would have to be taken seriously and which are capable of being subjected to quantitative analysis. Statements of the kind which he has made belong to the category of 'arm waving.'" After Fleischmann and Pons corrected Morrison's criticisms, they encouraged critics to return dignity to the scientific debate:

> It is our view that a return to this traditional pattern of communication will in due course eliminate the illogical and hysterical remarks which have been so evident in the messages on the electronic bulletins and in the scientific tabloid press. If this proves to be the case, we may yet be able to return to a reasoned discussion of new research. Indeed, critics may decide that the proper course of inquiry is to address a personal letter to authors of papers in the first place to seek clarification of inadequately explained sections of publications.
>
> Apart from the general description of [our experiment], we find that the comments made by Douglas Morrison are either irrelevant or inaccurate or both.

For Morrison, "cold fusion" became an obsession. He published his "cold fusion" newsletters and attended "cold fusion" conferences almost until his death, on April 29, 2001.

Morrison's First "Cold Fusion" Newsletter

Morrison's first "cold fusion" newsletter, on March 31, 1989, was the only one of his newsletters that was objective and accurately reflected the facts. Morrison arrived for Fleischmann's lecture at CERN 20 minutes early, but because of all the advance media reports, the auditorium was packed, and he had to sit on the steps.

Morrison wrote that Fleischmann had a reputation as an expert in his subject and was a first-class scientist. Morrison thought that Fleischmann had made a breakthrough, though it was a stumbling block for particle physicists. That stumbling block, as Garwin and many other physicists had already pointed out, was the billion-fold discrepancy between Fleischmann and Pons' reported neutron rate versus the rate predicted if nuclear fusion was in fact occurring.

Unlike Garwin, Morrison was momentarily willing to entertain the idea of a possible alternative nuclear process:

> A conclusion that can be drawn from Fleischmann's talk is that the heating is not due to the reactions $2d + 2d \longrightarrow 3He + n$ or $2d + 3t \longrightarrow 4He + n$. ... Instead of saying that there is a discrepancy between the number of neutrons produced and the heat produced, perhaps we should assume that all the results are correct and that the reactions occurring are different.

This was a stunning proposition, particularly coming from a physicist. Morrison had assumed good faith, that Fleischmann and Pons were competent and honest, and he was willing to consider the possibility of new science. Rather than doubting the data, Morrison thought that a new idea was needed:

> Maybe the dominant reaction is fusion, $d + d \longrightarrow Helium\text{-}4$, but we need something else to share the energy and momentum produced — this could be the close neighboring structure of the lattice.
>
> Thus, the dominant reaction is to produce heat! Of course, other reactions will also occur, which is why there is an observation of tritium, and one would expect some production of helium-3 and helium-4 and neutrons and gammas.
>
> If this were true, and again this is mainly a suggestion which needs experimental confirmation, then this would have tremendous social effects, as we would have a simple source of energy without the particulate matter, sulphur and other gases from coal- and oil-fired power stations that are killing so many today. Also, the radiation danger would be

very much less than with nuclear reactors (sell your coal and oil shares if you have any!)

In his lecture at CERN, Fleischmann discussed the amount of time it took to load the required amount of deuterium into palladium by electrolysis. This was a crucial fact that was missing from his and Pons' paper.

"In his lecture," Morrison wrote, "Fleischmann had said that an experiment takes months to perform. However, recent experiments have been performed in days. He explained that, to charge with deuterium, a 1 millimeter diameter rod of palladium takes about 2 days, a 2 mm rod 8 days, a 4 mm rod a month, and an 8 mm rod about 4 months."

But Fleischmann and Pons were inconsistent in the details they offered. Back in the United States on the same day, Stanley Pons gave a lecture in Dallas, Texas, and said that a 4 mm electrode takes "10 days or so" to charge. I do not know the explanation for the discrepancy.

The Problem With Normal Water

Nobody gave Fleischmann any trouble with the still-problematic gamma-ray peak at CERN. The most glaring problem was gone; Fleischmann displayed it with the new label, identified at 2.2 MeV.

As shown in Chapter 7, Laura Garwin, who saw the Fleischmann-Pons manuscript when they submitted it to *Nature*, was taken aback when she noticed that Fleischmann and Pons had claimed deuterium-deuterium fusion, using heavy water, because they apparently had done no control experiments using light water.

Heavy water, D_2O, is made of deuterium and oxygen, compared to light water, H_2O, which is composed of normal hydrogen (also called protium) and oxygen. Deuterium and hydrogen atoms each contain one proton, but deuterium also has a neutron, making it twice as heavy. The resulting heavy-water molecule is slightly heavier than normal water.

In thermonuclear fusion research, the required energies to create hydrogen-hydrogen fusion are so great that researchers do not even attempt it. Instead, they stick with deuterium alone.

Critics were bewildered: How could two prominent electrochemists fail to do what appeared to be the obvious control experiment? The experiments Fleischmann and Pons had done with a platinum "blank" satisfied the two chemists that their calorimetry system was working properly, but those experiments did not satisfy physicists who were much more concerned about the theoretical claim of D+D fusion.

The Fleischmann-Pons hypothesis was so preposterous that this apparent omission immediately cast doubt on the excess-heat claim. Critics assumed that, if Fleischmann and Pons used light water instead of heavy water, and if they still saw excess heat, then the "excess heat" must be the result of error or incompetence.

When Frank Close gave a lecture in 1991 on "cold fusion," he played a recording of part of the conversation that took place toward the end of Fleischmann's talk at CERN. Fleischmann had been asked a question by Carol Rubbia, the director of the lab. Close played only a few seconds of the audio recording. Rubbia had asked Fleischmann the question about normal water. Fleischmann had either refused to answer or evaded Rubbia's question. Rubbia, in the kindest voice, persisted:

> RUBBIA: Martin, If you don't allow me to do that by asking you a direct question, namely, as all this has been done with deuterium in the water, did you ever try to do the same experiment replacing deuterated water with ordinary water? And if you do so, what happens?"
>
> FLEISCHMANN: I must confess to you that those experiments are just going on now, and we will hope to give you that answer very shortly. The reason we haven't done so is that, if you do that experiment, you do, in fact, ruin the electrodes, and we have, so far, had very limited resources.
>
> We know that, when the system is dead, when there is no excess heat, it balances to 1 milliwatt, so we did not feel especially compelled to do the blank experiment. But we accept that we have to. We are, at the moment, running a whole new set of cells which include the thermal balances on ordinary water.

Fleischmann's statement about ruining the electrodes sounds dubious; I've never heard any other electrochemists in the field say that is unique to H_2O experiments. The part about not feeling "especially compelled to do the blank experiment" was evasive but truthful.

Vague responses were common with Fleischmann. He was very often vague with me, and his good friend John Bockris knew Fleischmann's tendency as well. Bockris called it "being wooly." Sometimes, he had his reasons. Sometimes, it was just arrogance; he didn't like explaining things that were, to him, elementary. In this case, he did not say that he and Pons did not do the blank experiment; he merely said they were not motivated to do it. I will explain why.

It was obvious to Fleischmann and Pons when their experiments were producing excess heat and when they were not. An experiment using a "dead" cathode would remain at equilibrium and give a precise thermal balance to 1/1000 of a watt. An experiment with a working cathode, while in the loading phase, behaved like a normal electrolytic experiment. It warmed up by Joule heating and showed the exact amount of heat evolution expected by the electrical input; that is, they saw no excess heat within 1/1000 of a watt. That's why they didn't feel "compelled" to do the light-water experiment. They knew how to measure heat, and there were few researchers who could do so better.

Therefore, their experiments which did not produce excess heat were, effectively but not intentionally, blank experiments. This proved to them that their excess heat, when observed, was measured correctly. It did not, however, prove that the heat was from deuterium-deuterium fusion, and this is what the light-water control test would have proved.

The reason that Fleischmann was evading Rubbia's question is that Fleischmann was hiding something. He was afraid to reveal to the packed auditorium that he and Pons had already performed light-water experiments and that they had seen a small amount of excess heat. Fleischmann knew that, if he had revealed this fact, the physicists would have laughed and walked out. Room-temperature deuterium fusion was a stretch; room-temperature hydrogen fusion was just inconceivable.

The other problem for Fleischmann was that he was extremely reluctant to give up his attachment to the idea of fusion. This was not true of Pons, who was far less attached to the concept.

Rubbia's question about light water, however unresolved, was probably the most significant part of the meeting. The question came up again when Fleischmann spoke on April 12 in Erice, Italy. And it came up when Pons spoke in Dallas, Texas, the same day.

Fusion Capitulation

On March 31, the same day that Fleischmann gave his lecture at CERN, Pons gave a lecture at the University of Utah. The announcement for the lecture, written by James Brophy, the University of Utah vice president for research, was "Background For Nuclear Fusion Seminar."

The thrust of Brophy's announcement was a partial retreat from the fusion claim. Brophy tried to place the emphasis where it should have been in the first place, on the experimental data, not the interpretation of fusion.

"There is not yet a complete understanding of where the heat is coming from," Brophy wrote. "Fusion occurs in the cells, but fusion reactions do not account for all the heat that is observed. As we stated at the press conference last week and on several occasions since then, the investigators believe that no chemical reaction can account for the heat output, so they attribute it to nuclear processes."

Brophy's backtracking was too little too late. It was a move in the right direction, but Brophy was still holding onto a partial-fusion interpretation. He also proposed a new definition for evidence of fusion, based on electrochemistry. "Evidence for nuclear fusion," Brophy wrote, "includes generation of heat over long periods that is proportional to the volume of the electrode and reactions that lead to the generation of neutrons and tritium which are expected by-products of nuclear fusion."

Evidence for nuclear fusion had never had anything to do with electrodes. Critics, rightly so, would have viewed Brophy's attempt to redefine nuclear fusion with contempt, at best.

Someone on the IBM newsgroup posted a first-person account of the Pons lecture written by Timothy K. Reynolds, apparently a student at Utah, who attended it. Two paragraphs are of particular interest.

"Pons admitted that the results were just as puzzling to him as they are to many others," Reynolds wrote. "He openly admits that much more work is needed to understand this phenomenon. He did not seem to resent any questions and was honest in his responses."

Robert Bazell, of NBC-TV, quoted Pons that day. Bazell said that Pons admitted that he did not know what was generating the heat.

"We don't know that it's fusion," Pons said, "but we cannot imagine any other process that can come up with that kind of energy in that small volume with the components [that are] in the system."

"Pons ended his talk," Reynolds wrote, "with a WARNING: 'Please do not, DO NOT, attempt to repeat this experiment until you have read the journal articles or have consulted with Drs. Pons or Fleischmann directly.' The initial experiment which vaporized is no joke. Please consult with them, or wait for the articles to appear before you begin a possibly dangerous experiment. Please act responsibly in this regard."

The Vaporization Incident

By now, March 31, 1989, the story of Fleischmann and Pons' dramatic 1985 experiment began to come out. For several months in late 1984, they had been loading deuterium into a 1 cm cube of palladium by electrolysis. One evening, in either January or February 1985, Pons asked his son Joey to go to the lab and turn down the electric current.

Fleischmann and Pons often talked about avoiding "sharp edges" in the operating parameters of their experiment. This is something they had learned only by experience, and this was about to become their first lesson. What they meant by "edges" was any abrupt changes or perturbations to the cell and its environment. Joey turned the current down too much, too quickly (from 1500 milliamps to 300 or 400 milliamps), Although Joey did not know it at the time, he had caused a sharp edge and had triggered a rapid rise in the cell temperature.

The next morning, Fleischmann and Pons came in and found that most of the palladium cube had vaporized. Kevin Ashley, a graduate student in the Chemistry Department, was walking past the room on the way to his classroom that morning when he noticed a commotion in the

Pons laboratory. At the time, Ashley had no idea what sort of work Fleischmann and Pons had been doing. I interviewed him in 2004.

"It was strange," Ashley said. "I came in the morning, and there was a huge mess in the lab. Pons and Fleischmann had these strange smirks on their faces, and I couldn't understand what they were so happy about when the experiment had obviously blown up overnight. They weren't exactly talking about it."

Charles Beaudette had interviewed Ashley for his 2000 book *Excess Heat & Why Cold Fusion Research Prevailed*. Beaudette wrote that Ashley was one of the first to witness the scene in Room 1113 in the Henry Eyring building that morning.

"The lab was a mess, and there was particulate dust in the air," Ashley said. "On their lab bench were the remnants of an experiment. The bench was one of those black-top benches that was made of very, very hard material. ... The experiment was near the middle of the bench where there was nothing underneath. I was astonished that there was a hole through the thing. The hole was about a foot in diameter. Under the hole was a pretty-good-size pit in the concrete floor. It may have been as much as 4 inches deep."

Ashley noticed that, far from looking devastated at their destroyed experiment, "Stan and Martin had these looks on their faces as though they were the cat that had just swallowed the canary."

They knew their experiment did something extraordinary, but they were worried that the university might shut them down if word leaked out. They managed to get the damage repaired without blowing their cover. They described this experiment in their preliminary note, using thermonuclear fusion terminology, as an "ignition" event.

They were very worried about people replicating their experiment without taking proper precautions. Even when Fleischmann and Pons appeared on the *MacNeil/Lehrer News Hour* the evening after the press conference, Fleischmann did not let correspondent Charlayne Hunter-Gault close the interview without interjecting a warning that people shouldn't try the experiment before reading the preliminary note.

Fleischmann told me in 2003 that he was not entirely proud of the accident. "This was a potentially lethal experiment," Fleischmann said. In hindsight, Fleischmann realized that he and Pons should have looked for evidence of transmutations, as he told me in an Oct. 21, 2003, letter.

"It was here that we made a crucial mistake," Fleischmann wrote. "We should have scraped out the inside of the fume hood housing the experiment and preserved the scrapings for future analysis. However, we said, 'Right. We'll do this experiment again in the future when we have a better understanding of the system.' But this has never happened!" In 2004, I interviewed him again and probed further.

KRIVIT: Did the school administrators know about the accident?

FLEISCHMANN: No!

KRIVIT: How did you cover up the hole in the floor?

FLEISCHMANN: Oh, that was easy.

KRIVIT: How did you avoid an investigation?

FLEISCHMANN: Look, we did it in America, you realize. Your country was very lax about security back then. I remember saying to Stan, "We can't do this in Southampton; we would be stopped." The question Stan asked was, "What do we do? Do we stop or carry on?" And I said, "We carry on but under very mild conditions. We change the design, and we keep the whole thing quiet."

FLEISCHMANN: If we had the resources we would investigate it using large cathodes again under a variety of conditions. It would take a lot of money to do that experiment. If you were trying to do this under controlled conditions, you'd want to do it in a secure environment. You'd want to do it in a laboratory devoted to explosives research. You'd want to be able to monitor it as it was going on. The thing is that other people have had this thing go out of control, too: the people at Lawrence Livermore, for example.

Meanwhile, as March turned to April, the cold fusion frenzy began to escalate.

CHAPTER 9

Cold Fusion Frenzy

Claims, Counterclaims, and Press Releases

On March 31, 1989, three "cold fusion" lectures took place. Martin Fleischmann gave a lecture at CERN, in Switzerland, followed by a press conference. Stan Pons spoke at the University of Utah, in Salt Lake City, Utah; cameras and recording devices were strictly forbidden by the university. Brigham Young University's Steven Jones gave a lecture at Columbia University in New York City, and spoke afterward at a press conference.

The same day, Pons released to five of his colleagues a preprint of his and Fleischmann's revised manuscript for the *Journal of Electroanalytical Chemistry*. Despite the fact that he marked them "confidential," the documents quickly made their way around the globe by fax machine. "Cold fusion" was in high gear.

Experimental physicists Gyula Csikai and Tibor Sztaricskai, at Lajos Kossuth University, part of the University of Debrecen, in eastern Hungary, had started their "cold fusion" experiments several days earlier. On March 31, they detected neutrons. Within hours, they reported the news to the Hungarian news agency MTI. They were the first scientists to claim a confirmation. Traditionally, scientists published results in scientific journals to establish their scientific priority. But this was "cold fusion." Claimants were instead issuing press releases or holding news conferences.

A few days later, on April 4, 1989, Pons gave another lecture, this time at Indiana University-Purdue University Indianapolis. He had been scheduled to speak on another topic long before the fusion announcement, but "cold fusion" took priority. Pons was now a science superstar, and, according to author Frank Close, the university decided that admission to the lecture would be by purchased ticket only. At the press conference following his lecture, Pons responded to two hours of questions from reporters, according to Valerie Kuck, the organizer of the American Chemical Society meeting that April.

This chapter reports on a few "cold fusion" events that took place in the first two weeks of April 1989 in the United States. Worldwide, there were many more — hundreds, probably — of attempts to replicate the Fleischmann-Pons experiment in April. The timeline in Appendix C shows claimed replications in Russia, Korea, India, Czechoslovakia, Italy, and Brazil, to name a few.

MIT Fails Fast

Within a few days of the University of Utah press conference, researchers at the Massachusetts Institute of Technology began their attempts at "cold fusion." MIT, with its Alcator series of fusion reactors, operating since 1973, had been one of the major recipients of U.S. government funding for thermonuclear fusion research. Professors in the MIT Plasma Fusion Center teamed with the MIT Chemistry Department to create tabletop fusion but allegedly failed.

Whereas Fleischmann and Pons had gone public after they had submitted and learned of the acceptance of their manuscript in a peer-reviewed journal, MIT researchers went first to the news media to report their failure.

On April 6, 1989, according to the Associated Press, MIT issued a press release and said that chemist Mark Wrighton failed to repeat the Fleischmann-Pons experiment. "We see no physical basis at the moment for thinking that nuclear fusion is going to occur," Wrighton said. "And we certainly have no evidence that there's a big effect, one that would have technological consequences in the near term." Nobody accused Wrighton of being unethical for announcing his results in a press release, let alone doing so three months before he had submitted a manuscript to a scientific journal.

By March 28, Wrighton had five cells running. Of the five, only Cell C could have given positive results by the time he spoke with the news media. The other cells would have required at least two weeks of preparation because of the large dimension of the cathodes. The larger the cathode, the longer it takes to load the required amount of deuterium into the palladium. Without sufficient loading, excess heat is not possible. Cell C, with a 1 mm diameter cathode, started on March 28 and finished on April 19. Wrighton publicly announced his failure after Cell C ran for one week.

Phase I Cell Parameters.				
Cell	Electrode Size (cm)	Electrolyte	Start (m/d/y)	Stop (m/d/y)
A	0.64 x 2.5	0.1 M LiOD	3/27/89	5/19/89
B	0.64 x 2.5	0.1 M LiOD^b	3/27/89	5/19/89
C	0.10 x 2.5	0.1 M LiOD	3/28/89	4/19/89
E	0.64 x 2.5	0.1 M LiOD	3/28/89	4/7/89
F	0.64 x 2.5	0.1 M LiOD	4/1/89	5/19/89

MIT Phase 1 "cold fusion" experimental data. (Albagli, 1990)

For most of that first week, as he told the news media, Wrighton had no scientific paper to use as a guide. He based his experiment on information he read in the newspapers. The situation placed science journalists in the odd role of being the primary conduits of scientific information between scientists.

Wrighton received a copy of the Fleischmann-Pons preliminary note several days later, possibly on March 31. He told the Associated Press, however, that the copy confirmed that he had been doing the experiment correctly.

But if Wrighton was expecting the cells with 0.64 cm cathodes to show excess heat by April 6, he was mistaken. Given the loading requirements, only experiment C — with a 1 mm cathode — assuming all other protocols were correct, had a chance of working.

Later that month, Pons gave a lecture at the Los Alamos National Laboratory. He gave a more technical explanation to someone in the audience who asked how long it took for the experiments to get to equilibrium and then to begin producing excess heat.

> There is a slow rise, an expected rise in temperature from just Joule heating in the cell over about several hours after you start the experiment. Then for an electrode, it takes, say, three weeks to charge up [with deuterium], and you will see not too much of a change over the next week and a half. Then in the last week, you will start to see the temperature rise from that up to, say, 6½ degrees that you saw here, so that cycle takes about a week. We usually wait about five times the diffusional relaxation time before we think the cell is at total equilibrium.

Wrighton implied but did not say to the Associated Press (AP) that his failure was evidence that Fleischmann and Pons were wrong. MIT provost John Deutch was quoted in the article and went further, judging the Fleischmann-Pons experiment false simply because Wrighton's half-completed experiments produced nothing.

Wrighton told the AP that he was doubtful. John Deutch, MIT's provost, said he didn't think the Utah claim was true. Wrighton told the AP that there were two possible reasons for his lack of results: 1) Fleischmann and Pons were wrong or 2) the effect is too small to be detected.

Wrighton omitted other possibilities: 3) as a chemist rather than an electrochemist, he may have been unfamiliar with certain skills familiar to electrochemists; 4) Fleischmann and Pons may not have communicated all the required details; 5) Fleischmann and Pons may not have known all the required details; or 6) Wrighton did not properly replicate all the material elements and procedures of the experiment.

Follow the Money

On April 6, the same day as the MIT news release, thermonuclear fusion research funding in the U.S. came under attack. The U.S. House of Representatives Energy, Research, and Development Subcommittee accepted an amendment from Robert Walker (R-PA) authorizing $5 million to be taken from the Magnetic Fusion Program and given to the Basic Energy Sciences activity, specifically for room-temperature fusion.

The source for this fact is the transcript of the hearing on "cold fusion," titled "Recent Developments in Fusion Energy Research," which took place on April 26, 1989, in the House of Representatives Committee on Science, Space and Technology, in Washington, D.C. As far as I know, this authorization was never reported by the news media, but thermonuclear fusion researchers likely knew about it.

Fusion workers in academia and industry had seen their budget drop each year since 1984. Roxane Arnold, writing for the *Los Angeles Times* on May 8, 1989, described the situation:

> What's at stake in fusion research is more than a matter of reputation. Fusion scientists are competing for shares of an ever-shrinking pie of government money. The U.S. Department of Energy has cut funds for fusion research from a peak of about $460 million a year in the early 1980s to about $350 million last year. At Princeton, where the largest share of the money goes, budget cutters have trimmed spending from about $140 million to $102 million during this period. In addition to Princeton, the government underwrites programs at the Massachusetts Institute of Technology, Oak Ridge and UCLA, among others.

Researchers at the Princeton Plasma Physics Laboratory also tried to repeat the Fleischmann-Pons experiment, but they, too, failed. Arnold quoted Don J. Grove, who had been a member of the Princeton team since 1954 and the project manager for the Tokamak Fusion Test Reactor. "I doubted from the beginning that 'cold fusion' was true,"

Grove said. "While it's unfortunate if it turns out that some segment of the scientific community may be caught with its pants down, at least it isn't physicists."

The Princeton fusion reactor shut down in 1997 after 15 years of experiments. In 2013, Congress sharply cut back funding for the MIT Alcator fusion reactor, and it was forced to shut down. After heavy lobbying, MIT was able to get funds to restart the reactor, but only temporarily. Stephen Dean, the director of Fusion Power Associates, reported in his newsletter on Dec. 14, 2014, that fiscal year 2016 will be the final year of funding for the MIT fusion reactor. Nevertheless, as of March 2016, the MIT Web site was still promoting thermonuclear fusion:

> Today, we are closer than ever to realizing the dream of harnessing the nuclear process that powers our sun. This stellar process, called fusion, produces minimal waste and offers the hope of an almost limitless supply of safe, dependable energy.

Sixty-four years after the first U.S. attempts to harness the power of the sun, MIT was nowhere close. Yet critics accused Fleischmann and Pons of wishful thinking, only two weeks after their announcement.

Bard Finds Nothing

By the beginning of April, scientists around the world were trying to confirm "cold fusion." Most failed. At the University of Texas, Austin, Allen J. Bard, a prominent electrochemist and editor of the *Journal of the American Chemical Society*, tried it albeit with lukewarm interest. Author Gary Taubes spoke with Bard and learned that he wasn't thrilled about following in Fleischmann and Pons' footsteps, as Bard told his researchers.

"It's not that we're going to make any big contribution now," Bard said. "It's done. If it's true, it's true, but it's important to reproduce this result. I don't think you guys ought to jump off and get into this unless you really feel very strongly motivated about it."

Bard, 56, and postdoctoral student Norman Schmidt attempted a replication around March 24, a week before the Fleischmann-Pons preprint made its rounds. Bard didn't know what electrolyte to use, he didn't have clear information about the loading times, and he didn't know the primary effect (heat rather than neutrons) to look for.

They used an ultra-small-diameter electrode — a 25 micron wire ultramicroelectrode of palladium — but, according to Taubes, they found no evidence of "fusion." Taubes was not specific about whether they searched for heat or neutrons. Bard, has been unable to remember any details about his experiment and has been too busy to locate and retrieve any of his original documents.

With such tiny electrodes, Bard had made it easy to load the palladium with deuterium, but such tiny electrodes also made it difficult to produce measurable products. Bard's electrode was 40 times smaller than the smallest electrode used by Fleischmann and Pons. But he tried.

Texas A&M: Charles Martin Claims Heat

At least for some scientists, the race to confirm "cold fusion" was on. They knew that the first person to do so could earn a coveted spot in the history of science, assuming the results were easily repeatable and unambiguous. One hundred miles east of Austin, in College Station, Texas, electrochemist Charles Martin, at Texas A&M University, was eager to make his mark.

Texas A&M was particularly active in "cold fusion" research in the weeks following the University of Utah announcement. This was not unexpected; it was one of the most well-equipped and expertly staffed universities in the country for this kind of electrolytic research. Four independent groups at the university dived into the research.

Martin's group seemed to have better success than Bard and appeared likely to be the one to report the first confirmation of the Fleischmann-Pons experiment in the U.S. and the first group worldwide to confirm excess heat.

Martin, 35, an associate professor of chemistry, was not only more enthusiastic but also in a better position than Bard to get inside information about the experiment. Martin was a close friend of Pons, his senior by 11 years, and thought highly of him. Martin was also much more enthusiastic than Bard, who, coincidently, was Martin's postdoctoral thesis advisor at Austin. Pons went out of his way to give Martin a few tips, as Taubes reported.

Martin set out to work with chemists Kenneth Marsh, director of the Texas A&M Thermodynamics Research Center, and Bruce Gammon, also at the center. By Friday night, April 7, 1989, they thought they saw evidence of excess heat. It continued through Sunday. They quickly prepared a manuscript and faxed it to Ronald Fawcett, the American editor of the *Journal of Electroanalytical Chemistry*. As was the norm for "cold fusion," they contacted the university public relations office to schedule a press conference for the next day, April 10. AP reporter Michael L. Graczyk covered the story and reported some of the technical details.

They used a small, 1mm-diameter electrode, the same size as the smallest used by Fleischmann and Pons. This gave them a quick loading time, and it was large enough to put the excess heat within range of conventional calorimetry. Their calorimeter was as sensitive as Fleischmann and Pons': 1/1000 of a watt. Graczyk wrote that Martin detected a constant 1.144-watt power gain in the experiment.

Thomas H. Maugh II, covering the story for the *Los Angeles Times*, reported Martin's percentages of excess heat. "This is in agreement with the findings of Pons and Fleischmann," Martin said. "We have run the experiment using four different amounts of electric current and have found that excess power varies between 60% and 80%."

As eager as Martin and his group were to report their experimental findings, they were cautious about making theoretical interpretations. They refrained from claiming that that the heat was from a fusion reaction, as Graczyk reported, because they did not see significant levels of neutrons. "This is only one aspect. The excess-energy aspect has been confirmed," Martin said. "I would feel a lot more comfortable if we detected fusion, and we have not yet."

Reporter JoAnn Jacobsen-Wells, writing for the Salt Lake City *Deseret News*, a newspaper owned by the Church of Jesus Christ of

Latter-day Saints (the Mormon church), was not as circumspect as the AP about the news. "The University of Utah's cold nuclear fusion breakthrough," Jacobsen-Wells wrote, "which ultimately could revolutionize world power production, has been confirmed."

The big secret, as only a few people knew, was that the previous evening Martin and his colleagues began to see excess heat with the light-water control cell they had started just that day. They said nothing about the light-water excess heat during the press conference. This perplexing result sent Martin into a tailspin. It caused him to doubt the validity of his heavy-water experiment even though his team had extensive expertise in electrochemistry. Their assumption, shared by everyone at that time, was that, if Fleischmann and Pons' deuterium-deuterium "cold fusion" was real, then a light-water experiment without deuterium should give no excess heat.

There is more to the Charles Martin story, but I have been unable to garner reliable information. In his book about "cold fusion," Gary Taubes went into great detail about Martin's subsequent distress and confusion. Taubes spoke at length with Martin's postdoctoral researchers — but not with Martin himself — about the group's follow-up experiments. There are some inconsistencies among the facts Martin gave to Taubes. There are also some inconsistencies between Martin's account in Taubes' book and Martin's audio interview by Bruce Lewenstein in the Cornell Cold Fusion Archive.

Martin declined my interview request about this matter.

Texas A&M: Appleby Sees Heat

Anthony John Appleby, 49, the director of the Center for Electrochemical Systems and Hydrogen Research, at Texas A&M University, obtained much clearer excess-heat results that month than Martin, but Appleby, a world-renowned electrochemist, was cautious. Perhaps with the benefit of his additional wisdom, he did not schedule a press release within hours of his first sign of excess heat.

Appleby and his group used what is known as microcalorimetry, giving them sensitivity down to one-millionth of a watt. Word of

Appleby's results leaked on May 3, 1989. JoAnn Jacobsen-Wells, with the *Deseret News*, spoke with Ed Walraven, the assistant director of public information at Texas A&M.

"Appleby's confirmation," Walraven said, "is no state secret, but he has only mentioned it to colleagues in a conservative way." Walraven told Jacobsen-Wells that Appleby's "formal announcement is pending publication of his data in a scientific journal."

The first written statement from Appleby appears to be one page he sent to the House Committee on Science, Space and Technology on April 25, 1989, for its April 26 hearing on "cold fusion."

A month later, Appleby and his colleague Supramaniam Srinivasan, also a highly respected electrochemist, first reported their results to the scientific community at the Electrochemical Society meeting in Los Angeles. According to his obituary, Srinivasan "worked on the kinetics of electrochemical reactions on porous electrodes." Thus, he likely understood better than most people the value of allowing time for the deuterium to soak into the palladium electrodes.

Appleby discussed their results again, at a Department of Energy-sponsored workshop on "cold fusion" in Santa Fe, New Mexico, later that month, and again in October at a meeting co-sponsored by the National Science Foundation in Washington, D.C. Toward the end of April 1989, a third group at Texas A&M confirmed the Fleischmann-Pons experiment.

Texas A&M: Bockris Finds Tritium

John O'Mara Bockris, 66, professor of electrochemistry at Texas A&M, was leading another independent group in an attempt to replicate "cold fusion." Bockris was recognized internationally among electrochemists. He too, received some inside tips. The day after the March 23, 1989, press conference, he spoke with Fleischmann, his old buddy from graduate school, as Bockris told me:

> When I heard about this the next day, I called Martin
> Fleischmann. Martin at this time was being called about 10

times a minute, and he had to shut off the phone. But when he knew it was me, it was different. We had been friends for years, so he accepted my call.

"You go along at low current density for many hours," Fleischmann said, "and then you jack it up! – 10 times, 100 times more current density, and then it happens!"

As we found out later, that's not enough information, because the "go on for a long time" is a *very, very long time,* 500 hours or so. That is the essence of it, and then you see the excess heat.

Also, I asked Martin another question, which was, "How did you get the lithium hydroxide?" He told me that he had got it by putting lithium [a soft metal] in the solution and letting it dissolve to form lithium hydroxide. Those were the two bits of information that he gave me, and in that sense, I had a flying start, with all my graduate students at Texas A&M. I had 20 of them working with me at the time.

Bockris confirmed part of the Fleischmann-Pons results on April 24. Rather than base his group's claim of a nuclear reaction on excess heat, they measured the production of tritium, a definitive nuclear byproduct. Unlike Charles Martin, Bockris did not call a press conference or issue a press release.

Vanishing Neutrons

On April 10, 1989, the same day that Charles Martin announced his results in a press conference, researchers at Georgia Institute of Technology, in Atlanta, joined the "cold fusion" media parade.

The leader of the group was James Mahaffey, 38, a senior research scientist with a doctorate in nuclear engineering. That weekend, Mahaffey thought he had seen neutrons. Following the "cold fusion" convention, he worked with the university media relations office to send out a press release and arrange a press conference for late Monday, April 10, 1989.

Graczyk reported the Georgia Tech news in the same article that he reported the Texas A&M news. "We think we've confirmed the Utah experiment to prove 'cold fusion,'" Mahaffey said.

James Mahaffey, Georgia Tech researcher

Mahaffey, thinking he had seen neutrons, was in a slightly better position than Martin to make a claim of nuclear fusion. But neutron detection can be tricky, and Mahaffey learned the hard way. Three days later, Mahaffey held another press release to announce some bad news.

"On Thursday," the AP reported, "Georgia Tech researchers disclosed that their experiment may have been skewed by inaccuracies in a key piece of equipment that measured an emission of neutrons, a sure sign of fusion. The researchers said their probe may give false readings caused by heat at low temperatures."

The AP reported that Mahaffey was getting nervous and that his results were not as sure a thing as he had thought. Thomas Stelson, the university's executive vice president, told AP writer Joseph B. Frazier that the scientists got their experimental protocol from press accounts and computer bulletin boards.

"We thought we had it all calibrated right," Stelson said, "but we found out that one instrument was more temperature-sensitive than we thought. We have new detectors. We're in the process of checking them out in some detail."

The Back Channel to Utah

Nathan (Nate) Lewis, 33, a hotshot associate professor of chemistry at Caltech, was neither an electrochemist nor a friend of Fleischmann or Pons. He was also in no rush to confirm the Fleischmann-Pons experiment. Years later, in 2006, I met Lewis at an energy conference in Los Angeles. He told me that he didn't speak about "cold fusion" anymore so I defer to the interview he gave to author Gary Taubes.

By the time Lewis got to his lab on Friday morning, March 24, two of his postdoctoral researchers, Mike Sailor and Reggie Penner, had a "cold fusion" experiment set up and about to run. Here's Taubes' account:

> They had bootlegged a strip of palladium and purchased the heavy water from the stockroom — $60 for 100 grams. They had assumed Lewis could afford it.
>
> Lewis was [doubtful]. He told Penner and Sailor that the experiment wasn't worth more than one day's effort. It was also possible that Lewis, having a high opinion of his own talents, did not want to sweep up after Pons and Fleischmann, even if they had just discovered the greatest thing since fire. Lewis was proud of his reputation.
>
> "You can go out and get a reading on me," Lewis said. "I did things and got the respect of a fair number of people by going into an area that people have studied already, and doing things meticulously and seeing things that other people didn't see. So I'm one of the people where you say, 'If Nate says it's right, it's right.'"
>
> Lewis had done his doctorate at MIT under Mark Wrighton who had been electrochemistry's resident prodigy before Lewis came along. Then Lewis went to Stanford as an

assistant professor without bothering with a postdoc. Now he was 33 and tenured at Caltech.

Fleischmann and Pons had gone out of their way to help a few of their fellow electrochemists. Lewis, a chemist, wasn't one of them. Pons didn't provide much assistance to Lewis in their one phone call. But there was a back-channel network that passed information between the University of Utah and Caltech — not that it seemed to do any good. The source of these e-mails is Bruce Lewenstein's Cornell Cold Fusion Archive.

John Gladysz, a chemist from the University of Utah who had known Pons for seven years, was on sabbatical for the semester at Caltech. Gladysz bridged the communication gap between the two labs. So did Gary Holland, an assistant professor of chemistry at the University of Utah. During April, Holland was the key liaison to Michael Sailor at Caltech. On March 31, 10 days before the Fleischmann-Pons preliminary note published, Holland sent Sailor the key details of the experimental protocol. A day later, Holland sent another e-mail with more information:

> I'm pretty sure that the electrode discharges electrochemically in D_2O; the catch seems to be that you need to go to very high D2 [concentration] within the Pd before anything happens — like near 1:1 Pd:D loadings. Current density is also important — 250 ma/cm² is too high. They observed meltdown between 250-125 ma/cm². So at the lower current densities, it takes a while to charge up the Pd to the appropriate loading, hence the volume dependence. Love and kisses to NSL. Good luck.

Lewis was missing one crucial detail, the need for a dynamic trigger. Holland had hinted at this clue in his warning to avoid sharp edges, that is, rapid changes in the operating parameters of the experiment. Fleischmann and Pons were very worried that people attempting their experiment might unintentionally duplicate their 1985 accident and cause injury.

In the following month, Lewis told many reporters that his Caltech team had precisely replicated the Fleischmann-Pons experiment but failed to replicate their results. Lewis also complained to reporters that he didn't know the exact protocol because Pons wouldn't speak with him and give him personal assistance. Reporters were apparently too distracted by Lewis' public drama to notice the contradiction: If Lewis didn't know key details, he couldn't possibly have precisely replicated the experiment according to the correct protocol.

MIT Claims "Cold Fusion" Explanation

Even though Mark Wrighton had dismissed "cold fusion" on April 6, MIT still had a "cold fusion" believer on campus. Peter Hagelstein, 34, an electrical engineer, computer scientist and associate professor in the MIT Department of Electrical Engineering and Computer Science, said that he could explain the mechanism of "cold fusion."

According to several scientists who have worked with Hagelstein, he didn't have "a theory" to explain "cold fusion." Instead, he had only had an assortment of ideas that, assuming fusion was taking place, could explain different aspects of the phenomena. At last count, Hagelstein said he had tried more than 150 models to explain "cold fusion."

Between April 5 and 12, 1989, Hagelstein submitted four theory papers for peer review. The first was "A Simple Model for Coherent D+D Fusion in the Presence of a Lattice," $(d + d \longrightarrow 4\text{-}He)$ which he sent to *Physical Review Letters.*

Even though Mark Wrighton told the news media on April 6 that he had failed to confirm "cold fusion," a week later the MIT news office announced in a press release that Hagelstein had figured out ways to explain how "cold fusion" worked. Not only that, but MIT filed patent applications for inventions based on Hagelstein's ideas.

Despite the bitter antagonism among people who dismissed the Fleischmann-Pons claims, academic freedom, at least momentarily, was alive and well at MIT. In the press release, the provost, John Deutch, who had dismissed the Fleischmann-Pons experiment as untrue on April 6, now endorsed Hagelstein's claim. "MIT is a place where creative individuals are encouraged to address scientific subjects of the greatest significance," Deutch said. "We are pleased to see professor Hagelstein proposing an explanation for 'cold fusion.'"

The next day, April 13, Ronald R. Parker, the head of the MIT Plasma Fusion Center, indirectly gave Hagelstein a whack in the *Washington Post*. "In fusion research," Parker said, "there are always crackpot claims to produce fusion in a simple way."

Lee Dye saw the irony of it all. "MIT did a back-flip," Dye wrote, "and announced that it was going to patent the theory for a process that other MIT scientists had concluded doesn't even work."

Pops Like a Firecracker

Meanwhile, Edward Teller's group at the U.S. Department of Energy's Lawrence Livermore National Laboratory, in Livermore, Calif., had been hard at work since getting one of the rare advance copies of the Fleischmann-Pons preliminary note, on March 24, 1989. On April 12, Dan Stober, writing for the *San Jose Mercury News*, reported that the Livermore researchers' experiment exploded.

"There was a pop as from a firecracker," Stober wrote, "and broken glass was scattered about, according to employees at Livermore and its sister lab in Los Alamos, N.M. However, there were apparently no injuries or serious damage. Livermore officials refuse to talk about the incident, but other scientists speculated that the explosion was caused not by nuclear fusion but by the ignition of hydrogen and oxygen gases produced by the experiment itself."

Stober tried to get more information from the lab's scientists, but they didn't return his phone calls, and the public affairs officer refused to comment. Somebody must have leaked the information to Stober. The lab clearly wanted none of this reported publicly. Author Frank Close wrote about this incident in his book.

"It had nothing to do with fusion," Close wrote. "It seems that a palladium rod had been loaded with hydrogen and then left out on some Handi Wipes. The hydrogen gas started leaking out into the air, and, releasing this strain, the palladium began to get hot, very hot. This is a phenomenon known since early in the 19th century, a chemical effect, not fusion, that has been used in the Döbereiner Cigarette Lighter. The temperature rises rapidly and is sufficient to light a cigarette."

Thanks to Richard Garwin's willingness to grant me access to his archives, I discovered an internal report on this accident that had been sent from the Livermore laboratory to the Department of Energy in the summer of 1989. The accident, discussed in Chapter 30, could not have been caused by the recombination of hydrogen and oxygen, and it had nothing to do with Close's Handi Wipe story.

Among Electrochemists

Electrochemists had the best success in replicating the excess-heat claim of Fleischmann and Pons. This was partly the result of the nature of scientific specialization and the use of terminology specific to electrochemistry practices. It was also partly the result of their fraternal nature. A few inside tips from Fleischmann paved the way for Bockris. Charles Martin, despite his ambiguous results, also directly received some private tips from Pons. Anthony John Appleby seemed to figure things out on his own.

The electrolytic experiment of Fleischmann Pons was not the domain of general chemists, let alone physicists. This fact was not entirely appreciated by non-electrochemists, particularly physicists, who previously held exclusive control in the court of nuclear science.

Certainly, Fleischmann, Pons and the University of Utah placed a tantalizing temptation in front of the world's scientists. The news media had depicted "cold fusion" as the discovery of the century, yet for an entire week, most scientists had nothing but newspapers and televisions from which to obtain the laboratory protocol.

Pons faxed copies of the preprint on March 31 to five people. Everyone else received faxes of faxes of faxes. Two and half weeks after the University of Utah press conference, the Fleischmann-Pons preliminary note published. With this, the scientific community finally got clean and legible copies of the document, but it was still very difficult for non-electrochemists to follow Fleischmann and Pons' cryptic and inadequate description. Many physicists grew resentful.

For example, Fleischmann and Pons expressed their values of excess heat not in "Watts" but in "Watts per cubic centimeter." Although this was more informative because it gave a sense of the power density, it also made it harder for non-electrochemists to get a sense of the power level. But it also may have introduced an error. Fleischmann and Pons assumed that power density varied with volume rather than surface area. In hindsight, this assumption seems wrong.

Paul Recer, writing for the AP, reported on March 31, 1989, the complaints from Anthony DeMeo, the head of public relations at the Princeton University Plasma Physics Laboratory, and John Soures, a fusion researcher at the University of Rochester.

"People have been throwing some equipment together and trying some things, but it's lousy science," DeMeo said. "It's not the type of science that could let you publish the confirmation or the negation of the Utah experiment because we don't know what they did."

"The major problem," Recer wrote, "is that nobody outside of Pons' laboratory seems to know the precise details of the Utah experiment. Until these details are published or shared with other labs, the Utah experiment cannot be tested. This frustration was felt at laboratories in California, Idaho, New York, Tennessee and Arizona."

"There are all kinds of questions in the fusion community, but nobody is getting any answers," Soures said. "Attempts to talk to scientists at Utah have met with polite silence."

The scientists in the academic sector — Caltech, Rochester and Princeton, all prominent institutions with fusion research programs — expected that, among the hundreds of phone calls and e-mails Pons received in the first few days, Pons should have placed a higher priority on their requests.

Other scientists, mostly chemists, booked themselves flights to Dallas, Texas, for the April 12, 1989, American Chemical Society meeting, where Pons was the featured speaker.

A different group of scientists, mostly physicists, headed to southern Italy for Zichichi and Garwin's Fusion Forum workshop on "cold fusion" at Erice.

CHAPTER 10

Meeting of the Minds

Chemists, Physicists and Cappuccino

On April 12, 1989, in a small pre-medieval town situated on a 2,500-foot mountaintop on the island of Sicily, Italy, about 25 hand-picked scientists from Italy, the United States, the Soviet Union, the United Kingdom, Germany, and the Far East gathered for a one-day workshop to discuss "cold fusion."

Fusion Forum, as it was called, was the first open scientific meeting to address the profound nuclear claims. And it was the only meeting in the history of the field — to this day — at which uninvolved physicists and chemists came together with open minds and a sincere desire to share ideas and solve the issues raised by the claims. It was an exemplary scientific meeting.

The scientists met at the Ettore Majorana Foundation and International Centre for Scientific Culture in Erice, Italy. According to legend, the town was founded by Erice, the son of Venus and Neptune, more than 3,000 years ago. The scientific center is named for Ettore Majorana, an Italian physicist, born in Sicily in 1906.

Every year since the center's establishment in 1962, it has hosted several workshops covering a broad range of scientific disciplines. It was founded by John Bell, Patrick M.S. Blackett, Isidor I. Rabi, Victor F. Weisskopf and Antonio Zichichi.

Antonio Zichichi, the director of the center, was one of the organizers of the Fusion Forum. He is an emeritus professor of advanced physics at the University of Bologna, past president of the

Italian National Institute for Nuclear and Subnuclear Physics, past president of the European Physical Society, past president of the NATO Science Committee for Disarmament Technology, president of the World Federation of Scientists, and president of the Enrico Fermi Centre.

Antonio Zichichi

Zichichi's friendship with Richard Garwin went back some years, and, as discussed in Chapter 7, Zichichi had asked Garwin to help him organize the Fusion Forum and bring some key scientists to the meeting. Zichichi, as he did then, still enjoys his role of popularizing science to the masses. Here's an excerpt from his opening remarks in 1989:

> The last point that I want to tell you is that in his lecture hall there are journalists, the reason being, the following. We have been contributing here, during all our life, to promote science without secrecy and without frontiers. So we cannot forbid anybody to listen to what has to be said.

They have no right to ask questions, of course, but they have the right to listen. We will try to organize a press conference later in the day.

As Garwin remembers it, in addition to the invited participants, 20 scientists came to observe, as well as a dozen journalists. The workshop was mentioned in a short report by Garwin and a few news articles in April and May 1989 and briefly in three books. No other records of the Erice Fusion Forum workshop show up in the public domain.

Richard Lawrence Garwin (1997)

The first time I met Richard Garwin in person, I was going down the escalator in the San Francisco Hilton on Feb. 15, 2007, although we had been in contact since 2004. Moments earlier, I was sitting in the back of a meeting hall among thousands of members of the American

Association for the Advancement of Science for its annual meeting.

Three giant projection screens were displaying photos of a dozen of America's greatest scientists. Many of them were no longer alive. Garwin's larger-than-life image flashed on the screens, and the moderator announced that Garwin was in the room. I wondered whether I might run into him. Seconds later, a man 10 rows in front of me got up from his seat and began walking toward me, toward the exit. I immediately recognized Garwin. Quickly, I gathered my stuff and chased after him.

Garwin has been one of my more interesting, albeit reluctant, sources on "cold fusion" and LENRs. Despite our differing views on "cold fusion" and LENRs, he has always been responsive and cordial to me. He's been generous with his time and has, on more than one occasion, written to other people that I am a journalist whom he respects.

In 2010, while I was passing through Westchester County, New York, he invited me into his home to talk shop. Over cups of jasmine tea and his wife's homemade cookies, we discussed research and debated science. I have always enjoyed our at-times contentious discussions. As a result of our discussions, I have learned aspects of the science that withstand tough scrutiny and other aspects that do not. Garwin has never conceded an argument with me, but I have seen him run out of bullets on several occasions.

In the fall of 2014, I asked Garwin, then 86, about his plans for preserving his "cold fusion" records in an archive. He told me that, many years earlier, he had loaned his six "boxfiles" of "cold fusion" papers to the Cornell Cold Fusion Archive. He said I should go there to get them. I told him I had already been there and seen what they had but that, to my recollection, only a small number of his papers were at Cornell. On hearing this, he agreed to send me his papers, and I agreed to scan and archive the full set, which comprised 5,000 pages.

I placed the entire digital archive into the public domain on the www.archive.org Web site and sent a copy to the Cornell Cold Fusion Archive. The archive contains the first and, so far, only public record of the inner workings of the 1989 Department of Energy cold fusion review panel.

A few months later, when I was midway through writing this book, I opened an e-mail from him and found a download link to eight hours of high-quality audio recordings of the entire Erice "cold fusion" workshop. I thank Garwin for providing this material.

The keynote speakers at the meeting were Martin Fleischmann, Steven Jones, and Bart Czirr, Jones' colleague who was his expert on neutron detection.

In the next few chapters, I will refer to my transcriptions of Fleischmann's talk and the question-and-answer session that followed, as well as Garwin's lecture and its question-and-answer session, and I will provide some excerpts from them. I will also refer to a few other excerpts from the meeting. The complete transcripts of the Fleischmann and Garwin presentations and question-and-answer sessions are available on the *New Energy Times* Web site.

Zichichi's enthusiasm for the topic was evident in his opening remarks, although he remained impartial about the validity of the research. His eagerness and commitment to pull the workshop together were direct evidence that he took the question of "cold fusion" seriously. He put together a thoughtful agenda that facilitated excellent, detailed discussions for such a new and ambiguous topic.

Here is the agenda, edited to reflect how the day actually progressed.

Part 1 — Plenary Talks
Part 2 — Status of New Experiments
Part 3 — Questions Raised by the Experiments
Part 4 — What Kind of Critical Experiments Can Be Done?
Part 5 — Theoretical Speculations
Part 6 — What Kind of Fusion Reactions Can Provide
    Energy Without Emission of Neutrons and Particles?
Part 7 — Non-Fusion Explanations for Source of Heat
Part 8 — Ancient History of Palladium-Catalyzed Fusion

Zichichi gave Garwin a 20-minute time slot to raise questions about the Fleischmann/Pons and Jones claims. Zichichi's introduction of Garwin established Garwin's stature and scientific authority far above anyone else's in the room.

"The [next] part of the program is very important, fascinating and disturbing," Zichichi said. "And this is 'Questions Raised by the Experiments.' If I had to write a list, this space would not be enough. So I have decided to ask one of the greatest experimental physicists living on this planet to speak on this topic himself — no one else — and this is Richard Garwin. He's not only my friend; he is the greatest experimental physicist living." Fleischmann, however, was the guest of honor at Erice and spoke first.

Guest of Honor

The Fusion Forum workshop recording shows that Fleischmann was his usual self at Erice: relaxed and poised. But he was evasive on three topics: tests performed with tritium as a starting material, results of light-water experiments, and potentially rapid releases of energy. He mentioned, on more than one occasion, that he was willing to be wrong about the whole thing. He encouraged the wider scientific community to prove or disprove his and Pons' claims.

"I know [our claims] have astonished many people," Fleischmann said. "I want to assure you they have astonished us, as well. So we are very pleased to be here, to take these first steps in the discussion, and we appreciate the fact that this is an international meeting devoted to this problem."

He began his lecture by discussing what was going on, based on his best understanding. He explained the general working hypothesis: The electrochemical environment produced localized conditions that led to very high pressures and high energies. He cited his and Pons' primary evidence: the calorimetrically measured excess heat. He explained how they confirmed the baseline by using dead (inactive) cathodes as controls. He said that they had the ability to measure heat to within one-tenth of a milliwatt. "When the rod is dead," Fleischmann said, "it balances to a milliwatt. The input and output enthalpy balances."

Nobody questioned his expertise in electrochemistry or calorimetry. He said that at least one experiment produced excess heat for 1,000 hours. He explained that, based on the levels and amounts of heat they had observed, they were convinced the phenomena had to be based on nuclear reactions. Fleischmann explained:

> Now the reason we think it is very likely to be a nuclear reaction is that we run these experiments typically on the order of 100 to 200 hours, in which, for example in this series here, you generate typically 5 MJ/cc of heat. If you can fix your mind on 5 MJ/cc, this is about two orders of magnitude above any conceivable chemical reaction. I'm convinced that, if we ran it for 1,000 hours, we would get 50 megajoules, but we haven't had time to do such long-term experiments.

Fleischmann talked about his and Pons' 1985 experiment in which they unintentionally vaporized most of a massive 1 cm^3 palladium cube that had been their cathode. He made it clear that, because nobody died from radiation exposure, the number of neutrons was not even in the ballpark for a fusion reaction:

> When we calculated back the enthalpy into a neutron flux, if we had the expected nuclear reaction, we decided that we would be dead. We did actually measure the radiation around us. We found a slight increase in the gamma radiation around the experiment, but we decided that this experiment was dangerous, so we discontinued it, and we went to using short rods of various lengths.

Fleischmann mentioned his and Pons' gamma-ray measurements of the water bath, which provided indirect evidence of neutron-capture reactions. This was supplementary evidence for the direct evidence of neutrons they collected using a BF_3 (boron trifluoride) counter, though Fleischmann didn't give much weight to the BF_3 detector data:

> Now, we do have some supplementary evidence. We have measured the gamma radiation. We are not able to measure the neutron radiation. We have only very crude

neutron measurements because we just have a neutron dosimeter to check on the safety of the experiment. So when we have a highly active electrode, we can actually measure a significant increase of the neutron level, above the background level.

This is the gamma ray spectrum which you would expect for an n-gamma reaction of thermal neutrons with the bath. This scale is wrong. This should be 1,000, and this is an accumulation time for about 10 hours in the experiment.

Fleischmann was not contradicting himself when he said they were not able to measure neutron radiation and he immediately followed that by saying they had measured it. What he meant was that they had not been able to perform defensible neutron measurements. Nevertheless, the crude neutron measurements provided for their safety.

When challenged by Garwin, Fleischmann acknowledged that his and Pons' gamma-ray spectrum was wrong. In fact, the gamma-ray spectrum still had problems even after scientists at Harwell had pointed out the first error to him.

Fleischmann and Pons faced many obstacles, including poor reproducibility and a challenge to nuclear theory going back 60 years. In his presentation, Fleischmann made it clear that the measured heat did not correlate with the predicted neutron flux if the effects were the result of fusion. Fleischmann also explained, using as an example the 1 cm^3 cube that partially vaporized, why he thought the reaction had the potential to be self-sustaining:

We're not talking about small gains. We are talking about very large gains in heat. This is an experimental observation. If you recalculate, if you predicted what the neutron rate should be, it's between 10^8 and 10^{10} times higher than the rate which we observe. If we are right about this, what is required is independent experimental confirmation, many times over. If we are right, and this is what we are going to no doubt discuss, then there must be another decay channel.

> There are good reasons for believing that, if there is a temperature runaway, the reaction will actually accelerate. ... If you disconnect the system from the heat source, it will still go on. This cube actually fused the platinum contact of the electrode and fell down and reacted away.
>
> We had this strong evidence that there was a heat source. We did not believe, of course, it was a fusion source. It might just possibly have been a fusion source, so we redesigned the experiments in several cycles. We are now at the position where we know we have this large excess enthalpy, maintained for a period of 100 to 1,000 hours, liberating an amount of excess heat, which is, to us, inconceivable unless it is a new nuclear process.

Fleischmann's statement that, at the time of the experiment, he and Pons hadn't believed that it was a fusion reaction contradicts their statements during the University of Utah press conference.

Garwin's Response

Garwin was first on the sign-up list to ask questions. Their exchange was terse. His first set of questions was about the Fleischmann-Pons gamma-ray graph. Garwin did not take Fleischmann to task for the depicted energy of the peak. It was too early in the conflict to understand this critical nuance. Instead, Garwin suggested that Fleischmann and Pons had not used the optimal configuration to detect gamma rays. Fleischmann explained the subtleties of the electrochemistry configuration that made Garwin's suggestion impractical, if not impossible.

After six rounds of questions and answers between the two men, Garwin had made no headway. Fleischmann explained that Garwin's idea wouldn't work, but Fleischmann revealed a new detail of the experiment.

FLEISCHMANN: That would not work. If you use a 4 mm electrode, you have to wait 10 days for the thermal effects to decay.

GARWIN: So the thermal effects persist?

FLEISCHMANN: The thermal effects persist.

GARWIN: When the electricity is off?

FLEISCHMANN: The thermal effects persist.

GARWIN: Do you have data? You know which days each of these runs was done on?

FLEISCHMANN: Yes, those are all in laboratory notebooks.

Garwin was incredulous. This may have been the first time that Fleischmann revealed that he and Pons had observed the self-heating aspect of the phenomenon. They had observed instances when the cells continued to release heat after the input current was turned off. If confirmed, this was not only another approach to validating the heat (with no input, all of it would be "excess") but also was indicative of a potential process that, much further down the road, could be used as a self-sustaining source of energy.

Garwin switched topics. He began talking about the current input. He wanted to know whether he and Pons underreported the input power. He didn't seem to know that Fleischmann and Pons were masters in calorimetry and could measure thermal balances with a precision of 1/1000 of a watt.

Fleischmann and Pons always used a constant-current power source provided by a precision instrument known as a potentiostat/galvanostat. Fleischmann, in fact, had invented the modern potentiostat, a faster and more accurate device than the original 1942 invention. (Williams, 2013)

Fleischmann and Pons allowed the cell voltage to vary, based on resistance changes that took place within it, in response to temperature changes. Garwin then asked questions about the accuracy of their neutron-gamma-ray measurements.

Garwin exposed problems with the gamma spectrum and with Fleischmann's lack of expertise in gamma measurements. Fleischmann and Pons had also taken direct neutron measurements using a Bonner-sphere BF_3 type instrument, and detected neutrons three times higher than background. But Fleischmann and Pons didn't hang their hat on the neutron-gamma or the direct neutron measurements. Their primary evidence was the heat measurements. Even the television-watching public knew this.

A caption from NBC-TV on April 13, 1989, summarizes the two key points about the Fleischmann-Pons experiment.

Wilkinson's and Koonin's Responses

Sir Denys Wilkinson, nuclear physicist and past vice chancellor of the University of Sussex, was next to ask questions of Fleischmann. The exchange between him and Fleischmann included the one question participants asked Fleischmann that he was unwilling to answer.

The question was about his and Pons' control tests, or lack thereof, with light water. For Fleischmann, and for most people who understood how he and Pons performed their calorimetry, the light-water control was not so important. As mentioned earlier, they had used inert cathodes to confirm the accuracy of their calorimetry and determine the zero baseline in their experiments before the onset of excess heat. Non-working duds made of palladium worked great for blanks, as did cathodes made from platinum.

Although the question about a light-water control was irrelevant to the potential validity of the excess heat, it was entirely relevant to the potential validity of deuterium-deuterium fusion. From this perspective, Wilkinson's final question was right on target. He began by asking about the self-heating phenomena.

> WILKINSON: Are you qualitatively convinced that there is continuing heat generation when there is no electrolytic action?
>
> FLEISCHMANN: Oh, yes. But it is also a little bit more complicated than that because, as deuterium is expelled from the lattice, when you switch the current off, you actually get absorption of heat. In fact, you can, for a certain length of time, eventually push the thing below bath temperature because of the reverse of the exothermicity of the dissolution of the deuterium. So it is very difficult to do the unambiguous experiment here. It is phenomenally difficult to devise an experiment which will answer all those questions.
>
> WILKINSON: Thank you. You know what the deuterium-palladium ratio is under the conditions of your experiment?
>
> FLEISCHMANN: We think, roughly, it is approaching about 1.0.

The 1.0 deuterium-palladium ratio, a fact that had been omitted by Fleischmann and Pons or excised by attorneys from the preliminary note, was a crucial piece of information. I don't have any information about whether the attorneys reviewed the preliminary note. I discuss this omission below in the Broer section. Wilkinson asked several more questions about electrochemistry, and Fleischmann gave clear, immediate answers that seemed to satisfy Wilkinson.

> WILKINSON: Good. The final question: What happens when you use ordinary water?
>
> FLEISCHMANN: I'm not prepared to answer that question at the moment.

Wilkinson did not respond. There was a lot of murmuring among the audience, but no words are clearly perceptible on the recording. After a 7-second silence, Zichichi called on Steven Koonin, who was next on the sign-up list.]

ZICHICHI: Professor Koonin.

FLEISCHMANN: I'm sorry, I do not wish to go into experiments with ordinary water or with tritiated water.

KOONIN: My first question was in fact the same one, and let me ask it again to make sure I've heard the answer correctly. I would have thought that standard scientific procedure would ask you to run a control experiment with light water to prove that the deuterium is really important. How can you make such a claim without running such a control?

FLEISCHMANN: I've said quite specifically I'm not prepared to answer that question at this moment.

KOONIN: OK.

FLEISCHMANN: You can read into that whatever you like.

Koonin was forceful and incisive, but he politely and respectfully thanked Fleischmann, who responded with more detail.

"I'm sure that everybody is worried about the blank experiment," Fleischmann said. "The best blank experiment you have is that, if a cell does not give out excess heat, then the heat balances to 1 milliwatt."

The questions about light water had begun at CERN and continued here, at Erice. On the same day as the Erice workshop, Pons was speaking at the American Chemical Society meeting in Dallas, Texas. (Chapter 12) Pons was asked the same question about light water tests. His responses were remarkably different and revealed new insights into the science as well as how he and Fleischmann differed in their perspectives about the potential theoretical mechanism.

Ziegler's Response

Next on the sign-up sheet was James (Jim) Ziegler, 53, an accomplished IBM physicist with extensive experience in electrical and electronic engineering. Ziegler had been involved in a "cold fusion" experiment for several days.

He asked Fleischmann many thoughtful and detailed technical questions about the configuration and operation of the experiment. He was sincerely interested in trying to repeat the experiment. The exchange between the two men went rapidly. It was friendly, even jovial at times, professional and productive. There was no hint of any cynicism or animosity from Ziegler. Fleischmann clearly enjoyed the discussion and was eager to share what he knew with a fellow scientist.

Broer's Response

Physicist Matthijs Broer, 33, was working for AT&T Bell Labs. Like Ziegler, Broer was sincerely interested in getting the experiment to work in his lab. The audio recording reveals him to be an enthusiastic, energetic and perhaps contentious man with a quick wit and masterful elocution. After a dozen specific technical questions, Broer asked the most important question that anybody could have asked. Here's the complete discussion, starting with that question. He was polite, but his frustration was evident.

> BROER: OK, let me finish with one question. I'm really intrigued, and I must compliment you on your results. This is mind-boggling. But the thing that bothers me, and I know I speak for many, many people, is there's no really firm confirmation. That really annoys me. Maybe you have to be patient. Is there anything in your experiment that you have left out, [or failed] to mention to us, for perfectly good reasons? For proprietary reasons? That's perfectly acceptable to me. But I would like to duplicate this, and I cannot do it at this time.
>
> FLEISCHMANN: Which experiment are you trying to duplicate?
>
> BROER: I'm trying to do the rods, the wires. I'll mention it later in my review here; like many people, we're trying to duplicate, to the best of our knowledge, what you have done, and so far it doesn't work.

FLEISCHMANN: What doesn't work?

BROER: We don't see the yields in heat. We don't see the neutrons. We don't see the gammas. Absolutely zilch. What's going on here?

FLEISCHMANN: Well, I don't know. You'll have to tell us what you have done.

BROER: OK. That's what my question is. Is there anything that you have left out, and am I wasting my time?

FLEISCHMANN: No.

GARWIN: Do all of the samples work?

FLEISCHMANN: I've told you, for example, that the 8 millimeter rod did not work. I would have to see what experiment you have done. The only [successful] experiment I know on the thermal measurements to date is the Texas A&M [Charles Martin] experiment.

BROER: Right, but the nuclear experiments have been done at my Bell Labs and many, many other places, with virtually the same results. We just don't see what you're seeing as of now.

FLEISCHMANN: I'm terribly sorry. I have no detailed information about those experiments whatsoever. So I cannot comment on that.

BROER: But you have not left anything out?

FLEISCHMANN: I have not left anything out.

But Fleischmann and Pons — or the attorneys — did leave at least three crucial requirements out:

1. The deuterium-to-palladium ratio needed to be near 1.0.
2. The minimum time required to allow that much deuterium to soak into palladium. Physicists were unlikely to have known this.
3. The requirement for a dynamic trigger.

The third requirement needs explanation. Fleischmann and Pons knew well that excess heat was never produced when the cell was left with all conditions, such as power and temperature, running steady.

Some sort of abrupt stimuli was required to perturb the cell to trigger excess heat. Fleischmann and Pons often referred to these as "sharp edges."

A common method used by Fleischmann and Pons was to suddenly increase the current. But quickly turning the current down worked as well. This is how their 1 cm³ palladium cube vaporized. I've spoken with other researchers who have also seen intense reactions occur in their experiments after turning the current down or completely off too quickly.

They also omitted a critically important analytical fact: Although the data in their manuscript showed that heat, rather than neutrons, was the dominant signature of the reaction, Fleischmann and Pons failed to make clear to other researchers that they should not expect to see large fluxes of neutrons and, therefore, should look for heat, instead. When some scientists went looking for neutrons that were nearly at background level, Fleischmann and Pons were partly to blame for not explicitly telling them to look for heat, instead.

These omissions, coupled with the fact that the two electrochemists appeared on national television and told the world that it was a "relatively simple experiment," were disastrous. The omissions interfered with most scientists' ability to repeat the experiment, they interfered with the proper operation of the scientific method, and they made Fleischmann and Pons more enemies than they could count.

Monti's Response

Roberto A. Monti, 44, a physicist with the Instituto TESRE of the National Research Council in Bologna, who trained as a biophysicist, also attended the workshop. Monti's interests were, and still are, in low-energy nuclear transmutations and neutron reactions. He was well aware of the $e + p \rightarrow n + \nu$ reaction that could produce neutrons in the stars. Here's the exchange between Fleischmann and Monti in Erice:

MONTI: I suppose that, in your apparatus, you have realized three different kinds of nuclear reactions. One is the fusion of deuterium, but I think that is not the dominant one, and I justify this by the lack of neutrons. The second is the nuclear reaction which may have happened between neutrons and gamma. Again, I don't think that it's the dominant one. I think that the dominant one may be what is known as the Kervran Effect, which constitutes the absorption by neutrons.

FLEISCHMANN: The dominant one is what?

MONTI: The Kervran Effect, which constitutes the absorption by palladium of neutrons, giving different isotopes of palladium. And the second aspect, the absorption of hydrogen by palladium, giving origin to silver atoms. So I want to ask you: Did you examine the initial composition in isotopes of your electrodes? And after the experiment, did you measure the variation of composition of isotopes and the search for silver atoms?

FLEISCHMANN: No. That is something which will have to be done. We have not been able to do an isotopic analysis on that. Our view is that the primary neutrons produced would be of such high energy that they would escape. The maximum cross-section for fast neutrons in palladium, as far as I know, is about 8 barns. So I think the chance of getting a fast-neutron reaction is small. If we have thermalized neutrons, I accept that.

MONTI: No, this is not a fast-neutron reaction. It's a slow neutron.

FLEISCHMANN: Yeah, but I don't see how you can get a slow —

MONTI: I only wanted to ask you, did you make an analysis of the composition of the —

FLEISCHMANN: No.

Corentin Louis Kervran (1901-1983) was a French scientist who is best-known for his controversial research in biological transmutation. Kervran's work didn't relate specifically to palladium, but Monti thought there was a connection to Fleischmann and Pons' work.

CHAPTER 11

Problems and Possibilities

Trying to Solve the "Cold Fusion" Riddle

F ollowing Fleischmann's presentation at the Erice Fusion Forum workshop, scientists attempting to replicate his results spoke. Garwin had a reserved place on the agenda, and he used his time to criticize the Fleischmann-Pons experiment. Following Garwin's talk, theorists pitched their ideas: some in support of and others in opposition to the idea of "cold fusion."

The experimentalists were physicist Matthijs Broer (AT&T Bell Labs), nuclear chemist Richard L. Hahn (Brookhaven National Laboratory), physicist James Ziegler (IBM), and physicist Francesco Celani (Italian National Institute of Nuclear Physics in Frascati).

Broer's Best Effort

Matthijs Broer described his group's unsuccessful efforts to duplicate the Fleischmann-Pons experiments. "What we have done," Broer said, "is concentrated on palladium only; we have used rods, wires, electro-deposited films, and foils. We have the advantage, purely coincidentally, that AT&T owns 75% of the palladium stock in the United States. So we have no shortage of palladium. We can make rods as big as we want and do anything we want."

At Bell Labs, he explained, he had access to virtually every sophisticated scientific instrument he might want. Broer said that his team had done very preliminary calorimetry but that the focus of their efforts was to detect radiation. They found nothing. Broer made it clear that their instruments could have detected neutrons, had they been present, even at the low rates claimed by Fleischmann and Pons.

"It's been just nothing," Broer said, "for the last two weeks. Our back-of-the-envelope calculation, in terms of an upper limit of decay rates, is less than 0.6 neutrons per second per cm^3."

Broer said they tried to analyze the amount of deuterium in the palladium, but he did not reference whether, during the two weeks of their experiments, they achieved the required loading ratio. Broer did not indicate whether he was aware of the required loading ratio, the time required to obtain that ratio, or the need for a dynamic trigger. It seems clear now that without the proper ratio and dynamic trigger, they would not have seen any positive results. Broer had a (correct) sense of the importance of materials preparation.

"I think an important issue here," he said, "is the materials characterization. We've just begun to do that. If any of this stuff is true, the metallurgy, defects, and surface chemistry will be key."

To this day, materials preparation, particularly surface geometry and treatment, appear to be key factors in obtaining fully reproducible excess-heat results.

Hahn's Outlook

Richard L. Hahn, a nuclear chemist, gave a 10-minute review of experiments taking place at Brookhaven. None had shown significant results, but one researcher measured some neutrons slightly above background levels. Hahn said that a *Wall Street Journal* reporter had talked with that researcher, and then the *Journal* published news that Brookhaven had tentatively confirmed the Fleischmann-Pons work.

"I commented to my wife," Hahn said, "that having a tentative confirmation was similar to a woman being slightly pregnant, and she agreed with me." He got a few laughs.

Hahn was aware of the "really long initiation periods in terms of loading up the electrode and getting the effect to go." He said the other researchers at Brookhaven expected to get results right away. Hahn seemed to have the patience and perspective needed for the difficult problem that lay before them:

> Over the years, I've been involved in a variety of researches in nuclear chemistry and physics, among them trying to discover new elements at the top of the periodic table at accelerators, like element 106 for example, searches for super-heavy elements. Now, I'm involved in looking for neutrinos from the sun in a massive 30-ton gallium detector in a big collaboration here at the Gran Sasso lab.
>
> I mentioned that because low-level counting is a very special game in itself. Very often, when you build an instrument, and you want to do nuclear physics and count neutrons, you know that you are going to count neutrons. You're going to see them, and you're worried about resolution and efficiency and things of that sort, but background is perhaps a few percent of your signal. So you worry a little bit about it, but you don't spend a big part of your effort really trying to suppress the background instrumentally by shielding or things of that sort. In this [Fleischmann-Pons] kind of game, this may be critical. So I think one has to worry about that.
>
> But there have been a couple of other things that have been, perhaps, alluded to that I want to specifically point out. As part of the overall effort to find a new phenomenon, you have to address a couple of fundamental questions. If there are different characteristics of the process, if energy is coming out, if various particles are coming out, are these things really correlated? If you're going to talk about a mechanism, can you really demonstrate correlations between the emissions that reinforce your interpretation? That's extremely important from a positive viewpoint, but by the same token, are there things that should not happen if the effect is turned off? If you throw your switch and the current is off, does all of the radiation die away in a certain time, for example? It's a negative result, but it's very

supportive of the thing you are arguing about.

Another thing was brought to my mind from the Jones data: Is your effect reproducible in the sense that, if you have an electrode that is your best case, can you make 10 of those and do 10 experiments identically and in 10 different experiments always see the effect you're looking for? If you only see it in two or three of the cases, you start worrying about things that you have not considered yet.

The bottom line is, Are we seeing a nuclear process or a chemical process? Many of us think that whatever is happening is very exciting and interesting, but clearly a lot more has to be done before all of us really understand what's going on. I hope you'll be hearing more from Brookhaven in the weeks ahead.

Hahn had put his finger on the mode of thinking required to solve the problems presented by these experiments.

Dictum of Reproducibility

Hahn's presentation drew an important philosophical point about science from Zichichi and a quarrel from Garwin.

Hahn told the workshop attendees about the neutron rates the Brookhaven researchers had seen. He also mentioned that they could not repeat the results very often. Both Zichichi and Garwin said this meant that the results were not scientifically valid.

"If an experiment gives sporadic reproducibility, it is to be considered of a nonscientific nature," Zichichi said. "So the experiment ought to be reproducible and should always give the same results; it cannot be sporadic."

This is an incorrect interpretation of the scientific method. There is no such rule in the scientific method. Nothing in the scientific method says that a carefully performed and well-measured experiment that sometimes succeeds is unscientific. In fact, it may mean that one or more necessary factors or parameters are not well understood.

There is also no rule that says every scientist who attempts someone else's experiment must succeed in order for the original experiment to be valid.

Ziegler's and Celani's Attempts

Physicist James Ziegler gave a talk demonstrating, as Broer had, that although he had made a sincere effort to replicate the Fleischmann-Pons experiment, given the limited details that he had, Ziegler, too, saw no positive results.

Physicist Francesco Celani, who was working at the Italian National Institute of Nuclear Physics in Frascati, started his replication attempt on March 27, 1989. At Erice, he reported that he had not found any convincing results.

On the recording, Fleischmann didn't seem perturbed by Ziegler's, Broer's, or Celani's failures and he didn't ask them any questions to determine whether they had applied the necessary experimental parameters.

The experimentalists were obviously frustrated; they had expected the experiment to be easy to replicate. They blamed Fleischmann and Pons because the pair had said on television that it was "relatively easy" to do the experiment yet it was turning out to be anything but easy.

Garwin's Problems

Garwin's turn at the podium was next, but he didn't talk about IBM's experiments. Rather, he spoke for 20 minutes, complaining about "problems" with the Fleischmann-Pons and the Jones experiments.

Like Zichichi, Garwin brought up the Reproducibility Dictum. "Before we can have physics," he said, "we have to have a reproducible effect." Garwin highlighted Jones' neutron counts. "Even though one can argue that it's 170 counts, plus or minus 23, that is not persuasive," he said. Garwin criticized Jones because Jones had not obtained more data in the preceding three weeks and suggested that the lack of new results meant that something was wrong with Jones' first set of results.

During the question-and-answer session, Garwin's complaint came up again.

> GARWIN: "OK, then, let me ask this question. There's a lot of time since March 23, 1989, when your pre-print was concluded. Why are there no more data like run number 6?"
>
> JONES: We've actually done work on the experiment, not on taking data but on improving the electronics and preparing for future experiments.
>
> GARWIN: OK, that would not be my priority.
>
> JONES: Well, that is our priority, thank you.

Later, during the questions, another participant minimized the criticism from Garwin, the "greatest experimental physicist living." "One short comment concerning the last question from professor Garwin to professor Jones," the participant said. "Our impression is that, in this type of measurement, measuring backgrounds and accurately calibrating electronics may turn out to be, if not as important as measuring new data, even more important."

Garwin quickly moved on to the Fleischmann-Pons experiments and offered a laundry-list of complaints. "Now, on the University of Utah results, Fleischmann-Pons results," Garwin said, "there are more questions because they have not been content with finding fusion as detected by the neutrons and the gamma rays; they have insisted on finding a billion times as much heat. And this is a totally different problem, as I will explain."

In contrast, Monti, another physicist, immediately recognized that, regardless of what Fleischmann and Pons called their work, it was clear that it wasn't fusion; the experimental data proved it. In fact, many people who spoke that day acknowledged that the low rates of neutrons were, in fact, a *signature* — however inexplicable — of the phenomena. Garwin elected to portray that signature as a serious problem.

Garwin discussed some real problems with the Fleischmann-Pons gamma-ray spectrum, and he was justified to tear apart that data. Garwin then moved on to the excess heat and conveyed his disbelief. "The heat, of course," Garwin said, "is the phenomena-squared discovery; and there are two aspects to this heat. There is the explosive release of heat, and that resulted in partial vaporization of the palladium cathode and destruction of the apparatus. There is no reason to believe that that is nuclear at all, let alone fusion."

"Let me point out what I think may be going on," Garwin said, "and since this is at least as plausible, it is required that it be negated by argument or experiment."

But Garwin was not in a position to argue about any specific data from the vaporization incident: There were none, apart from a recognition of the minimum energies that would have been required. Fleischmann and Pons had only anecdotal evidence from that incident. Garwin made his case for the "required negation." He imagined an alternative, non-nuclear explanation. He explained what could have happened and what would have happened, based on an *ad hoc* speculation, without any physical test. "So I believe that is the more likely explanation of what is going on in the explosive release," Garwin said.

At the end of Garwin's talk, Fleischmann addressed only a couple of the "problems" that Garwin presented. Fleischmann argued against Garwin's hypothetical explanation for the vaporization incident. There is a break in the original recording, but Fleischmann's point still comes across clearly:

> The only point I'll make about the disintegration of the palladium electrode is this. I've asked many people whether that has ever been observed, and all the people I have asked have said, "No, that has not been observed." And I'll just make one other observation: Electrochemical equivalence of palladium diffusion tubes have been used forever. You polarize one side of a palladium tube, and you accumulate hydrogen on the inside. To my knowledge, nobody has ever [25-second break in recording]. It's a perfectly standard piece of equipment. If there was any account of this thing disintegrating or melting or anything like that, it would surely not be a technologically useful device.

Coincidentally, as I report in Chapters 9 and 30, on the same day in California, Edward Teller's researchers at Lawrence Livermore National Laboratory, hard at work to replicate the Fleischmann-Pons experiment, observed an uncontrolled heating event that they said "may be the same phenomenon reported by Fleischmann and Pons."

In his talk, Garwin began discussing the Fleischmann-Pons experiments that, instead, released heat more gradually. "The steady release of heat," Garwin said, "is another problem. In order to persuade a cynic — and we must all be cynics when we are provided with such a new and far-reaching phenomenon — we need run-by-run details."

But Fleischmann and Pons had given run-by-run details on pages 3 and 4 of their paper. Garwin later said that Fleischmann and Pons needed to give him their raw data in order to convince him.

Despite a long list of issues (such as faulty thermistors or uneven electrolytic mixing) that he portrayed as serious problems (gamma peak notwithstanding) and his insinuation that Fleischmann and Pons didn't know how to do calorimetry, most of his arguments were irrelevant. In short, the experiments that Fleischmann and Pons performed with blank cathodes, which resulted in steady, continuous thermally neutral levels of heat (power in = heat out, with no excess), nullify most of Garwin's objections. For example, if a thermistor was unexpectedly sensitive to an accumulation of deuterium during an experiment that produced excess heat with an active cathode in heavy water, it should show the same response with an inactive cathode in heavy water.

Ponomarev's Limited Possibilities

The next part of the program featured several people who attempted to explain the Fleischmann-Pons and Jones results as nuclear processes.

Leonid I. Ponomarev, a theoretical physicist at the Soviet Union's famous Kurchatov Institute (where the first Russian nuclear reactor, the first Russian A-bomb and the first Russian H-bomb were developed), participated in the workshop. Also present was his colleague Semen S. Gershtein, a theoretical physicist at the Institute of High Energy Physics in Protvino. Both men were well-recognized for their pioneering work in muon-catalyzed fusion.

In Ponomarev's talk, he wasn't able to make any definitive proposition or come to any conclusion, but he did his best on such short notice. "During this day," Ponomarev said, "our plans have changed several times because of so much information, so many different people and opinions. Nevertheless, the main line was the same."

He reminded participants of the custom regarding theoreticians and experimentalists. As the saying goes, a theorist always believes his or her own theory, but nobody else does. An experimentalist never believes his or her own work, but everyone else does.

Ponomarev didn't believe that the Fleischmann-Pons reactions were nuclear because the experimentalists were still alive. It was a reasonable perspective based on the knowledge of physics at the time. He tried to suggest ways to overcome the Coulomb barrier. In the end, he could only shrug and offer a lighthearted resolution. "You can try one miracle to explain another miracle," Ponomarev said. It got a good laugh.

Koonin's Calculations

Koonin and his colleague Michael Nauenberg had, as I mentioned in Chapter 7, submitted a manuscript to *Nature* on April 7, in which they attempted to provide a theoretical explanation for Jones' and Fleischmann-Pons' "cold fusion." Koonin's presentation summarized the two primary points of the paper. Here is an excerpt from Koonin's talk:

> As a theorist, it's a little bit disconcerting to have so many experimentalists come up and tell you that they cannot reproduce the effects that are the subject of this workshop, and to know that there are so many people around the world trying very hard to do the same without any reported success yet. On the other hand, because of the potential importance of the two positive experiments that we have, in terms of both basic science and potential applications, I think they have to be taken seriously.
>
> So we have to ask, as theorists, What is really going on?

I think we have two choices, or at least there are two modes of attack. One is to do a kind of benchmark calculation to take a situation that you think you understand and calculate what the fusion rates could be. The other is to speculate about new mechanisms or new phenomena that might happen to explain the experimental results. I'd like to offer you a little bit of each in the next few minutes.

For those of you who are not nuclear physicists, we believe that there are no mysteries in the structure of light nuclei at this point. This is a well-studied subject. We've been doing it for over 50 years, and we think we understand the physics pretty well.

Koonin did not share Garwin's cynicism. He had assumed good faith on the parts of Fleischmann and Pons, and Jones. Koonin assumed that Fleischmann and Pons knew how to measure temperature accurately and that they understood how the stirring or sparging should take place. Koonin also assumed that Fleischmann and Pons had correctly converted their raw data into the published data. Koonin's last sentence is telling; it reveals how confident he and other physicists were that they knew everything of significance about the scope of nuclear physics.

Although Koonin was enthusiastic and respectful of Fleischmann and Pons at this meeting, he is best remembered for publicly accusing them of incompetence and delusion three weeks later.

In Erice, Koonin was willing to consider that Fleischmann and Pons had discovered something new in physics. He had tried his best to understand it. Koonin's presentation inspired a lively, interested and optimistic discussion among several of the participants, including electrochemists Gerischer and Fleischmann.

Gerischer spoke about the high loading of deuterium into palladium and how unusual it is. "It will only happen under dynamic conditions, and those diffusion coefficients decrease very rapidly," Gerischer said.

Koonin responded with an insightful comment: "Everything we know about the experiments is consistent with the phenomenon happening under unusual conditions, since it's not easy to reproduce." Clearly, Koonin did not subscribe to the Reproducibility Dictum.

At that point, Fleischmann responded to Koonin. The two men went back and forth pleasantly discussing aspects of physics, such as the Coulomb barrier, the Boltzmann factor and the concept of breaking symmetry. The productive and pleasant exchange between the electrochemist and the theoretical nuclear physicist ended in both men joking and laughing.

Wilkinson Wonders

Sir Denys Wilkinson was next to wonder how "cold fusion" might be possible.

> My task, which was handed to me by Nino Zichichi shortly before midnight last night is, roughly speaking, to ask if we can understand, in shorthand, how the heat generation can be very much greater than the neutron emission. I'm going to discuss this, to your surprise, in terms of the D+D reaction, and I'm going to conclude, to your astonishment, that we can, indeed, understand that inequality.

The most astonishing part of his presentation, however, was that he, like so many other scientists at the meeting, was sincerely trying to understand the reported anomalies.

Maiani Invokes Mozart

Luciano Maiani, a physicist at the University La Sapienza, reluctantly tried his hand at an explanation but couldn't come up with much:

> [This subtitle] summarizes my opinion about the discussion that we have had so far today. It is the same phrase that Don Giovanni [from Mozart's opera] says when the marble statue that he has invited to dinner actually shows up at the table. "I would have never believed that, but

I'll see what I can do." [audience laughter]

So what I can do — as you can see, [laughs] will be something which can be encouraging but not completely. In a way, many of these things have already been said, especially by professor Ponomarev. I share his views to quite a considerable degree.

Maiani's summary included a suggestion that the gamma radiation might somehow be suppressed.

If you see neutrons but don't see gamma rays, at least at the same level, then that would suggest some form of quenching, due to the surroundings. So the final question is whether cold fusion is here to stay. Maybe. I'm moderately optimistic; however, something mysterious must still be happening, so somebody's going to eat his hat in the end.

A week later, when Garwin reported on the workshop for *Nature*, he quoted Maiani's comment about somebody eating his hat.

Breaking News From Soviet Union

Before the question-and-answer session for Maiani began, Zichichi grabbed the microphone to announce breaking news.

ZICHICHI: Keep quiet! This is news coming from — apparently cold fusion has been confirmed in the Soviet Union.
[Someone in the background shouts, "Ha, ha!" There is a lot of chatter. Another person asks, "Which lab?"]
ZICHICHI: Nothing is known, but it is an announcement given by TASS — for you to know everything.
MAIANI: Can you get the announcement? Can you give more details?
ZICHICHI: Just nothing. There is nothing more.

Everyone was talking. This went on for half a minute until Garwin redirected everyone's attention to the sobering matters at hand. He began to ask the first question of Maiani, but he had to talk over the background chatter about the exciting Russian news.

Back to the Future: Ancient History

Wrapping up the day, Heinz Gerischer, a prominent electrochemist from Germany, delivered some forgotten news that he had just learned, thanks to colleagues of his who faxed him a copy of the 1926 paper "On the Conversion of Hydrogen into Helium," by Fritz Paneth and Kurt Peters. Gerischer briefly explained the paper, and it is clear from the way he described the paper that it was news to him. This suggests that, by the time Gerischer began studying chemistry at the University of Leipzig in 1937, the Paneth and Peters work had already been forgotten.

Gerischer had not yet done the research to learn more about the ensuing controversy and Paneth and Peters' attempts to discredit and retract their own work.

On April 27, 1989, Steven Dickman wrote an article in *Nature* (Dickman, 1989) and cited Paneth's 1927 retraction in, among other places, *Nature.* (Paneth, 1927)

I found it curious that, in his retraction, Paneth wrote that "the liberation of helium from glass (and from asbestos) is dependent on the presence of hydrogen [and that] glass tubes which gave off no detectable quantities of helium when they were heated in a vacuum or in oxygen were found to yield helium ... when they were heated in an atmosphere of hydrogen."

That 1926 paper led me to discover not only that controversy but also a lost era of nuclear transmutation research, which I report in *Lost History: Explorations in Nuclear Research, Vol. 3.*

End of the Day at the Fusion Forum

The Erice Fusion Forum workshop recordings provide a unique look at this history. They show that the real problem with "cold fusion" had nothing to do with Fleischmann and Pons' gamma measurements, even though they were wrong. At Erice, only Garwin was bothered by the faulty gamma-ray spectrum.

The greatest problem, initially, with "cold fusion" was not the discrepancy between the observed experimental data and the prevailing theory of nuclear fusion. Clearly, most of the participants at Erice accepted that as a *signature* of the phenomena.

The larger problem with "cold fusion" was that it was difficult to repeat. It was even difficult for Fleischmann and Pons to repeat 100% of the time.

The tone of the discussions at the workshop was optimistic. The behavior was professional and productive. Despite Ziegler's failure to repeat the experiment at IBM, his demeanor and his approach to Fleischmann showed no cynicism or hostility, in contrast to that of Garwin.

As discussed in Chapter 10, Wilkinson and Koonin had hit Fleischmann hard with the question on light water; nevertheless, the conversation remained scientific, and both men attempted to explain the reported phenomena in good faith.

An effective two-way scientific exchange on this topic between involved and uninvolved parties never recurred. Within three weeks, the topic degenerated into an ugly display of unscientific behavior that few of the original participants would ever want to remember.

Because these details of the Erice Fusion Forum workshop have never been published before, the 1989 "cold fusion" conflict has been primarily characterized by the vicious events that took place three weeks later at the American Physical Society in Baltimore, Maryland.

History has forgotten that Fleischmann, at CERN and at Erice, was welcomed and treated with respect and dignity and like an honest, ethical scientist.

On that same day, separated by a few thousand miles and a few time zones, chemists and electrochemists had gathered in Dallas, Texas, for the American Chemical Society national meeting and welcomed Stanley Pons like a hero.

CHAPTER 12

The Woodstock of Chemistry

7,000 Gather for Dallas Symposium

On April 12, 1989, an unprecedented event in American chemistry took place: 7,000 chemists gathered in Dallas, Texas, to listen to one man speak — Stanley Pons.

The American Chemical Society Cold Fusion Symposium, held in conjunction with the organization's annual meeting, was dubbed "The Woodstock of Chemistry" because it was, for chemists, the once-in-a-lifetime spectacular event that the Woodstock Music & Art Fair was for music fans in 1969. The people I spoke with who went to this ACS meeting and to the corollary American Physical Society "Woodstock of Physics" meeting a few weeks later were all certain they would never again attend such exciting and significant science meetings.

This ACS meeting was the first of three U.S. science meetings at which "cold fusion" was the main attraction. These meetings had major impacts on the new field. The first one, the ACS meeting in Dallas, accelerated interest and activity. The latter two, the American Physical Society meeting in Baltimore and the Electrochemical Society meeting in Los Angeles, had a chilling effect.

Enthusiasm and Optimism

Scientists around the world were scrambling to replicate the Fleischmann-Pons experiment. The only laboratory at that time to publicly give an official negative verdict on "cold fusion" was Mark

Wrighton's, at MIT. European scientists had just finished listening to Martin Fleischmann and Steven Jones at the Erice Fusion Forum workshop. In Dallas, Pons was welcomed like a rock star. Enthusiasm and optimism for "cold fusion" were at an all-time high.

Partial view of the Dallas Convention Center arena, April 12, 1989. Image: James Krieger/Chemical & Engineering News

The official announcement for the ACS symposium went out to the news media only three days in advance, but by then the news had traveled far by word of mouth. The headline on the flyer for the circus-like symposium read "You Read About It in the Paper. Come Hear the Experts Discuss It." The symposium was called "Nuclear Fusion in a Test Tube." It included a panel of scientific talks, a question-and-answer session, and the obligatory press conference. The panelists were Harold Furth, Allen Bard, Ernest Yeager, Stanley Pons, and K. Birgitta Whaley.

An Unprecedented Event

The impact of this meeting was immense: 7,000 chemists newly interested in fusion, previously the exclusive domain of physicists.

The impact of the meeting was also felt by the U.S. Energy Secretary as well as by staff members in the White House. Excitement was in the air when ACS President Clayton Callis took the podium. His prepared comments were thoughtful, and he was enthusiastic:

Good afternoon! Welcome to the president's special event. I'm Clayton Callis, president of the American Chemical Society.

This scientific meeting has to be a precedent-setting event for the American Chemical Society, both in attendance and in general interest. The potential benefits for all humanity of the use of controlled nuclear fusion as a source of electric power have been recognized in the scientific community since the days of the Manhattan Project at the end of World War II. Starting with Project Sherwood in the early 1950s, the federal government and other countries have funded what was and is hoped to lead to a method of initiating, sustaining, and controlling the fusion of deuterium and tritium nuclei.

Many billions of dollars have been invested in attempting to reach a net energy gain through magnetic or inertial — that is, laser or particle beam — confinement, heating and compression of fusion fuel. This research generally involves temperatures of hundreds of millions of degrees and, in some instances, comparable pressures. While much has been learned about plasma physics, and while much progress has been made, the goal has remained elusive, and the large, complicated machines that are involved appear to be too expensive and inefficient to lead to practical power.

Now, it appears that chemists may have come to the rescue. [Applause] I first learned about cold fusion when I read of Dr. Pons' work in the *Wall Street Journal* on March 24. The next day, I discussed it with ACS board chairman Ernie Eliel, and we agreed it would be a good idea to see if Dr. Pons would be willing to address this ACS meeting. In the meantime, Valerie Kuck, chairman of meetings and expositions, was checking into the possibility of such a program. As president, I authorized her to proceed, and this session is the result of her efforts and arrangements. Of course, you wouldn't be here unless you shared this excitement.

When, in 1980, Congressman Mike McCormick introduced what was to become the landmark fusion law that bears his name, he stated, "The practical development of nuclear fusion power will be the most important energy-related event in human history since the first controlled use of fire." Today, we are discussing what may be the first step of that development, and I am proud that this is happening at the American Chemical Society meeting and that, once again, a chemist is reporting on the results of his work, which may be of great service to all mankind.

In the book, *Science, Reason, and Rhetoric,* Trevor Pinch perceived Callis' comment about chemists coming to the rescue as a humorous jab at fusion physicists. It was not. Callis was serious. Pinch also wrote that the audience laughed. The videotape shows that they applauded, but no laughter is audible. (Pinch, 1995)

I had the opportunity to interview Valerie Kuck about this meeting on Oct. 29, 2014. The interview transcript is available on the *New Energy Times* Web site.

Science as Usual, Almost

None of the speakers on the panel who followed Callis echoed his enthusiasm. They all spoke in the dispassionate tone expected of scientists. Even Harold Furth, the director of the Princeton University Plasma Physics Laboratory, representing traditional fusion research, was polite and refrained from denouncing the claims of "cold fusion."

Furth talked about fundamental concepts of fusion. He also made it clear that he, like all physicists, was puzzled about the lack of large neutron fluxes in the Fleischmann-Pons experiments. The talks by Bard and Yeager seemed to provide useful information. Kuck had invited Fleischmann as her first choice. Fleischmann, who was a Fellow of the Royal Society and former president of the International Society of Electrochemistry, had greater name recognition than Pons. Fleischmann, instead, spoke in Erice, Italy, that day.

Harold P. Furth, nuclear physicist

"We did ask Dr. Martin Fleischmann to be with us today," Kuck said. "For a brief moment, he was going to be with us, and then things changed. He [Fleischmann told Kuck] is visiting today with the prime minister of Italy, so he will not be with us. But we did extend an invitation; he did accept, and then there was a change."

Before introducing Pons, Kuck passed on the same news that Antonio Zichichi had announced in Erice just a couple of hours earlier.

"Before I have Stanley come up," Kuck said, sounding like a TV announcer, "I'd like to comment that we were just informed before we came on that a Dallas radio station has reported that the University of Moscow has just announced that it has successfully repeated the Pons-Fleischmann experiment." A wave of applause rolled through the arena.

When Pons took the podium, he explained the mundane details of his and Fleischmann's experiment, the results, and how they came up with the idea. He seemed relaxed and at home among fellow chemists. Overall, he appeared humble but didn't miss the chance to throw in a jab at plasma physicists. He got a roar from the audience when he displayed a photo of his "cold fusion" cell and called it the "U-1 Utah Tokamak."

The question-and-answer session went for more than an hour, with

Pons responding to most of the questions. He answered question after question, directly, without hesitation, and he had immediate, clear answers to every question except two. This contrasts with the statement that fusion physicist John Soures gave to the Associated Press, that "nobody is getting any answers" and that all attempts were met with "polite silence."

A member of the audience asked Pons, "Prometheus, Pandora or Piltdown Man: Which are you, and do you have any thoughts?" Pons chuckled and said, "No comment."

Joe Templeton, from the University of North Carolina, commented on the character of Pons' responses. "The ACS has organized an excellent forum for discussing this objectively," Templeton said, "and I'm amazed that professor Pons has been able to respond so accurately to all these details."

The affiliations of the questioners reflected widespread interest. They included Lithium Corp., Westinghouse Savannah River, Brookhaven National Laboratory, Lawrence Berkeley National Laboratory, Dow Chemical, Ralston Purina, Kodak Research, National Institute of Standards and Technology, General Motors Research Laboratory, Advanced Fuel Research Corp., Chevron, AT&T Bell Labs, Air Force Academy, Lawrence Livermore National Laboratories, and many U.S. and some foreign universities.

Notably absent among the chemists who took the opportunity to ask questions was Nathan Lewis of Caltech, who later complained to the *New York Times,* "Pons would never answer any of our questions."

A scientist from Westinghouse Savannah River asked Pons about the working parameters of his and Fleischmann's 1985 vaporization incident. It was one of the few times that he or Fleischmann provided these parametric details. "1.5 amperes for about seven months," Pons said, "then dropping the current too quickly, to about 0.3 or 0.4 amps."

Light-Water Problem

A scientist from MIT gently asked Pons about light water: "Have you thought about or tried any control experiments with H_2O instead of

D₂O?" Pons' amiable but ambiguous response was drastically different from that of Fleischmann when he had been asked a similar question several hours earlier in Erice.

"Yes," Pons said. "Several people are looking at this right now, including us. With the confinement parameter which we have, that sort of reaction might be interesting."

Pons knew a little more information than he was revealing. Obviously, so did Fleischmann, who had refused to respond to the question. In Dallas, this was the only question from the 7,000 chemists about a light-water control experiment. None of the other attendees of the meeting pushed the question. However, reporters, who were not allowed to ask questions during the scientific session, pressed the issue later, at the press conference.

Furth told the audience that, as soon as someone could demonstrate that a light-water experiment produced the theoretically expected null result, nuclear physicists would take "cold fusion" seriously. But that proposed test was predicated on three assumptions. The first assumption was that Fleischmann and Pons had not accurately measured the excess heat in their heavy-water experiments. The second was that Fleischmann and Pons had not performed any control experiments. And the third was that the heat effects were exclusive to deuterium. All three assumptions were wrong.

Storage Mechanism Concern

Fleischmann explained in his talk at CERN that, before they triggered the cell to get excess heat, the cells were running at a thermal-neutral baseline: No power was gained, and no power was lost. Because the cells were thermal-neutral during the loading phase, accounting for the minor amount of heat released by the absorption of deuterium into palladium, there was no energy storage during the loading phase.

Therefore, any periods of subsequent excess heat were also periods of excess energy. But it would take a year before Fleischmann and Pons showed this in a paper. Without understanding the nature of the thermal balance during the loading period, some critics were unwilling to accept the validity of Fleischmann and Pons' excess-heat claim based on power alone. They asked to see data representing the total integrated energy for the duration of the experiments. Pons responded to a similar question about this from an MIT chemistry graduate student.

"The most recent experiment that has run," Pons said, "for approximately 800 hours, has now 50 MJ of energy. There is no conceivable known chemical reaction [that would produce this]. Even if you consumed all of the matter in the cell by any known chemical reaction — the palladium, the glass, and everything else — you could not generate, within orders of magnitude, that much energy."

Reproducibility Challenges

Someone from the University of Illinois asked a key question: "Why is it so difficult to reproduce your experiment? It should be trivial." Pons explained that the loading time, or charging time, was a crucial factor. The experiment would not produce any excess heat without the required amount of deuterium loaded into the palladium cathode:

> The reproduction of the thermal data simply takes time. It takes a fairly long time to charge these palladium lattices. If you make measurements on the smaller electrodes, then the heat effects are very small, and they're quite difficult to measure. A 4 mm electrode — where you get degrees of difference in temperature — takes 10 days or so. And you really should wait a couple of hundred hours before you start seeing the heating effect. For instance, a 20 mm rod, which we want to do, will take about 18 months to charge up.

The target zone (or ratio) for loading deuterium into palladium was about 1.0 (also written as a 1:1 ratio). Later research showed that D:Pd loadings of 0.9 generally work as well. Nathan Lewis' team at Caltech achieved loadings of only 0.77, 0.79 and 0.80. Lewis may have had a huge team of skilled researchers, the world's best detection equipment, and dozens of cells running simultaneously, but none of that mattered.

He and his team didn't wait long enough for the electrodes to load enough deuterium. (Lewis, 1989)

Fleischmann and Pons bore much of the responsibility for this failure. They did not specify the required loading ratio in their preliminary note. However, as shown in Chapter 9, Gary Holland, from the University of Utah Chemistry Department, told the Caltech team on April 1 that "the catch seems to be that you need to go to very high D_2 within the Pd before anything happens — like near 1:1 D:Pd loadings."

Other factors were required before "cold fusion" cells began producing excess heat, but the loading was a critical one. It was also a relatively manageable parameter; researchers just had to be patient. In contrast, perhaps the most unmanageable parameter was the metallurgy of the palladium cathodes. For researchers experimenting with the Fleischmann-Pons electrolytic method, even today, all the metallurgical characteristics that lead to excess heat are not generally known.

I do not know whether Fleischmann and Pons ever knew the specific metallurgical characteristics that were required. But their supplier, Johnson Matthey, claimed to know important characteristics about the working cathodes. According to a statement by a Johnson Matthey representative in a 1993 Japanese documentary by NHK-TV, the company considered the metallurgical details a trade secret. That secrecy, of course, was of no help to the general scientific community.

Press Conference

The ACS press conference started around 3 in the afternoon. John Harrington, of KTVX-TV, in Salt Lake City, began. "Dr. Pons," Harrington said, "a question more on the human level rather than the scientific level: Describe in a way as *unscientific* as possible how you personally felt when you first realized your observations might be what you think they are now. What were you thinking of the implications of your experiment at the time?"

"We had results very early on," Pons said. "We were quite excited and saw high temperature vaporization of the metal. We were quite excited at that point. We continued making measurements for the next five years." Pons' response to an achievement that *Time* magazine called "the most important discovery since fire" was remarkably dull. But this was Pons; he was not a showman or an entertainer, save for a few jokes here and there about physicists and tokamaks.

Then the hardball questions came, starting with Lawrence Framburg of the *New York Tribune*. "Dr. Pons," Framburg asked, "have you had an opportunity to make observations of the branching ratio between the tritium and neutrons?"

"To the extent that we see the conventional deuterium-deuterium reaction," Pons said, "we see possibly equal amounts, 50-50, of the tritium and neutrons."

That wasn't correct. I can't explain why Pons said that. Neutron production was much lower than tritium production.

Light-Water Problem, Again

Three reporters, the first one unidentified, then asked pointed questions about light-water (H_2O) versus heavy-water (D_2O) results.

"Dr. Pons," the reporter said, "Dr. Furth raised — and I've heard many other people raise — the issue that these results would be much more convincing if they saw the results of what seems to be the obvious control experiment of running the experiment with all conditions identical but with H_2O rather than D_2O. Why wasn't that something that would've been done before your initial announcement? Are you doing it now, and how soon do you expect results from that?"

Pons knew that he couldn't answer the reporters' question satisfactorily. He knew that he and Fleischmann had seen light-water excess heat in recent tests they had performed. Pons did not explain that the light-water excess-heat response was much smaller than the heavy-water excess-heat response, but he may not have known this with certainty. He and Fleischmann appear to have performed only one light-water experiment, started in January 1989. Pons was more forthright about the light-water question than Fleischmann.

"I'm aware of many groups carrying out that reaction," Pons said, "and I'm not convinced that that's a baseline reaction, that that is a control reaction." Jeffrey White with the *Dallas Morning News* asked Pons about light-water control experiments. "Does that mean," White said, "you have some reason to believe that an H_2O reaction would also produce excess heat as well as the D_2O reaction?"

"It's not out of the question," Pons said. His answer took courage, because he knew that critics would erroneously interpret light-water excess heat as evidence that heavy-water excess-heat measurements were a mistake.

Fleischmann and Pons had put the idea into everyone's minds during the University of Utah press conference that the duo was seeing deuterium-deuterium nuclear fusion. Adding to the confusion, Fleischmann and Pons had not communicated clearly how their blank cathodes worked as controls.

After Jeffrey White, his colleague Tom Siegfried, also with the *Dallas Morning News*, pushed further on the light-water issue. "The question of the reaction in normal water?" Siegfried asked. "It came up in the meeting, and we've been hearing rumors today that other groups have in fact been seeing heat production in the regular H_2O reaction, and your answers, before and now, have been less than completely direct."

"It is just too early to speculate," Pons said. "We have very little data accumulated."

"What does the data you have accumulated so far suggest?"

"That a baseline reaction run with normal water is not necessarily a good baseline reaction," Pons said.

Pons knew that physicists had wanted to see results from a light-water experiment as a control. Pons knew that physicists were assuming that, if the "incompetent chemists" (as a few of them said) found excess heat in light water, then it was proof that the electrochemists didn't know how to measure heat and somehow had made a mistake, because deuterium-deuterium fusion could not happen without deuterium.

The question of light-water versus heavy water continued to be a growing concern and source of confusion as well as mockery in the next few weeks. Cartoonist Brian Basset precisely captured the physicists' attitude toward the idea of light-water excess heat.

Physicists rolling on the floor laughing at the idea of light-water excess heat.
Cartoon by Brian Basset

When researchers later began performing gas-loading experiments, similar confusion developed when they used hydrogen gas rather than deuterium gas. The question of light- versus heavy-hydrogen continued to provoke confusion and debate for years.

In April, researchers at an Italian national laboratory made headlines with a deuterium-gas and titanium experiment.

Francesco Scaramuzzi with ENEA administrators during April 18, 1989, press conference (Caricature by Arturo Aguirre based on Reuters photo)

CHAPTER 13

The Bets Are Placed

Rush to Pronounce or Denounce

In early April, "cold fusion" gained momentum. Experimental confirmations had come in from Texas A&M University, University of Moscow, and Georgia Institute of Technology. (Georgia Tech scientists withdrew their claim several days later, citing an instrument error.) Many other attempts failed, but they did not make the headlines. Such failures led to frustration among many researchers.

"Cold fusion" experiments were being conducted in dozens of U.S. government and university labs, in Brazil, Korea, India, Czechoslovakia, Mexico, Italy, Germany, Canada, Argentina, and other nations. A congressional subcommittee agreed to redirect $5 million from the Department of Energy's magnetic fusion program to "cold fusion." The Utah state legislature approved a $5 million appropriation for "cold fusion" research. Korea would soon commit $1.5 million for "cold fusion" research.

Two major science conferences featuring "cold fusion" had occurred: one in Erice, Italy, and the other in Dallas, Texas. The April 12 American Chemical Society meeting in Dallas, filling an arena with 7,000 scientists, gave the appearance of a victory lap for "cold fusion."

Meeting in the White House

The following day, April 13, Glenn Seaborg's telephone rang. Seaborg was a preeminent nuclear chemist and former chairman of the

U.S. Atomic Energy Commission with a long history of advising the federal government on important science matters. He worked at the University of California, Berkeley, and was an associate director-at-large at the Lawrence Berkeley Laboratory. Seaborg also shared the 1951 Nobel Prize in chemistry.

The Department of Energy wanted him in Washington, D.C., immediately. On Oct. 6, 1992, Seaborg provided excerpts from his journal to Bruce Lewenstein at the Cornell Cold Fusion Archive.

Before leaving for Washington, Seaborg made a few phone calls to get the latest information on "cold fusion." One call was to John Nuckolls, the director of the Lawrence Livermore National Laboratory. A day earlier, a "cold fusion" experiment at Livermore had blown up in a way that defied conventional explanations. This had been reported the same day by Dan Stober in the *San Jose Mercury News*.

When Seaborg arrived in Washington, he was picked up at the airport by Robert Hunter, director of the Department of Energy's Office of Energy Research. Hunter handed him a draft of a DOE summary of the status of "cold fusion."

Seaborg checked in at the Capitol Hilton and called his colleague Bogdan Maglich, who gave him an update. Seaborg learned that "people at Princeton, Livermore, Brookhaven and Harwell were failing to find the neutrons." Maglich also told Seaborg that he should call Art Kerman, at MIT, to get an opinion about Hagelstein's ideas. "Kerman expressed extreme [doubts] about the cold fusion work," Seaborg wrote, "and said that the physicists at MIT have disowned Hagelstein because they have such little faith in his calculations."

To the physicists at MIT, Hagelstein was a disgrace because of his support of "cold fusion." Whatever open-mindedness about "cold fusion" existed among scientists at Erice didn't make it back to MIT.

The next morning, Seaborg met with Hunter and, Seaborg wrote, he "suggested to [Hunter] that there should be an impartial panel created for the purpose of making an assessment of this cold fusion situation, and Hunter immediately agreed." Hunter and Seaborg then attended a meeting in the office of U.S. Energy Secretary Admiral James Watkins. Two other DOE staff members, John Tuck and Steve May, were there. Together, they planned and prepared for their 11:15 a.m. meeting with

John Sununu, White House chief of staff for President George H.W. Bush.

Seaborg said he summarized the Fleischmann-Pons and the Jones work, but he did not record in his journal whether he mentioned the Texas A&M or University of Moscow claims or the Livermore explosion. He told Sununu that there was a lot of doubt about the claims even though there had been a "tremendous reception in favor of the work" at the annual ACS meeting. Seaborg also said that, "on the other hand, a number of nuclear physicists take a very dim view of this development and, in effect, do not believe it."

Seaborg wrote that he told Sununu that he was highly doubtful but that it should be investigated by a panel. "Watkins indicated that the appointment of such a panel is under way," Seaborg wrote. The panel was not solely Seaborg's idea, but there is little doubt that Seaborg was the primary technical advisor during the meeting. When the meeting was over, Sununu checked with President Bush to see whether he was available to meet Seaborg. He was. They exchanged pleasantries, and a photographer captured the moment.

When Seaborg told this story in a 1995 lecture at Lawrence Berkeley National Laboratory called "My Service With Ten Presidents of the United States of America," took some poetic license. He made it seem that the president did what he told him to do and that the panel did what he predicted it would do:

> I was called to Washington on April 14, 1989, to brief George H.W. Bush on cold fusion. I don't know whether you know what cold fusion is, but it was the idea that you could fuse nuclei very easily and get a lot of energy just by passing electric current through heavy water, whereas, of course, physicists had built huge machines and worked for decades trying to do this, spending billions of dollars.
>
> The chemists thought they'd really stolen a march on them. The idea swept the country, and I was called to Washington to brief President Bush on it. It was a real dilemma. What should I do?
>
> I decided to take my background as a nuclear scientist

and really come to the sensible conclusion that this work was not right, that it was really cold. You couldn't do it. So that's what I told him at that time. I said, "You can't just go out and say this is not valid. You're going to have to create a high-level panel that will study it for six months, and then they'll come out and tell you it's not valid," and that's what he did.

Although the DOE panel eventually said it wasn't valid, it is not because of what Seaborg told Bush to do, according to internal documents that were part of the review (Chapters 27-32).

Admiral Watkins, who directed the DOE to conduct the review, and his staff were far more open-minded about "cold fusion" than Seaborg. The review took place under Watkins' command, not Seaborg's.

There are other discrepancies: Seaborg's earlier version of this history, which he sent to Lewenstein in 1992, presents him as advising Sununu and his aides, then speaking with Bush only as a social formality, not as a direct advisor to the president.

Birth of the 24 MeV "Cold Fusion" Concept

In early April, the concept of the hypothetical "cold fusion" reaction $d + d \longrightarrow Helium\text{-}4 + 23.8$ MeV, emerged. The idea went back at least to March 31, 1989, when Douglas Morrison proposed the concept in his newsletter:

> Maybe the dominant reaction is fusion, $d + d \longrightarrow He\text{-}4$, but we need something else to share the energy and momentum produced — this could be the close neighboring structure of the lattice. Thus, the dominant reaction is to produce heat!

Peter Hagelstein, as discussed in Chapter 9, went further with the idea. Between April 5 and April 12, he had submitted four theory papers to a journal for peer review. On April 14, when the only available experimental data for "cold fusion" was the Fleischmann-Pons preliminary note, Hagelstein gave a lecture at MIT before hundreds of faculty members, researchers and students.

He expanded on Morrison's very general idea and proposed that the reaction $d + d \longrightarrow Helium\text{-}4 + 23.8 \ MeV$ could explain "cold fusion" — if it existed. According to an April 21, 1989, article by Eugene Mallove, a staff writer at the MIT news office, Hagelstein acknowledged the limitations of his model:

> [This model is] highly speculative; little hard experimental evidence currently exists supporting the claim of cold fusion, and no support for any aspects of the present model have been demonstrated experimentally. The model does not predict that cold fusion should or should not exist, but if it does, the model proposes how it might work.

Hagelstein's idea was that two deuterium nuclei react and form one helium-4 nucleus and that 23.8 MeV of energy, rather than emitting as a dangerous gamma-ray, instead stay inside the palladium lattice and somehow come out as heat. The equation does not exist in accepted physics and is presumptive. It is a hypothetical variant of a real equation that describes the well-understood thermonuclear fusion reaction $d + d \longrightarrow Helium\text{-}4 + 23.8 \ MeV \ gamma \ ray.$

Mallove explained to the Associated Press on April 14 that Hagelstein "was in a hurry to get credit for his ideas, and that's why he didn't want to wait to try them out in the laboratory."

Helium-4 in Utah

On April 17, 1989, Pons participated in the first of what would become weekly news conferences as an alternative to spending day and night responding to repetitive questions from the news media.

During this news conference, Pons said that two professors in the University of Utah Chemistry Department, Cheves Walling and John Simons, could explain the mechanism for "cold fusion" and submitted a manuscript to a journal by April 14. (Walling and Simons, 1989)

In a follow-up telephone interview, Walling told reporter Lee Dye, at the *Los Angeles Times,* that he and Simons used one of the same experiments that produced heat for Fleischmann and Pons and that he (Walling) and Simons had detected helium-4 when they put the experiment into a mass spectrometer. Dye wrote that Walling and Simons' helium-4 observation offered the most compelling evidence supporting Fleischmann and Pons.

In his 1991 book, on Page 140, Frank Close mistakenly reversed the facts of this story and attributed the helium-4 measurement and claim to Pons. Charles Beaudette also reversed these facts in his 2000 book, on Page 223. I notified both authors of the error on July 13, 2015. Beaudette did not reply, and Close could not remember any other details. I tried to contact Simons by e-mail on July 13, 2015, but he did not reply.

Reporter Anne Burnett, at Salt Lake City's KUER-FM public radio station, gave her report on April 21, 1989. Her "Fusion Update" broadcasts are available in the Cornell Cold Fusion Archive.

"The news that has Pons smiling is the theory that is being published by his chemist colleagues Walling and Simons," Burnett said. "Their theory says that the high heat output being generated by Pons and Fleischmann's experiment is the result of an accelerated fusion process that releases most of its energy as heat instead of radioactivity."

Walling and Simons were the first researchers to report helium-4 production from "cold fusion" experiments. Unlike excess heat, helium-4 was definitive evidence of a nuclear reaction. Dye interviewed Walling — one of the most respected chemists at the time — by phone. "The amount of helium-4 corresponds to the amount that should have been there if the heat was coming from nuclear fusion," Walling said.

Heat in California

The following day, April 18, Robert Huggins, a prominent materials scientist at Stanford University, invited the news media to his lab. Huggins and his team performed side-by-side light and heavy water Fleischmann-Pons-type experiments. Huggins and his colleagues told the news media that their D_2O experiment was producing excess heat and that their H_2O experiment was not.

The difference, Huggins told me, was proof-positive of excess heat in heavy-water. But an implicit assumption in this side-by-side test was that deuterium fusion was responsible for the excess heat.

Huggins, sticking closely to experimental evidence rather than theory, was careful not to claim "cold fusion." He merely insisted that his group's results confirmed Fleischmann and Pons' excess heat.

The side-by-side experiment should have satisfied people who didn't trust that Fleischmann and Pons had the skill to accurately measure heat.

Differential Versus Absolute Calorimetry

Pons had seen low levels of excess heat in his and Fleischmann's light-water experiments. Huggins' claim of zero excess heat in light water seemed contradictory. But it wasn't. Many years later, I asked Huggins detailed questions about his group's experiments. Unlike Fleischmann and Pons, Huggins said, he used differential calorimetry rather than absolute calorimetry.

Huggins did not independently determine thermal balances in each of his cells. Instead, he simply compared the temperature difference between them. The heavy-water cell became hotter than the light-water cell. When he told the news media that his light-water cell showed no excess heat, he did not have data to support that statement. He had assumed that, because he was observing heat from D+D fusion, a light-water cell without deuterium was a good reference for a thermal-neutral baseline. As Pons said in Dallas, light water was not a good baseline.

A 1990 paper written by SRI International electrochemist Michael McKubre and colleagues includes a useful explanation of differential calorimetry. As the authors wrote, differential calorimetry is a convenient way to detect anomalous heat by "comparing the temperature or heat flux from identical cells, where one cell is restricted to producing Joule heat only." As electrochemists knew well, the best blanks to use were inert cathode materials, because switching heavy water with light water introduced other differences, as McKubre explained in a 1990 paper. (McKubre, 1990)

Pons Speaks at Los Alamos

Also on April 18, Pons spoke at the Los Alamos National Laboratory, reiterating the talk he gave in Dallas at the ACS meeting. Pons seemed calm and confident in the video of his talk, and he appeared appreciative of the chance to discuss his work with fellow scientists.

The first question, not surprisingly, was about light water. Someone asked Pons to respond to a news report about a duplication of the experiments with light water. Pons answered in several parts. First, he said that Walling and Simons had developed a theory to explain light-water excess heat. Fleischmann and Pons didn't think to look for helium-4 in their data until after Walling and Simons reported that they had seen helium-4 in mid-April.

Second, Pons said that he and Fleischmann had, without knowing it at the time, collected data for helium-4 back in January:

> We went back to the old mass spectral data for January, and sure enough, there's two very-well-defined peaks at mass 4: one for deuterium, and one for helium-4, which we substantiate. We went back to the lab last Thursday and Friday and indeed have seen a large amount of helium.
>
> We tried, in January, to get a blank experiment in regular water with a palladium electrode that we knew worked. We did not get the heat-in/heat-out balance like we expected; we did get some excess heat. So that still is open for investigation. That experiment is set up and running again right now. The baseline experiment that we've done the most with has been a dead palladium rod, one that never got started because of some metallurgical reasons, probably, or some surface effects. That rod gives absolutely — in the same cells, under all the same conditions — the same amount of heat out that you put in. On the other hand, the palladium rod that had been active in D_2O does not give total thermal balance in plain water; it gives a slightly higher amount of heat out.

Helium-4 was not an obvious product to look for because, in thermonuclear fusion, which Fleischmann and Pons thought they were emulating, helium-4 is extremely rare. Only 1 in 1 million deuterium-deuterium fusion reactions produces helium-4.

Alternate Helium-4 History

Electrochemist Melvin Miles, whose group I consider the first to publish the observation of helium-4 data in "cold fusion" experiments, gives the credit to Fleischmann and Pons. (Bush et al., 1991)

In a paper Miles presented in 2015, he wrote, "Fleischmann and Pons were actually the first to observe that helium-4 was produced in the Pd/D system." However, Miles' only evidence is a Sept. 21, 1993, letter from Fleischmann stating "We had our first indication of helium-4 in December of 1988!" (Miles, 2015)

Fleischmann and Pons made very limited attempts to look for helium. In contrast to stories that have been written by some of their followers, Fleischmann and Pons never published any observation of helium-4 in their "cold fusion" experiments. In a memoir, Fleischmann wrote that their helium data were "un-publishable." (Fleischmann, 2000) They also never claimed that helium-4 was the dominant nuclear product in "cold fusion," let alone proposed that the amount of helium-4 was directly associated with excess heat. The idea of $d + d \rightarrow Helium\text{-}4 + 23.8\ MeV$ did not come from Fleischmann and Pons. All of these ideas about helium-4 came later from other researchers many of whom revised history and attributed the idea of $d + d \rightarrow Helium\text{-}4 + 23.8\ MeV$ "cold fusion" to them.

Very Cold Fusion Confirmed in Italy

On April 18, management of ENEA, Energia Nucleare e Energia Alternative, the Italian equivalent to the U.S. Department of Energy, held a press conference to announce a confirmation of "cold fusion."

Shortly after the University of Utah announcement on March 23, three physicists at ENEA quietly began their own attempts at "fusione fredda," as it is called in Italian. The physicists, Francesco Scaramuzzi, Antonella De Ninno and Antonio Frattolillo, in order to avoid the complexities of electrolysis, came up with the idea of gas-loading deuterium into titanium shavings. This "dry" deuterium gas method was the first alternative method to the Fleischmann-Pons "wet" heavy-water electrolytic method. Using shavings rather than bulk metal increased surface area, which then improved deuterium absorption. They had obtained their first successful results on April 7.

The ENEA physicists cycled the experiments through changes in pressure and temperature: going down to liquid-nitrogen temperature and then back up to room temperature. They didn't observe any excess heat, but they reported finding excess neutrons. At the time, they did not know the importance of a high loading ratio; however, they tried to induce a high hydrogen absorption in the metal by playing with thermodynamic cycling. De Ninno explained the early days of their "cold fusion" research to me in an e-mail:

> Antonio Frattolillo and I proposed to Franco Scaramuzzi (at that time, the head of the cryogenics laboratory) trying a dry version of the Fleischmann and Pons experiment. Scaramuzzi had had the same idea. We were quite doubtful, but our idea was that, if Fleischmann and Pons' claims were true, it should work in a dry version, too. It was a very simple idea; however, it worked. The three of us, plus five other researchers in our lab who worked on the experiments, prepared a manuscript which we submitted to a journal. (De Ninno, 1989)
>
> Suddenly, when the management of ENEA heard about our discovery, they got very excited, and they informed the press and the TV and scheduled a press conference. The general manager at that time was Fabio Pistella, and the ENEA director was Umberto Colombo. The next day, the news was in the *New York Times* and everywhere else. In a few weeks, our laboratory became very popular in many newspapers and magazines, and Scaramuzzi was invited to the Italian Parliament for an interview.

The news was distributed worldwide by Reuters. The director of the thermonuclear fusion department personally reported the news, according to the *New York Times* on April 19:

> Dr. Roberto Andreani, director of the Department of Fusion at the agency's research facility at Frascati, said that, in the first experiment, 20 to 40 neutrons were measured in successive 10-minute periods. This compares with a background level of 2 neutrons that are naturally present from cosmic rays.
>
> In the second experiment, researchers measured flows of 200 to 300 neutrons every 10 minutes for more than 12 hours, which was more than 100 times the background level, he said. Despite the high neutron yield, the team members said they had obtained virtually no [heat].

Scaramuzzi submitted the group's manuscript to *Europhysics Letters* on April 24. Word traveled fast. Almost immediately, Carlo Salvetti, a senior scientist, who had been the vice president of ENEA from 1963 to 1980, faxed a copy of the manuscript to Edoardo Amaldi.

"Salvetti," De Ninno wrote, "was one of the fathers of the Italian nuclear energy. He was a very respected scientist in ENEA and all over the world of nuclear energy research."

Amaldi was another scientist of immense stature in Italy and in Europe in general.

"For immediate delivery to Prof. Amaldi," Salvetti wrote. "The manuscript is going to be sent to *Europhysics Letters* today. Please inform professor Amaldi that Balbi from *La Repubblica* [a major Italian newspaper] will contact him this morning."

Somehow, a copy of that fax was shared with Richard Garwin, and I found it in his archive. Another document in his archive showed that the news hit the internal IBM message network on April 19, thanks to someone named Carlo Jacob. On April 21, somebody in the Frascati fusion division faxed a report of the Scaramuzzi group's results to Steven Koonin, according to another fax in Garwin's archive.

The ENEA result was a real nuclear signature that confirmed not only the general nature of Fleischmann and Pons' claim (ignoring, for the moment, the fact that they called it fusion) but also that the phenomena were much broader and not limited to experiments with heavy water and palladium.

Confirmations in India

The next day, April 19, the Press Trust of India news agency distributed a short article announcing another confirmation. The Indian news was reported in the *New York Times* on April 20:

> An Indian scientist, C. V. Sundaram, who is the director of the Indian government's Indira Gandhi Center for Atomic Research near Madras, reported a "30 percent increase in neutrons over the background level" but added that the energy produced had not been measured.

Two prominent international government nuclear laboratories had now reported confirmations. Although they didn't confirm Fleischmann and Pons' excess heat, neutrons were a much stronger confirmation, from the perspective of physics, than excess heat. *Nature* published detailed news of the Indian confirmations a week later, on April 27.

C.K. Matthews and his colleagues at the Indira Gandhi Atomic Research Centre (IGARC) at Kalpakkam near Madras had observed a 30 percent increase in neutrons over background from an electrolytic cell using heavy water. Researchers at the Tata Institute of Fundamental Research in Mumbai observed excess heat but didn't try to measure neutrons.

According to *Nature*, "Professor K.S.V. Santhanam and his colleagues in the chemical physics group said that passage of 0.25 watts of electrical power through the cell produced 1 watt of thermal output at the titanium cathode, whose temperature rose to 80° C in a sustained reaction." The experiments were not always reproducible, but that did not discourage the researchers from reporting their findings. These included the IGARC scientists as well as researchers at the Tata Institute of Fundamental Research in Bombay (now called Mumbai).

On April 21, 1989, researchers at India's largest nuclear research facility, Bhabha Atomic Research Center, in Trombay, found their first significant confirmation of neutron bursts and tritium. They did not publicize their results.

On April 25, 1989, Bhagwan Singh, writing for the Associated Press, reported that Matthews and his two associates at IGARC not only saw excess neutrons but also replicated the Fleischmann-Pons excess-heat claim using differential calorimetry.

DOE Shifts Into High Gear

On April 18, 1989 — the same day that the Frascati news broke — Robert Hunter, the director of the Office of Energy Research of the Department of Energy, called an emergency meeting.

On behalf of Admiral James Watkins, Secretary of the U.S. Department of Energy, Hunter told directors of 10 DOE national labs to report to Washington, D.C., the following day to find out what was going on in their labs with "cold fusion" research.

Gerd M. Rosenblatt, the deputy director for Lawrence Berkeley Laboratory, attended the meeting and summarized the discussions in an April 20, 1989, letter to his division directors.

"The meeting arose because DOE, from the Secretary down," Rosenblatt wrote, "is under tremendous pressure to be knowledgeable and to show leadership and progress in this area. Hunter is directing the labs to intensify their efforts to resolve the technical questions raised by the reports of fusion in hydrogen-adsorbing metals." (Rosenblatt, 1989)

Two days later, U.S. Energy Secretary Admiral James Watkins announced three initiatives, which he explained in a press release. First, he told the DOE lab directors, in an April 24 letter, that he wanted weekly reports on their "cold fusion" research and results.

Second, he announced that he was going to ask the DOE's Energy Research Advisory Board to "establish a panel to conduct an independent review of the entire situation," and third, he announced that "the Los Alamos National Laboratory, under the auspices of the

DOE, will sponsor a scientific workshop on the subject" in late May. Watkins was careful not to state that DOE believed fusion was occurring. His actions, nevertheless, were significant because Los Alamos and the other DOE national laboratories are "crown jewels" of physical science and engineering research in U.S.

"The main reason for the research by DOE laboratories is the potential for a new energy source," the press release said. "However, the origin of any heat released has not been established, be it nuclear, chemical, mechanical or another process."

Watkins' press release said that the ERAB review was "to provide DOE with an assessment of this new area of research" and that he wanted an interim report by July. The DOE invited 2,000 scientists from around the world to come to the Santa Fe "Workshop on Cold Fusion Phenomena."

Watkins' April 24 letter to the heads of the DOE laboratories was unambiguous about his position on the matter: He tentatively accepted the claim of "cold fusion" and took it seriously. "Because of the great scientific interest in this phenomena," Watkins wrote, "and the enormous potential benefits from practical fusion energy, I encourage you to continue and even intensify research efforts at your laboratories to more clearly understand the phenomena. ... Your aggressive and responsible pursuit of clarification of the situation is very important to the department." (Watkins, 1989)

He asked lab directors to provide summaries of the following activities in their weekly reports:

> Confirmation of experimental results claimed by others.
> Chemical assays for chemical reactants and products.
> Control experiments run with light water.
> Determination of mechanisms and rates for fusion reactions in solids.
> Verification of neutron and/or heat releases observed in the University of Utah experiments.

Garwin Places His Bet

The editors of *Nature* had asked Garwin to write a short report on the Erice Fusion Forum workshop. Garwin's report was laden with his concerns about violations of known theory. It was published on Thursday, April 20. Garwin surmised that the "cold fusion" conflict would be settled within a few weeks. He placed his bet against "cold fusion" and was thus the first prominent scientist to go on record suggesting that Fleischmann and Pons had made a mistake of major proportions and that "cold fusion" was not real.

Garwin tried to ascribe the neutrons he had heard about in Erice to spurious signals from electrical instrument failure:

> A few neutrons each second (or a few thousand) from an electrolytic cell may be cold nuclear fusion or may be due to an "arcs and sparks" origin. Within the next few weeks, experiments will surely show whether cold nuclear fusion is taking place; if so, it will teach us much besides humility and may indeed provide insight into significant geophysical puzzles. Large heat release from fusion at room temperature would be a multi-dimensional revolution. I bet against its confirmation.

The timing could not have been worse for Garwin. He had placed his "bet"; that is, he submitted the draft of his Erice report on April 17 and got the galley proof within 24 hours. The Italian and Indian confirmations had been reported the next day.

Lewis Complains

On April 21, 1989, Nathan Lewis and his colleagues gave a seminar at Caltech on "cold fusion. Until that moment, few people outside Lewis' lab knew that Caltech researchers had found nothing in their "cold fusion" experiments. Cold fusion fever was still running high. Lee Dye wrote about it in the *Los Angeles Times*.

"The seminar had to be moved from a classroom to the campus' largest auditorium to accommodate a crowd of several hundred," Dye wrote. "Many of those who showed up rushed to the front rows of the auditorium, expecting to see a demonstration of the fusion experiment, but [chemist] Nathan Lewis shattered that dream in an instant."

Lewis told the packed auditorium he had found nothing. In the absence of any positive results of his own to report, Lewis used the opportunity to criticize Pons. Lewis insinuated that there was something wrong with Fleischmann and Pons' claim because Pons was unwilling to submit to external validation as dictated by Lewis. Dye described the drama:

> At one point, Lewis asked in exasperation why Pons has not taken one of his experiments to one of the major national laboratories and asked scientists there to test it with their sophisticated equipment. Why not, he asked, when "everybody knows they were first now."
>
> During the Caltech seminar, speakers were frequently interrupted by applause and cheers when they suggested repeatedly that Pons and Fleischmann were so eager to claim credit for their work that they made their announcement prematurely and that their work was seriously flawed.

Physicists Compare Notes

On Sunday and Monday, April 23-24, physicists from around the U.S. were gathering at national laboratories, according to Frank Close. "The meetings had nothing to do with cold fusion," Close wrote, "and had been arranged weeks earlier to discuss the planning of the next five years of research in various aspects of nuclear and high-energy physics."

But as Close, who attended one of those meetings, explained, the chance to meet face to face provided an ideal opportunity for physicists to talk about the "cold fusion" problem.

It was there, for the first time, that Close began to piece together the way things were developing. So did the U.S. fusion physics researchers.

CHAPTER 14

Chemists Shill for Utah

Enthusiastic Congressional Committee

Decades of unfulfilled promises from thermonuclear fusion now placed fusion physicists in an uncomfortable position when Congress asked about the potentially competing yet unproven idea of "cold fusion."

On April 6, 1989, a congressional subcommittee had authorized the redirection of $5 million from the Department of Energy's magnetic fusion program to "cold fusion." Then, on April 21, Congress made it known that it was looking at "cold fusion" more carefully. Rep. Robert A. Roe, (D-NJ), the chairman of the House Committee on Science, Space and Technology, issued a press release announcing that, on April 26, his committee would hold a "Hearing on Recent Developments in Fusion Energy Research."

On April 26, Chairman Roe convened his hearing. Most prior accounts of this hearing depict it as a selfish attempt by Fleischmann and Pons to get money for their research and to circumvent the normal peer-review process associated with research proposals. Neither is true.

The previous month, on March 2, the Department of Energy had approved their 1988 proposal and offered them a $322,000 grant. They turned it down; they had enough money. According to journalist Ann Burnett of KUER-FM, Pons was given $1.2 million by the Office of Naval Research, a Navy branch with which Pons had a long-standing relationship. The two chemists were also getting offers, as Pons told me, from well-known industrialists.

Instead, University of Utah President Chase Peterson wanted the money to establish an off-campus "cold fusion" research institute. Fleischmann and Pons had no interest in such an institute and privately advised the university against it. (Chapter 26). In Washington, Fleischmann and Pons were shills for Utah, though they may not have done so gladly. On Feb. 6, 1991, Pons told the *Deseret News*, "We are made to be circus performers."

Stick in a Hornet's Nest

Diane B. Weisz, who worked at the National Science Foundation, wrote in a letter to her colleagues, "The hearing was packed — standing room only, lights and cameras. Committee members turned out in great numbers. The main question was, What should Congress do now?"

Prior accounts of this hearing have omitted something significant. Unlike the physicists, who were feeling growing pessimism toward "cold fusion" in late April, the committee members were inspired and enthusiastic about a potentially cheaper alternative to the still-unfulfilled promise of thermonuclear fusion research. They showed no particular allegiance to the prevailing fusion research. Some of the members asked critical questions of Fleischmann and Pons and with the exception of one question about light-water experiments, the members seemed to be satisfied with the answers.

Rep. Robert S. Walker, (R-PA), had already persuaded Rep. Marilyn Lloyd, (D-TN), the chairwoman of the Subcommittee on Energy Research and Development, to take $5 million from the magnetic fusion budget and give it to "cold fusion."

Walker, the ranking Republican member of the committee and a strong supporter of thermonuclear fusion research, was now also enthusiastic about "cold fusion." He was quoted enthusiastically in the press release that had announced the hearing:

> The committee is extremely excited by the opportunity to examine the promise of this new potential source of energy. Cold fusion may eventually turn out to be one of the

most significant sources of energy for the world in the 1990s and beyond, and I look forward to hearing from its discoverers and other experts on the implications of this intriguing breakthrough.

That hope-filled press release had gone out on April 21. Physicists at Department of Energy laboratories were meeting on April 23 and 24 to discuss plans and budgets for the next five years. For the thermonuclear fusion research community, "cold fusion" was a problem. A big problem.

Star Witnesses

The University of Utah contingent included Fleischmann, Pons, university President Chase Peterson, and Ira Magaziner, a technology lobbyist who was helping the university. Two Utah congressmen sat with them. The speaker's panel table had the usual items: microphones, pitchers of water and glasses. But the Utahns had brought one item that, on any other day, would have seemed out of place: a laboratory stand holding a glass "cold fusion" cell.

Pons spoke first, at a highly technical level. He offered three types of evidence to support his and Fleischmann's claim of nuclear reactions: measurements of gamma rays coming from secondary neutron-capture reactions in the water bath; neutrons directly measured with dosimeters; and their measurements of tritium. But Pons was clear about the heat. "While we do measure very low levels of these nuclear reaction products," he said, "we make a much more significant measurement with our calorimetric data."

Pons also explained, twice, very clearly, that the neutron flux and the amount of tritium were a billion times too small to account for the heat they measured if the reaction had been from fusion. "So, apparently, Pons said, "there is another nuclear reaction or another branch to the deuterium-deuterium fusion reaction that heretofore has not been considered, and it is that which we propose is, indeed, the mechanism of the excess-heat generation."

Fleischmann spoke at a broader conceptual level about the

experiment and discussed his and Pons' work in the context of thermonuclear fusion research.

Although Congress was eager for a new energy solution, Fleischmann downplayed that prospect when asked by Rep. Sid Morrison, (R-WA). "You are suggesting," Morrison said, "that we retain and maintain our investment in — let's use the term '*hot fusion*'." This appears to be the first public use of the term "hot fusion."

It's Fusion and It's Not Fusion

When Rep. H. Schiff, (R-NM), pressed Fleischmann and Pons harder to defend how they knew it was fusion rather than a chemical process, they didn't talk about nuclear products such as helium or neutrons. Instead, they defended their claim on the basis of the amount of heat they had measured, 100 times greater than any known chemical reaction.

Fleischmann and Pons gave mixed messages. On one hand, they were using the term fusion to describe their work. On the other hand, they were saying that their excess heat was the result of some previously unrecognized nuclear reaction. Fleischmann and Pons themselves did not agree on how to represent their work. Fleischmann was far more interested than Pons in proving the concept of room-temperature deuterium-deuterium fusion. Pons seemed to be more of a pragmatist: If they could make nuclear-scale heat without hard radiation, who cares whether it's fusion or some previously unrecognized nuclear process?

Physicists didn't care either way; the two electrochemists didn't have bona fide nuclear evidence, despite the chemists' assertion that their heat was evidence of nuclear fusion.

Pons had been asked two weeks earlier, at the American Chemical Society meeting and at Los Alamos, about his experiments performed without deuterium. He said that he and Fleischmann had seen low levels of excess heat in ordinary water.

Fleischmann, on the other hand, at CERN and at Erice, had refused to talk about light-water experiments. When asked directly by a legislator during the hearing, he also refused to answer. In a follow-up letter he sent to the committee, Fleischmann directly contradicted Pons' statements and denied any light-water excess heat. "As I have implied," Fleischmann wrote, "I am not able at this stage to comment about work in ordinary water, but I will certainly follow up with this letter when I am able to do so. It is, however, well-known that others have shown that there is no generation of excess heat when using light water."

Fleischmann contradicted Pons about the light-water excess heat more than once. Plasma physicist Howard P. Furth, who spoke at the hearing, wrote about one such interaction:

> I went up to Pons and Fleischmann and said to Pons, "Didn't you report an 'excess heat' release from ordinary water when you were at Los Alamos?" Fleischmann answered, "That's not what he said." Pons agreed with that.

The videotape made by Los Alamos of Pons' lecture there on April 18, however, shows Pons clearly stating that he had produced excess heat in light water.

Money and Innovation

When Rep. George Brown Jr., (D-CA), asked Fleischmann how much money he estimated the basic research would cost, Fleischmann explained that giving a qualified answer at such an early stage was very difficult. He said that the experiments take a very long time: months rather than days to carry out each individual experiment.

When Brown asked about the projected time required to commercialize the process, he responded conservatively. "Congressman," Fleischmann said, "I think that the normal timescale one thinks of in terms of commercial development is 10 to 20 years."

Eventually, Fleischmann said that the next immediate phase of the work would cost in the millions of dollars, but he said that Peterson was the designated member of the Utah group on the topic of money.

"I would really like to pass that topic over," Fleischmann said, "because the president of the university, Dr. Chase Peterson, will be talking to you about that, and I trust that he will be willing to quantify that in his presentation."

Brown, like the other members of the committee, was optimistic. "We are looking at something that appears to be a very-low-budget kind of an item," Brown said. "You know, what you spent on this experiment — I doubt if it's more than a few tens or hundreds of thousands of dollars. You contrast that with the half a billion a year that we are spending on other kinds of fusion research, and it represents quite a marked disparity. It would indicate, obviously, that we could proceed rather rapidly with this if it has the promise that you seem to indicate."

Those words were certain to send a chill through any scientist working in the thermonuclear fusion research field. Representatives of three of the largest and best-funded U.S. laboratories for thermonuclear fusion research were sitting in the room. One was Michael J. Saltmarsh, the associate director of the Fusion Energy Division at the Oak Ridge National Laboratory. Another was Howard P. Furth, the director of the Princeton Plasma Physics Laboratory. The third was Ronald G. Ballinger, a professor at MIT and a member of its Plasma Fusion Center team. They got their brief turns to testify much later in the day when the number of committee members in the room had dwindled.

Some scientists had publicly questioned Fleischmann and Pons' ethics for doing what the critics called "science by press conference." When committee members grilled them about this, Fleischmann and Pons made it clear that they had followed normal protocol and that their manuscript had been accepted in a peer-reviewed journal in advance of the press conference. Committee members appeared satisfied.

Rep. Don Ritter, (R-PA), was impressed that Fleischmann and Pons paid for the research out of their own pockets. "Mr. Chairman," Ritter said, "you're witnessing an example here where not all of the sweetness and light and new discoveries are going to come from the top five research universities in America. There's quite a bit out there in Utah."

Ritter had poured gasoline on a burning political fire. A comment from Rep. Christopher Shays, (R-CT), reflected the enthusiasm and optimism among public observers in the room. "If your discovery is verified," Shays said, "what you have done has obviously changed the course of mankind. There must have been a moment when you said,

'My God, we may have changed the course of mankind.' Did that happen? Was there a moment like that?"

Pons responded to the dramatic question with a single undramatic word, "No," which immediately evoked laughter from everyone. After the chuckling died down, he added, "I mean, it sure changed our lives, I'll tell you that."

"You don't have the advantage I do," Shays replied. "I see a lot of smiling faces behind you, [in reaction to] some of your responses."

For the physicists in the audience, there was no ambiguity, on this day, the chemists were being treated like kings.

When it was Peterson's turn to speak, the members of the committee asked him how much money he envisioned would be necessary to get a "cold fusion" program going." The figure that comes to mind is $25 million from the federal government," Peterson said. "Maybe that needs to be $125 million someday, but that's of not any importance right now. Twenty-five million dollars would allow us to start the 'onion' growing, with state and private sources. Ultimately, I would imagine it would be ... a hundred million dollars ... we would expect to raise, with a minority portion of that being from the federal government."

To my knowledge, the federal government never appropriated the money, but the state of Utah did allocate $5 million.

Vilified in His Lifetime

Rep. Dana Rohrabacher, (R-CA), was sympathetic to Fleischmann and Pons.

First of all, congratulations to both of you for maintaining your composure at what must be a tumultuous time in both of your lives, especially in front of a hearing like this. Sometimes, it gets a little difficult to express yourselves, and you've done very well today. I appreciate it. I'm sure the rest of us appreciate it, as well.

We all know that you've created a lot of heat, not only in the beaker but outside the beaker, as well. Do you think that some of this heat is being generated by the fact that there are a lot of people in the scientific community who are dependent on hundreds and millions of dollars worth of government grants that may not be open-minded toward the type of change you're suggesting is possible?

As Pons well knew, some of those people — Saltmarsh, Furth, Ballinger — were sitting right behind him. He spoke cautiously:

> The only comment I would make there is that I think it's always dangerous [for our critics] to point at incorrect experimental data based on [their] theory. I think theory must be used to explain experimental data, not to criticize experimental data. ... I think that you need to consider first that the experimental data must be duplicated and explained, and then a theory put forth, rather than just saying our data must be wrong because their theory doesn't predict that.

Fleischmann, ever the diplomat, quickly jumped in and — putting words in Pons' mouth — gently lent a softer edge to the discussion. Rohrabacher, however, didn't buy it.

FLEISCHMANN: I think Professor Pons is alluding to the nature of the criticism which has been leveled by people who are working in those areas of research. I don't really see that our work impacts too much on that work. It's another line to pursue and should be seen as that.

ROHRABACHER: But you're going to put some of these people out of business, aren't you, if you're successful?

FLEISCHMANN: Well, no. I think we will put them out of — if we are successful in demonstrating the science, and if we go to the point of technology, then the members of this committee and the scientific community at large will start to make a choice about whether to develop this technology.

ROHRABACHER: But contrary to public opinion or perception, isn't it true that many major scientific breakthroughs in human history have not been greeted by the professionals of the day with open arms?

FLEISCHMANN: How can you expect it? I think that a strange piece of research will strike people as being strange. You have to get used to it. You have to live with it. It's like an old bicycle. You have to grow old with it.

ROHRABACHER: And perhaps the fact that so many people in the scientific community are now dependent on government grants, that perhaps are heading in totally the opposite direction to achieve the same results, might actually make this problem even worse.

FLEISCHMANN: I hope not. I think that, in the end, all the people working in this area [thermonuclear fusion] will come to see this as just another arm of the research, one they will wish to be involved in, rather than one they wish to stand aside from. I think that, if we are correct, if we are opening up this gray area between physics and chemistry, where there is this strong overlap, then the people who have got the big experience in the high-energy-physics end will have an absolutely vital contribution to make. I think they will come to see that very shortly.

ROHRABACHER: I hope you're right. I would like to note that Jonas Salk, in my own time, was not greeted with open arms and was vilified for a certain period in his life, and there was a lot of confusion about that. I think he probably saved a lot of young people's lives. Mr. Chairman, if I could have one more question: If this is, indeed, the opening of a new door, what do you think mankind is going to see as we walk through that door? Just a very brief summary of the new potential that this may unleash.

FLEISCHMANN: Well, of course, as I said when I made my presentation, our motivation was social. If this is correct, then we have a source of energy which is clean, which avoids the pitfalls of generating carbon dioxide and sulfur dioxide. However, let's not again have too rosy a view. It will have a destabilizing effect initially as it is put into practice. Hopefully, eventually it will have a stabilizing effect on world economies. But the adoption of such an energy scenario would not be without difficulties for the developed and the developing countries of the world. I think those raise very profound political questions, which I'm sure this committee and other committees of Congress will wish to address.

Those in Favor

Huggins

Robert Huggins, the director of the materials science laboratory at Stanford University, spoke in support of Fleischmann and Pons' work at the hearing. A week earlier, he and his staff had reported a confirmation of Fleischmann-Pons' excess-heat phenomenon, though his experiments didn't seem to convince many people, perhaps because they did them in an Igloo lunch cooler.

Miley

George Miley, a professor of nuclear and electrical engineering and the director of the fusion studies laboratory at the University of Illinois, was moderately supportive. He was also the editor of the *Journal of Fusion Technology.* Miley wasn't a conventional fusion researcher; he had a long history of activity in nonconventional fusion approaches. He had his own "cold fusion" experiments running but no results yet.

Those Opposed

Saltmarsh

Four hours into the hearing, with only a couple of committee members remaining in the room, Michael J. Saltmarsh, the associate director of the Oak Ridge Fusion Energy Division, told the hearing panel that his lab had seen no positive results. Saltmarsh displayed a table of "cold fusion" experiments that had been performed and were continuing at the lab. Three experiments ran for just one day. Another three had run for two days. Six more were in progress. As far as he knew, he told the committee, nobody at any other labs had seen any positive results, either.

Saltmarsh was one of several scientists who began to deny any validity to "cold fusion" based on a tally of failures and successes. "Nationally and internationally," Saltmarsh said, "the vast majority of experiments have failed to duplicate the reported results. ... It would be a real mistake to try and draw firm conclusions at this point."

Furth

In his testimony, Howard P. Furth, the director of the Princeton University Plasma Physics Laboratory, analyzed the weaknesses in the Fleischmann-Pons work. He encouraged the legislators to force Fleischmann and Pons to reveal more of what they knew about the light-water tests.

The video of the hearing shows that Furth was very worried. He said that, if Congress funded "cold fusion" prematurely, it could lead to a regretful embarrassment. Furth also told the committee members that they should push Fleischmann and Pons to assay their cathodes for helium. He told the panel that this was a very simple, quick way to determine the validity of their "fusion" claim.

His well-intentioned admonition, however, was irrelevant, and his proposed test wouldn't have worked. He didn't know enough about the nature of the experiment; for that matter, neither did Fleischmann or Pons at the time.

Furth's test would have given a false negative. Years later, it became clear that the small amounts of helium from these experiments were generated at the surface of the cathodes and released in the gas stream. Surface analyses showed that reactions took place only in the outer layers of the cathode materials, contrary to what Fleischmann and Pons had presumed about a volume (bulk) effect. Other research — some of it from within the field — showed that helium does not enter or permeate intact, defect-free metals. (See Appendix E)

In a follow-up letter, the committee asked Furth about his views on future funding for fusion. He did not share the committee's enthusiasm for "cold fusion." "Continued reprogramming of funds from the magnetic-confinement effort into other areas is likely to have a significantly damaging net effect on the strength of the U.S. fusion effort," Furth said. Concurrently, the Princeton laboratory was fighting its own battle to maintain the flow of federal funds for its fusion project.

Jones

Jones was next to speak. He gulped, took a deep breath and pulled his chair closer to the table as he was introduced. He told the panel that he had been working on "cold fusion" since 1985. Thus, he implied that he was a qualified expert on Fleischmann and Pons' electrolytic work. This was not true; he had been working on muon-catalyzed and piezonuclear fusion. He had planned his presentation well and brought with him props, parables and thematic allegories. One prop was a small glass jar containing a wilting blade of grass, which he used to make a metaphor.

First, he displayed a hand-drawn cartoon depicting the mountains to climb, in order to get to the pot of gold at the end of the rainbow. He depicted someone representing the Department of Energy, floating in a hot air balloon, looking down on the mountains of research. Jones had put himself in the cartoon, as well.

After that, Jones borrowed the phrase, "a new door," that Rohrabacher had used when he spoke with enthusiasm moments earlier about the Fleischmann-Pons claim:

> This is nothing to get excited about from an energy-production point of view at the moment. Yes, a new door has been opened. But the gap between the bona-fide fusion yield and energy production by fusion is roughly equivalent to that which separates the dollar bill from the federal national debt, a factor of about a trillion to 1.
>
> First, cold nuclear fusion does not offer a shortcut to fusion energy. Second, based on my work of 10 years in fusion, particularly in cold fusion, magnetic and inertial approaches currently represent the best paths to achieving controlled fusion energy.
>
> Finally, I would like to emphasize that cold nuclear fusion is an exciting scientific discovery. Let us appreciate it for what it is and not decry it for what it is not. I would like

to compare cold nuclear fusion to this little plant, which is starting to wither — that may have some significance, as well. [Jones chuckled.]

Now, this is a tender shoot, as you can tell. It is difficult to say what it will become. Some think and suggest strongly that this is a tree, and it will grow up very quickly and provide us enough wood for all our energy needs for generations. I do not think it is. Let's give it a chance to grow. I think adding too much fertilizer at this stage will be detrimental.

I think we need to give it time, at least a couple of months, please, to see whether this is something that's a rose or a tree. If it should turn out to be a rose, we can then admire it for its beauty, even if we are a bit disappointed it was not a tree.

From the broad strokes to the subtle, his elocution was masterful. His fertilizer metaphor got the laughs he expected.

Decker

Jones' colleague Daniel L. Decker spoke next. Decker was the chairman of the Department of Physics and Astronomy at Brigham Young. He was fidgety, his delivery was irregular and it was often accompanied by a nervous smile and nervous laughter.

Nevertheless, and as a result, he seemed honest and transparent. Decker's testimony provides a rare glimpse of the frustration and anger felt by physicists about the seemingly unbelievable claims:

So now we have to make a 10^{12} jump in physics and a factor of a 100-fold jump in chemistry to try to explain the origin of this energy source.

I think that gives you an idea of what we feel, at least as physicists: Maybe the chemists should also look very seriously into possible chemical reactions and not tell us physicists that we need to change our physics to explain the process.

Ballinger

Ronald G. Ballinger, a professor in the departments of Nuclear Engineering and Materials Science and Engineering at MIT, was the last to testify that day. Ballinger had little to say that hadn't been said. By this time, attendance by members of the subcommittee had dwindled from a dozen to three. "I'm afraid," Ballinger said, "if there is another vote, that I'll be the only person here. The subcommittee is down to 1."

Unknown Source of Energy

On April 27, the day after the congressional hearing, Hans Bethe, a nuclear physicist who played a key role in the development of American nuclear weapons sent Chairman Roe a stern warning letter.

Bethe had tremendous stature in the scientific community. In 1939, he made three novel proposals: 1) hydrogen-hydrogen fusion is the process that powers the stars, 2) no elements heavier than helium-4 could have been formed in stars, and 3) the production of neutrons in stars is negligible. (Bethe, 1939)

His first proposal, for his theoretical work on the theory of nuclear reactions, earned him a Nobel Prize in physics in 1967. But his second and third proposals were wrong. As mentioned in Chapter 7, Bethe was friends with Richard Garwin, one of the key developers of the hydrogen (fusion) bomb. Garwin had been sending "cold fusion" information to Bethe.

"I am alarmed by the idea," Bethe wrote to Roe, "that your committee may authorize large funds for a commercialization of 'room-temperature fusion' on the basis of experiments which are highly controversial."

Of course, Congress had been authorizing large funds for the failed commercialization of thermonuclear fusion. That research wasn't controversial; it simply didn't work. Bethe didn't say in his letter to Roe that the "cold fusion" experiments were wrong. He didn't identify any errors with the Fleischmann-Pons procedure or data analysis.

Rather, Bethe expressed his concern based on three points: 1) many labs had failed to repeat the experiment, 2) the theoretical process of "cold fusion" was "very unclear," and 3) the theory of fusion had been well understood since 1933.

"Drs. Pons and Fleischmann of the University of Utah," Bethe wrote, "claim that they observe energy evolution about a billion times greater than would be produced by the nuclear reactions producing neutrons. Where this energy comes from is, at present, totally unknown, but it cannot come from fusion. The fusion reaction is far too well-studied and understood for this to be the case. ... It would be a black mark against nuclear science and against your committee, if you rushed into this unproven technology prematurely."

Bethe was correct, of course, that "cold fusion" didn't look anything like fusion. His true concern, however, was not that the experiments were "controversial." Rather, he thought it was a complete mistake. As he wrote in a May 19, 1989, letter responding to a letter from Michael Ravnitzky, coincidentally one of the editors of this book, "I strongly believe that [cold fusion] does not exist."

Doubts and Denial

By the end of April, battle lines were drawn. While many scientists applied legitimate scientific skepticism to analyze the "cold fusion" claims, others resorted to denial and accusations of fraud. Some of Fleischmann and Pons' critics said at the end of April that no one had successfully repeated any part of the Fleischmann-Pons experiment. In fact, as Appendix C shows, more than a dozen groups reported confirmations.

On April 27, two editorials were published by the 130-year-old peer-reviewed journal *Nature*, which served the scientific community as a weekly science news magazine, as well. The first editorial had no byline but was apparently written by the Washington, D.C.-based assistant physics editor David Lindley. He reported that claims of successful "cold fusion" had been announced in California, Brazil, and two Indian labs. For some reason, he omitted to mention the ENEA Frascati confirmation by Francesco Scaramuzzi's group. The group's work had been reported in the *New York Times* on April 19.

The second editorial was written by John Maddox, *Nature's* editor-in-chief. He complained that Fleischmann and Pons' research had been reported in newspapers before their preliminary note published in the *Journal of Electroanalytical Chemistry*. Maddox chastised them for their "astonishing oversight" of failing to perform tests with light water. He overlooked the fact that Fleischmann and Pons had used platinum and nonworking palladium cathodes as blanks, for control experiments. By now, in their lectures, Fleischmann and Pons had each explained his choice of controls.

"Fleischmann and Pons," Maddox wrote, "have done at least one great service for the common cause: They have kindled public curiosity in science to a degree unknown since the Apollo landings on the moon." Maddox placed his bet against the new science: "The Utah phenomenon is literally unsupported by the evidence, could be an artifact and, given its improbability, is most likely to be one."

Lee Dye, writing for the *Los Angeles Times* on April 23, 1989, understood the magnitude of the controversy:

> At this point, no matter which way it turns out, this scientific mystery will go down as one of the most unusual chapters in the history of science.
>
> Either Pons and Fleischmann, both of whom have solid reputations in the field of electrochemistry, have committed a blunder of extraordinary proportions, or the best and the brightest minds in energy research have been befuddled by an experiment that looks so simple it would not be a serious challenge for a freshman chemistry student.

The experiment did look ridiculously simple; however, it was anything but simple. And although Fleischmann and Pons' calorimetry was solid, they did commit a major blunder with their gamma-ray data.

CHAPTER 15

Inside the Atom

Tutorial on Atoms and Elements

It may be useful at this point to offer a brief tutorial on atomic structure and basic concepts in radioactivity. For nearly a century, scientists have known that all matter is composed of elements, elements are composed of atoms, and atoms are composed of three primary subatomic particles: the proton, the neutron and the electron.

Protons have a positive electrical charge, electrons have a negative charge, and neutrons have no charge. Protons and neutrons sit close together in the center of the atom and compose the nucleus. (See diagram of Carbon-12) Electrons, which have a negative electrical charge, remain outside of and orbit the nucleus.

Different Elements

All matter exists in the form of specific elements — for example, hydrogen, oxygen, and carbon. Each element is distinguished by the number of protons in its nucleus. At present, 98 elements are known to exist in nature, and a few others have been synthesized in laboratories.

When the number of protons inside the nucleus increases or decreases, the atom changes from one kind of element to another. For example, a proton added to a nitrogen atom changes it to an oxygen atom. This is called a nuclear transmutation. The first diagram below shows, on the left, one of the simplest atoms, a form of hydrogen called deuterium. Normal hydrogen has only one proton in its nucleus, but this variety of hydrogen, called deuterium, has a neutron, as well. On the right, the diagram shows a nucleus with two protons; this is the element helium. In this case, it's a variety of helium called helium-3, with has three particles in its nucleus: two protons and one neutron.

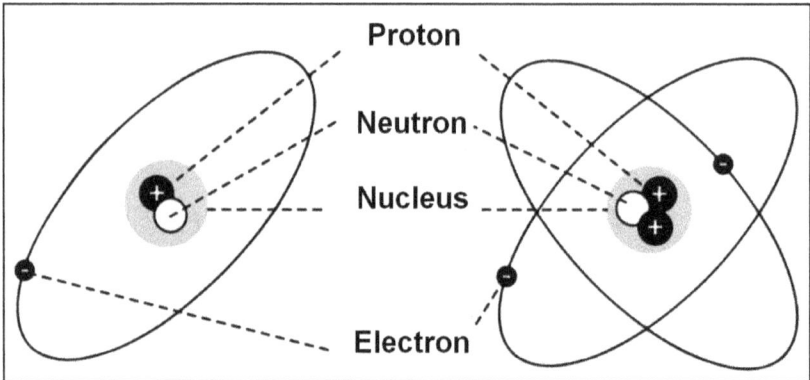

Deuterium atom (left): one proton and one neutron in the nucleus.
Helium-3 atom (right): two protons and one neutron in the nucleus.

Different Isotopes

Most elements exist in a variety of forms. Just as chocolate comes in different varieties, so do elements. However, different varieties of chocolate are still chocolate. A variation of an element is called an isotope. Each isotope is slightly different from other varieties of the same element. The difference between isotopes is that they have different numbers of neutrons in each nucleus, but the number of protons in each nucleus stays the same. An isotope of helium is still helium; an isotope of hydrogen is still hydrogen.

The pair of diagrams below provides an example. The one on the left shows a variety of helium called helium-3. It has three particles inside its nucleus: two protons and one neutron. The one on the right shows helium-4, which has two protons and two neutrons in its nucleus. Both are different isotopes of the same element, helium.

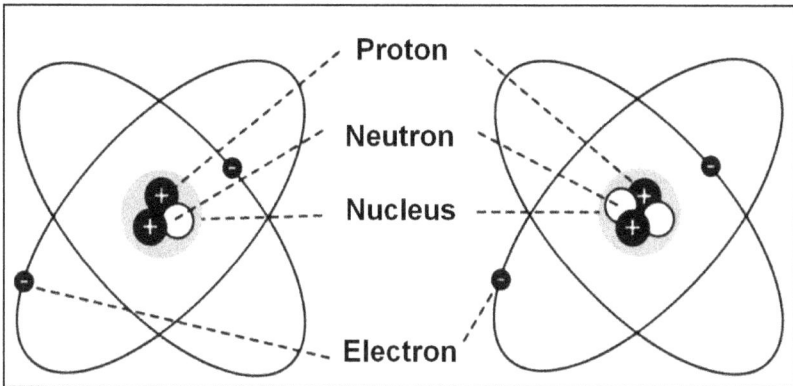

Helium-3 isotope (left): two protons and one neutron in the nucleus. Helium-4 isotope (right): two protons and two neutrons in the nucleus.

Nearly all elements have a variety of isotopes. Some elements have many isotopes; some have only a few. Some isotopes of a given element are more abundant than the other isotopes. For example, a lump of coal is mostly carbon. Isotopic analysis of the carbon reveals that most of it exists as the carbon-12 isotope.

Carbon-12 has six protons and six neutrons in its nucleus. Carbon has a total of 15 isotopes. Stable carbon-12 usually makes up 98.93% of the total amount of any carbon sample. Another stable isotope, carbon-13, for example, makes up only 1.07% of naturally occurring carbon.

The ratio between carbon-12 and carbon-13 is normally very nearly the same, whether it is measured in Colorado or Kiev. This phenomenon applies to all elements, not just carbon. The percentage of each isotopic abundance does not usually vary from its natural state. Because of this, these ratios act like scientific fingerprints. If scientists find a sample of an element that contains abnormal isotopic ratios, they know that an unusual event has occurred.

Generally, there are three types of events. First, environmental and biological factors can segregate some of the isotopes and cause minor shifts in the ratios between isotopes at certain locations over long periods, and forensic scientists can use this data to correlate biological samples to specific geographical locations. Second, a wide variety of man-made processes can be used to separate and concentrate isotopes. Methods include diffusion mechanisms, centrifuges, electromagnets, or lasers. Isotopic separation is the way low-grade uranium (containing very little U-235) is enriched to make high-grade uranium (containing much more U-235). Third, nuclear reactions can cause isotopic shifts that add or remove neutrons from isotopes. Although the first two type of events cause isotopic fractionation, only nuclear reactions change the number of protons or neutrons in the nuclei.

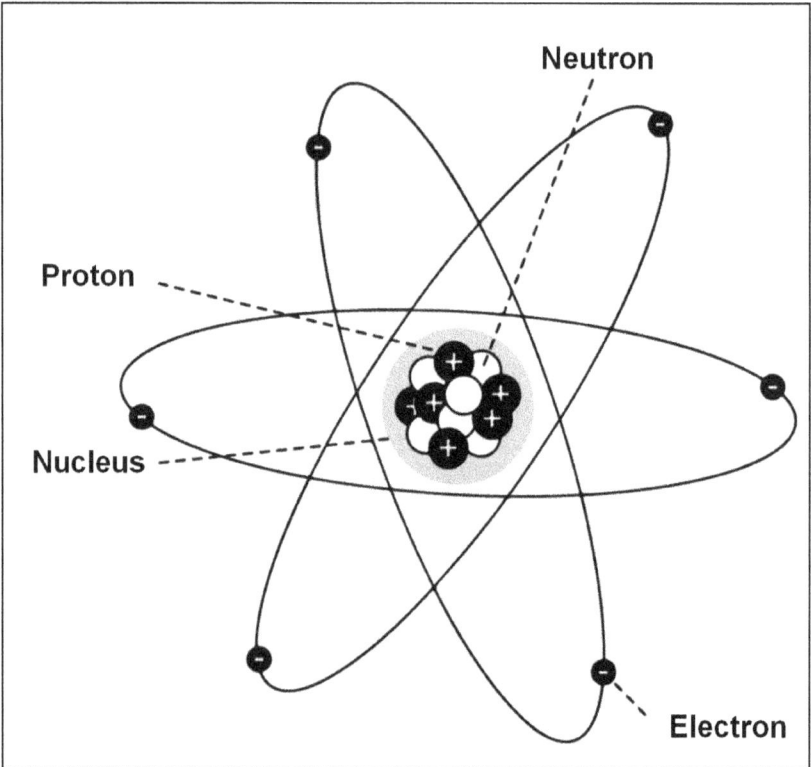

Basic diagram of a carbon-12 atom. The diagram isn't to scale; the actual distance between the orbiting electrons and the nucleus is much greater than shown here. Also, the specific arrangement of electrons in their valence shells is not depicted here.

A Matter of Power

Elements change, or transmute, into other elements by the addition or subtraction of protons. Isotopes change into other isotopes by the addition or subtraction of neutrons. Both kinds of changes are nuclear. Chemical changes involve either the addition or subtraction of electrons of a single atom, or the regrouping of atoms among themselves; they do not involve changes inside the nucleus. Not all reactions that occur in nature or science are equal. There is a big difference between reactions that can cause a chemical change and reactions that can cause a nuclear reaction. One of the biggest differences is the amount of energy required to initiate a reaction. A nuclear reaction typically requires one thousand to one million times more energy than a chemical change.

Primarily, two fundamental physics forces affect protons: the electromagnetic force and the strong force. Neither relinquishes its power and control over protons without a fight.

These forces, the electromagnetic and the strong, prevent protons from jumping from one atom to another and help provide stability for matter.

Under most circumstances, the electromagnetic force repels protons from each other like the north poles of magnets repel each other. It is no easier to squeeze two protons together than it is to try to press the north poles of two magnets together with your bare hands and expect them to stick. With protons, the exception is that, if they are squeezed incredibly close to each other, the strong force overcomes the electromagnetic force, which then slams the protons together. The strong force works only at very short distances. Conversely, to separate protons from each other, an immense force is required to free protons from the stranglehold of the strong force.

But no chemical process ordinarily has enough energy to bring protons together or to separate them. For a century, scientists have known that the sheer muscle required to make these kinds of changes can be triggered only by high-energy physics. This typically takes place through the use of particles that are emitted with high levels of energy. A moving projectile such as a bullet may be small, but the rate at which it is traveling gives it enough power to shatter dense material.

At the turn of the last century, chemists and physicists began to understand the nature and types of radioactivity.

Types of Radiation

The first radioactivity that scientists observed was two particles; which were initially labeled with the Greek characters "alpha" and "beta." Researchers later figured out that alpha particles are helium nuclei and beta particles are energetic electrons. The third type of radiation was the gamma ray. Gamma rays are highly penetrating forms of electromagnetic radiation emitted from nuclear reactions. On Earth, they are emitted by devices or radioactive materials and a few rare terrestrial events. Gamma rays are a class of photons (a larger group of massless entities) that, according to quantum mechanics, behave both as waves and as particles. A range of various-energy gamma rays can be depicted in a gamma spectrum.

Researchers also understood the concept of radioactive decay, which describes how radioactive elements spontaneously disintegrate and emit alpha particles or beta particles. As radioactive elements decay, the alphas and betas fly from the parent elements into the surrounding space. In their place, they leave smaller and slightly different child elements or isotopes. Gamma radiation is different in nature and is not a mechanism for radioactive decay, as alpha and beta particles are. For more information on the types of radiation and their characteristics, see Appendix D, Basic Types of Radioactive Emissions.

CHAPTER 16

Accused of Fraud

MIT Scientists Plant Story

Two days after the April 26, 1989, congressional hearing on "cold fusion," Fleischmann and Pons were vilified, just as Congressman Dana Rohrabacher had feared. Ronald G. Ballinger, a professor in the departments of Nuclear Engineering and Materials Science and Engineering at MIT, had just testified before Congress. Ballinger had been the last to speak that day.

Ballinger was also a member of the MIT Plasma Fusion Center team, one of the largest and most well-funded fusion research centers in the U.S. The news he brought back to MIT was not good. Members of Congress were enthusiastic about supporting Fleischmann and Pons, they were unhappy about the lack of progress in conventional fusion research, and they had treated the electrochemists like national heroes.

Ballinger and Ronald R. Parker, the director of the MIT Plasma Fusion Center, took matters into their own hands. Parker invited journalist Nick Tate, at the *Boston Herald,* to come to his office to speak with him and Ballinger. The meeting took place on Friday, April 28, 1989, three days before the start of the American Physical Society yearly meeting. Tate had the foresight to audio record the meeting. Gene Mallove, editor of *Infinite Energy,* got a copy of the tape from Tate a year later. Here are key excerpts:

"We're beginning to get a very short fuse on this whole issue," Parker said, "as you can tell, because, for example, these guys were down in Congress when Ron was down there on Wednesday asking for twenty-five million bucks. ... It's one thing when they come out with something that's potentially interesting scientifically. It's quite another thing when they're out there trying to fleece the public to push something that has no credibility at this point."

Tate got the sense that Ballinger and Parker had doubts and concerns about the Fleischmann-Pons claims, but Parker said that it was much more serious. "We can go beyond the concerns and questions to say that what they have reported is not true," Parker said.

Tate asked Ballinger his impression of the congressional hearing. "It was a fairly well-orchestrated attempt to short-circuit the well-established and well-recognized review process for any kind of research," Ballinger said, "much less this kind of research, and to divert government funds from other projects, presumably to the University of Utah." Tate pressed them to be specific about their concerns:

TATE: Back up a step. I presume you're talking about traditional scientific controls and traditional scientific methods that have not been observed in this particular situation.

PARKER: I'll give you a quote: This is scientific schlock.

TATE: Tell me specifically what they've done.

PARKER: [laughs] I'll just tell you about the neutrons. They've taken some data. They didn't even take it themselves; they had people take it for them. They published it in their paper, and they claimed that it showed the presence of neutrons from their experiment. The data has been patently falsely interpreted. Neutrons are not present at anywhere near the level their own data shows. They're not there. They've misinterpreted their results. They falsely interpreted their results. Whether they did this intentionally or not I don't know, but they did not interpret their results correctly. It's a key point in their paper.

TATE: Specifically, what were they claiming? That it was neutrons they were creating?

PARKER: That they were creating neutrons from their experiment. Their documentation unfortunately shows that not only was it falsely interpreted but also there were no neutrons at anywhere near the level they claimed. You can use the data in two ways, to show that they falsely interpreted it and that there weren't neutrons at the level they claimed.

TATE: So at best it's misinterpretation, and at worst it's — as you were saying?

PARKER: It's fraud.

TATE: How do you know this? From studying their research? From reviewing their information? I presume you've, in addition, attempted to parallel what they've done?

PARKER: We reproduced their [experiment], so we completely understand why they misinterpreted. Let me put it a different way. We don't see why they misinterpreted. We don't understand what they should have seen and didn't.

TATE: So you've reproduced their experiment?

PARKER: We've simulated the neutrons. We've said, Suppose there were no neutrons; what would it have looked like? And we find something quite different from what they claim.

BALLINGER: We find what we should expect.

TATE: Would you care to speculate on their intent?

PARKER: I think Ron made it perfectly clear that, when you're asking for [$25] million for the university —

During the interview, Parker's telephone rang. It was Richard Petrasso, a senior physicist with the MIT Plasma Fusion Center. Parker told Petrasso about a conversation he had just had with Richard Garwin:

PARKER: I just talked to Richard, who wrote the *Nature* piece. I don't know if you saw that. But he and I basically write [the gamma-ray] off. I said his piece was the best thing written so far. He told me he did see the original submission, and it did have the line at 2.5 MeV. The original submission had the line at 2.5, so that's the smoking gun with fingerprints. The original submission to the journal had 2.5, just as the 2.5 in the equation. It transcends the question of whether they misinterpreted to the question of whether there was deliberate fraud. All your detective work was correct, but now [Garwin] has the smoking gun with the fingerprints on it, right? [laughs]

The next two statements during the meeting with Tate are critical:

PARKER: I thought Garwin would be good to talk to, and he just volunteered. He'd seen the original submission to the journal. The line was at 2.5.
BALLINGER: That's what we suspected.

Some explanation is required. Fleischmann and Pons, in their original manuscript sent to the *Journal of Electroanalytical Chemistry,* included a graph, supposedly of a gamma-ray peak, shown at an energy of 2.5 MeV. They had submitted the same graph in the manuscript they submitted to *Nature,* for which Garwin was one of the reviewers. When their preliminary note published on April 10 in *JEAC,* the energy of the peak had been shifted to 2.2 MeV. How and why the value was shifted are important parts of the story, which I will discuss in the next chapter. But here's the important point: Ballinger and Petrasso had not known for certain that the peak had been shifted until now. Depending on exactly when Parker spoke with Garwin, Parker, too, may not have known about the "smoking gun" at the time he invited Tate over. Still, Parker had not personally seen the graph depicting a peak at 2.5 MeV.

They had called in the reporter. They had accused Fleischmann and Pons of fraud. And all they had at that moment was hearsay from Garwin and circumstantial evidence (discussed in the next chapter) that Petrasso had uncovered.

They were missing two crucial things. First, they were missing motive. Second, they didn't have sufficient information to determine whether the shift was the result of mistakes, sloppiness, or intent to deceive.

Parker told Tate that he and his MIT colleagues thought it was a scam and that they were preparing to "blast" Fleischmann and Pons at the American Physical Society meeting. Parker finished the meeting and asked Tate to hold the story until Monday, May 1, to coincide with the first day of the APS meeting.

Parker Frantic

Before Monday arrived, Parker became frantic. Late Sunday night, he received a call from a reporter at CBS, according to author Gary Taubes. "Dr. Parker," the reporter said, "we see that you've accused the Utah people of fraud and scientific schlock. We'd like to know if you'd elaborate on this." (Taubes, 263)

The next part of the story was told by Gene Mallove, who was the chief science writer at the MIT news office at the time. On Sunday night, after getting the call from the reporter, Parker called Mallove at home. The *Boston Herald* story was scheduled to publish the next (Monday) morning, but advance copies had apparently gone out to members of the news media. Here is Mallove's account:

> A frantic Ronald Parker, perhaps fearing that he would be sued by Pons and Fleischmann for the harsh words that were quoted a bit too explicitly for his taste, called me late on the night of April 30, 1989. He had me dispatch a press release to the wire services denying the impending *Boston Herald* story, the exact nature of which he had learned from a call from CBS television. ...
>
> I stayed up into the wee hours of the night of April 30–May 1, 1989, sending a press release dictated to me over the telephone at my home in Bow, New Hampshire, by Professor Parker. I telephoned it to UPI, Reuters, and the Associated Press, and it denied what Parker had said in the interview with *Boston Herald's* Nick Tate.
>
> When I arrived at the MIT News Office early that morning after a sleepless night, we hastily put together a printed form of the press release to handle the approaching storm.

Mallove provided an excerpt of the press release, in which he had quoted Parker:

The article erroneously characterizes remarks that I made regarding the cold fusion experiments done at the University of Utah. Specifically, I did not: (1) Deride the University of Utah experiments as "scientific schlock" or (2) Accuse Drs. Fleischmann and Pons of "misrepresentation and maybe fraud."

The headline in the *Boston Herald* took up the full width of Page 1: "MIT BOMBSHELL KNOCKS FUSION 'BREAKTHROUGH' COLD." Television footage shows that Parker also held a press conference that day while at MIT. Mallove later revealed that Parker lied to him:

Of course, I had at that time no reason to doubt what Parker was telling me, that his story was a distortion. I would learn the stark truth about this deception only over a year later when Tate allowed me to listen to the actual tape.
...
On June 7, 1991, I resigned from the MIT News Office to protest the outrageous behavior of the PFC and others at MIT against cold fusion. On the day of my resignation, I publically disavowed this press release — an unintended falsification of the truth in which I was used as a dupe in part of an orchestrated campaign against cold fusion.

As the next chapter shows, the MIT researchers had valid reasons to be concerned about the scientific integrity of the graph Fleischmann and Pons presented as a gamma-ray peak.

CHAPTER 17

Twin Peaks Over Salt Lake City

Shifted Gamma-Ray Graph

On March 28, 1989, MIT researchers used the news media to accuse Fleischmann and Pons of fraud, with liberal use of hearsay and circumstantial evidence. Some damning evidence existed, but the MIT researchers didn't have it — yet.

Fleischmann and Pons made several unexplained and improper changes to a graph that they thought, and claimed, was a measurement of a gamma-ray peak. This was secondary evidence they offered to assert their claim of fusion. They included this graph in the manuscripts they submitted to both the *Journal of Electroanalytical Chemistry* and to *Nature*.

For physicists, however, the claimed gamma-ray peak (assuming it had been real) would be extremely significant. Physicists didn't accept Fleischmann and Pons' excess heat as *bone fide* nuclear evidence. But gamma radiation would have been strong evidence of a nuclear reaction.

In the manuscripts, the graph displayed a curve with a peak at 2.5 MeV. Before the *JEAC* preliminary note was published, Fleischmann and Pons learned that a correct gamma-ray peak for that experiment should have been located at 2.2 MeV. On learning this, they changed the graph to 2.2 MeV and submitted the revision to both *JEAC* and to *Nature*.

There were other problems with and changes to the graph, in addition to the shifted energy value. Fleischmann and Pons should have withdrawn the graph as soon as Fleischmann learned of the first problem, which happened when he spoke at Harwell on March 28.

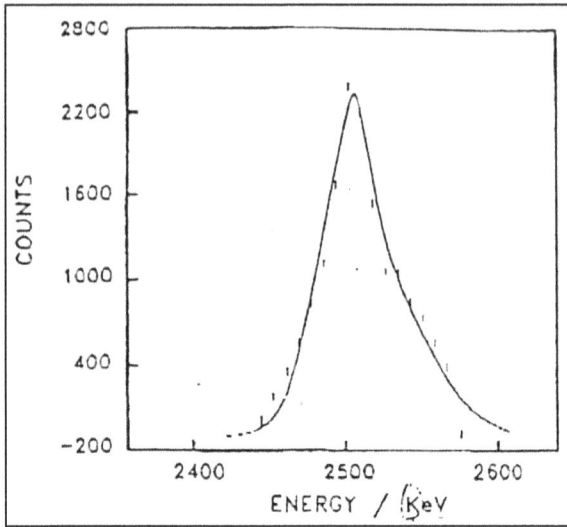

"Before" — *The graph as of March 11, 1989, when Fleischmann and Pons submitted their manuscript to the* Journal of Electroanalytical Chemistry.

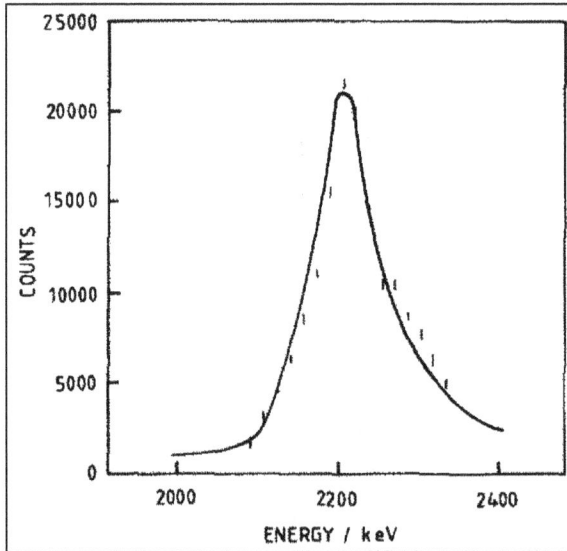

"After" — *The graph as of April 10, 1989, when Fleischmann and Pons' preliminary note published in the* Journal of Electroanalytical Chemistry.

The alterations to the graph were improper, as Fleischmann acknowledged two years later in an interview with Gene Mallove. "I don't think the MIT scientists behaved very well," Fleischmann said, "but I think we were also a bit stupid over those gamma-ray spectra."

Fleischmann and Pons' actions made them vulnerable to accusations of fraud. To begin with, Fleischmann and Pons did not know enough about gamma measurements to report them competently. The dialogue between Fleischmann and Richard Garwin at Erice proved this. Second, Fleischmann and Pons didn't collect the data themselves. They asked Robert Hoffman, a radiation health physicist, to record it for them.

Scientific Sleuthing

In late April 1989, the scientists at MIT had a hunch that there were serious problems with the Fleischmann-Pons gamma-ray graph. One of those scientists, Richard Petrasso, began his own investigation to try to uncover the facts. Garwin, a reviewer of the Fleischmann-Pons *Nature* manuscript, had seen the original Fleischmann-Pons graph with a peak at 2.5 MeV and passed that information — but not a copy of the graph — to the MIT researchers. This, the MIT researchers believed, was the "smoking gun," as they called it.

Garwin, as a reviewer, was unable to breach ethics protocols and give a copy of the "smoking gun" to a third party, such as the MIT researchers. As a result, the manipulated graph remained buried — until British theoretical particle physicist and author Frank Close picked up the trail in the fall of 1989. At the time, Close was a distinguished scientist at the Oak Ridge National Laboratory and a distinguished professor at the University of Tennessee. In 1991, he was appointed head of theoretical physics at the U.K. Rutherford Appleton Laboratory as well as the deputy chief scientist of the lab.

Most people in April 1989 were not aware of much of the following information. It did not become public until the first edition of Close's book, *Too Hot to Handle: The Race for Cold Fusion,* was published in January 1991. Close's book provides extensive details on the gamma-ray matter. Some of the information in this chapter, including the original graph purporting to be a 2.5 MeV gamma-ray peak, has never before been published, though Close has displayed it in many talks that he has given on the topic. In his book, Close redrew his own version of the peak to avoid possible litigation from Fleischmann and Pons.

The news of Close's exposé even made the front page of the *New York Times*. Close keeps a framed copy of that page of the newspaper on his bathroom wall above his toilet. The location, he told me, reflects his feelings on the conflict.

But some people, such as Garwin and John Maddox, the editor-in-chief of *Nature*, knew about the "smoking gun" by the end of March 1989. Theoretical physicist Steven Koonin likely was aware of it in mid-April because he was in close communication with Garwin that month and he noticeably altered his behavior toward Fleischmann and Pons after the Erice Fusion Forum workshop.

Measuring the Gamma-Ray Signal

Like most physicists, the MIT scientists didn't think "cold fusion" was likely, to begin with, as Philip J. Hilts quoted them in the *Washington Post:*

> "My first reaction was that it was incredible," said Ronald R. Parker, director of the Plasma Fusion Center at MIT. "In fusion research, there are always crackpot claims to produce fusion in a simple way. It always turns out that a little green man from Mars told them how to do it. When I heard of this, I thought — here's another one, but for some odd reason the *Wall Street Journal* bit on this one."
>
> "It's got to be wrong," Stanley Luckhardt, an MIT fusion physicist, told himself. "I'm afraid we'll look like idiots if we are seen trying this thing."

When Fleischmann and Pons described their experimental results in their preliminary note, they devoted the first two pages to the excess-heat data from their experiments Nos. 1 and 2. For their experiment No. 3, they displayed two graphs of a purported peak from a gamma-ray spectrum. The rest of the page was a discussion of tritium.

Fleischmann and Pons had taken some direct neutron measurements using a Bonner-sphere BF_3 type instrument, but they had only one such detector, so they didn't think that data was convincing. They also tried to measure neutron emissions indirectly by looking for gamma rays.

Here's what they did: Their cells were immersed in a water bath. This bath provided a constant temperature that enabled accurate heat measurements. The bath also served as a moderator in which neutrons emitted from the cell, if any, would be captured by the bath water.

Neutrons that hit protons in the water cause a secondary nuclear reaction, which emit gamma-rays. Gamma-rays emitted from a neutron capture by protons in the water would have an energy of 2.2 MeV.

Sometime later, Close gave a lecture on "cold fusion" at the National Institute of Standards and Technology, in Gaithersburg, Maryland. During his talk, he gave a technical explanation of the process:

> When the neutrons emitted from the cell meet the water bath, they bounce around and get slowed down until they come to rest, or are "thermalized," at which point they can be captured by protons in the water.
>
> The crucial thing is that the neutrons are captured at rest. All memory of their incident energy has been lost. Now, this is the thing that many of us did not know at the time; in particular, Fleischmann and Pons did not know it at that time. All memory of their original energy has been lost; they are captured at rest, and they form a deuteron. But the mass of a deuteron is less than the combined masses of a neutron and a proton, and that mass-excess is radiated off as energy in the form of a gamma-ray. The gamma-ray that comes off carries the binding energy of a deuteron, which is 2.2 MeV, nothing at all to do with 2.5 MeV, [as claimed in the March 11 Fleischmann-Pons graph].
>
> So Fleischmann and Pons were looking for gamma rays around the cell. If they found gamma rays at 2.2 MeV, that suggests that perhaps neutrons were present; therefore,

perhaps, fusion was taking place. Indeed, in the paper that was produced and shown to the world, this peak was presented, centered at 2.2 MeV, as evidence supporting their claims that this was fusion, as opposed to an ordinary chemical reaction.

The Persons of Interest

There were four people involved in the research and preparation of the graph that appeared in the Fleischmann and Pons preliminary note. The third person was Marvin Hawkins, a graduate student who was doing most of the hands-on laboratory work at the time. The fourth person was Robert Hoffman, a radiation health physicist, who, at Pons' request, brought in a portable gamma-ray detection system and took data. Hawkins may have recorded some of the gamma-ray data, as well.

By the time I investigated this topic in 2014, Fleischmann was dead. Pons, as he told me a decade earlier, had stopped talking about the past because it brought him too much pain. I spoke with Hawkins in August 2014 about Pons' accusations that Hawkins had stolen his and Fleischmann's lab books. (Chapter 6), However, Hawkins' inconsistent and unreliable memory discouraged me from asking him to recall details about the 1989 gamma-ray matter.

Hoffman had publicly stated in 1989 that the gamma-ray data was worthless, though I did speak with him in September 2014. He recalled that, after the press conference on March 23, 1989, he had done no further work with Fleischmann and Pons.

The Peak Labeled 2.5 MeV

On March 11, Fleischmann and Pons submitted the first version of their manuscript to the *Journal of Electroanalytical Chemistry*. The peak in this version was marked at 2.5 MeV. Not only was this peak at the wrong energy for neutrons captured in water, but, according to Close, no deuterium fusion reaction generates gamma rays with 2.5 MeV.

On March 13, the University of Utah fusion press conference was 10 days away. Fleischmann was worried. He knew there was an inconsistency between the neutron data and the tritium data. According to the branching ratio of deuterium-deuterium fusion, such reactions should produce equal amounts of neutrons and tritium. The ratio of these two products that Fleischmann and Pons were seeing in their lab was way off. More significant, the quantities of each of these products were much too low for fusion.

Fleischmann sent a fax to his friend and former student David Williams at the Harwell laboratory in the UK. Fleischmann told Williams about the excess heat they had measured. "The data is still very incomplete," Fleischmann said, "so you can imagine how annoyed we are to be rushed into premature publication." (Taubes, 89)

Fleischmann told Williams that he was bewildered by the ratio of the products, based on his assumption that the products were from a fusion reaction. "The intriguing thing," Fleischmann wrote, "is that the rate of tritium generation is much less than corresponds to the heat production. Neutron production is still lower. What on earth is going on?"

Michael Ravnitzky, an editor of this book, noted the relevance of a quote by Isaac Asimov: "The most exciting phrase to hear in science, the one that heralds new discoveries, is not 'Eureka!' but 'That's funny ... '"

Regardless of the discrepancy, Fleischmann and Pons were confident about their heat measurements, so they soldiered on. On March 20, Fleischmann and Pons submitted a revised version of the manuscript to *JEAC*, but it still had the graph depicting a 2.5 MeV peak. The cover page stated that this was the final version of the manuscript. It was not; Fleischmann and Pons submitted corrections to the proof on April 4.

On March 23, the University of Utah held the fusion press conference. Fleischmann and Pons did not release any preprint of their *JEAC* preliminary note at this time; thus, the graph showing a 2.5 MeV peak never entered public circulation.

On March 24, Fleischmann and Pons submitted a slightly modified version of their *JEAC* manuscript to *Nature*. It, too, contained the graph showing a 2.5 MeV peak. The manuscript was handled by Editor Laura Garwin. (Close, 281; Mallove, 44)

What about Hawkins? Pons discovered that his lab books were missing on March 26, and he suspected that Hawkins had removed them without permission. Pons then invited a postdoctoral researcher named

Mark Anderson to take over the fusion experiments. This means that Pons was unlikely to have requested or obtained any further help from Hawkins on the gamma-ray data.

Bad News at Harwell

On March 28, Fleischmann gave his lecture at Harwell. Physicists in the audience immediately recognized that the depicted gamma-ray peak at 2.5 MeV was at the wrong energy for that reaction, and they told Fleischmann that it should have been at 2.2 MeV. While he was still at Harwell, Fleischmann called Pons in Utah. Pons faxed a "corrected" version back to Fleischmann. (Close, 113)

On March 30, Pons faxed the replacement 2.2 MeV graph to the Washington office of *Nature*, to the attention of Editor David Lindley. Lindley, in turn, faxed the revision to *Nature* Editor Laura Garwin in London. According to Close, who obtained a copy of Pons' fax from Maddox, Pons provided no explanation for the change.

The Peak Labeled 2.2 MeV

On March 30 or 31, Pons released a pre-print of the manuscript with the revised 2.2 MeV graph to five of his close colleagues. Despite the fact that Pons clearly marked the preprint with the word "CONFIDENTIAL," within hours that version was distributed worldwide by fax machine as well as transcribed and sent through computer networks. (*Wall Street Journal*, April 3, 1989)

On March 31, 1989, Fleischmann spoke at CERN. I presume that he displayed the new 2.2 MeV graph. Had he displayed the 2.5 MeV graph, people would have jumped on the error, and Morrison made no such reference in his newsletter.

On April 4, 1989, Fleischmann and Pons received the galley proof from *JEAC*, and Fleischmann sent his and Pons' corrections back the same day. A copy of the fax that Close obtained shows that Fleischmann and Pons crossed out the graph depicting a peak at 2.5 MeV and submitted a revised graph depicting a 2.2 MeV peak.

On April 10, 1989, Fleischmann and Pons' preliminary note with the 2.2 MeV peak published in *JEAC*. If anybody realized at the time that a switch had taken place, they did not reveal it publicly. But the text in the preliminary note still showed 2.5 MeV in equation vii for what should have been: 2.2 MeV.

The general understanding in the nuclear community was that the 2.5 MeV value in equation vii was simply a typographical error, rather than the leftover detail it was, from the earlier 2.5 MeV graph. But the published graph was still missing important features for a legitimate gamma-ray spectrum. For example, Fleischmann and Pons published only a narrow portion of the apparent peak rather than a full spectrum.

Petrasso's Investigation

By April 27, MIT researchers had suspected something wrong with the graph, but they didn't have any hard facts. Richard Petrasso, a physicist with the MIT Plasma Fusion Center, began investigating. He called Hoffman to better understand the data he had taken. Petrasso did not have Hoffman's raw data at the time and wasn't able to come to a definitive conclusion. Petrasso then obtained an image of what he believed to the gamma spectrum in Pons' lab that was broadcast on a CNN news story. (Close, 170)

But when Petrasso examined the spectrum, he saw that the signal, whatever the source, was not at 2.2 MeV but at 2.5 MeV. Petrasso couldn't find any peak in that spectrum at 2.2 MeV, as he said in his May 2 presentation at Baltimore. "We conjecture that their purported signal is actually at 2.5 MeV," Petrasso said.

This means that a) Fleischmann and Pons had originally reported the data just as it had been measured by Hoffman and b) it was an artifact.

The first offense committed by Fleischmann and Pons was to include data that they did not understand well. Their second and more egregious offense was changing the graph to 2.2 MeV.

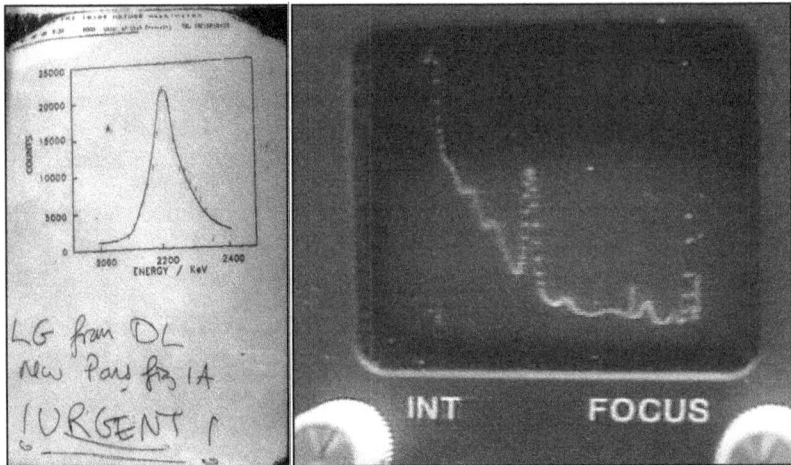

Left: Revised graph sent by Stanley Pons to Nature; Right: Screenshot from Fleischmann-Pons laboratory as displayed on television by CNN-TV. Image illustrates complete spectrum rather than just a single peak.

On April 28, as discussed in the previous chapter, MIT scientists Ronald R. Parker and Ronald G. Ballinger delivered the "fraud" story into the hands of Nick Tate of the Boston Herald. Parker, however, didn't have evidence that Fleischmann and Pons had changed the value of the measurement before he accused the chemists of fraud. He had only hearsay from Garwin. On April 30, Parker had the MIT press officer, Eugene Mallove, issue a press release denying what he had told Tate.

Slinging Mud

On May 1, the Boston Herald published Parker's accusation of fraud. The Herald included additional vicious comments from Parker:

> "Everything I've been able to track down has been bogus, and I think we owe it to the community of scientists to begin to smoke these guys out," said Parker, who will present the explosive findings to the American Physical Society in Baltimore tonight.
>
> Parker and other top nuclear experts at the prestigious

Cambridge institute also criticized Pons and Fleischmann for making a push in congressional hearings last week to grab millions in grant money and accused the University of Utah of trying to "fleece" the government.

A More Scientific Attack

Petrasso had been working overtime to prepare a paper on what Fleischmann and Pons had claimed was a gamma-ray spectrum. Because Fleischmann and Pons hypothesized that they were seeing fusion, they expected to see gamma-rays, normally associated with the third branch of deuterium-deuterium fusion. (Chapter 5) *Hacking the Atom*, the first book in this series, discusses the infrequent observations of gamma-rays in LENRs and offers an explanation for why the gamma-rays appear to be largely suppressed. On May 2, in the evening, Petrasso presented his analysis at the American Physical Society meeting in Baltimore. He also submitted it in a manuscript to *Nature*. He identified four major problems with the peak as published in *JEAC*. The line width was two times smaller than it should have been for a real gamma-ray peak, the spectral shape was incorrect, the neutron rate was off by a factor of 40, and the signal looked more like 2.5 MeV than 2.2 MeV. (Source: APS Video of Petrasso presentation) Petrasso's critique was accurate and scientific.

"Fleischmann and Pons make quite a point out of this," Petrasso said, "and it's on this basis that they are compelled to say and believe that they have seen fusion processes in their cells." But that part was not accurate. Fleischmann and Pons had made it clear that their primary basis for claiming fusion was their interpretation of the excess heat they had measured.

On May 8, 1989, Fleischmann and Pons faced the news media at a press conference following the Electrochemical Society meeting in Los Angeles. Reporters asked them about the gamma-ray data, and Fleischmann effectively but not explicitly retracted that data. "Stan Pons and I had a long correspondence about the deficiencies of our gamma-ray measurements," Fleischmann said. "I'm well aware of the criticism which was made today. We are well aware that those measurements are deficient. I think it is not meaningful to continue with those measurements in that present form."

On May 18, 1989, *Nature* published Petrasso and his colleagues' paper. It contained critiques similar to those Petrasso had presented during the Baltimore meeting. Petrasso concluded that the Fleischmann-Pons "purported gamma-ray line actually resides at 2.5 MeV, rather than 2.22 MeV," which is what their original data had shown. Nevertheless, Petrasso knew of no reaction in that cell that could have produced a 2.5 MeV gamma-ray peak. "We can offer no plausible explanation for the feature," Petrasso wrote, "other than it is possibly an instrumental artifact, with no relation to a gamma-ray interaction." (Petrasso et al., 1989)

Flip-Flop

On June 29, 1989, Fleischmann, Pons, Hawkins and Hoffman replied to Petrasso's paper. They reversed themselves and wrote that their gamma-ray peak was actually at 2.5 MeV. This is despite the fact that there was no known reaction in that environment that could produce a 2.5 MeV gamma-ray peak, according to Close and Petrasso.

In that same issue of *Nature*, Maddox gave Petrasso the opportunity to respond to the Utah group. Petrasso and his colleagues pointed out the obvious problems and demonstrated that Fleischmann and Pons were out of their league on this matter. Without making any direct accusations, they made it clear that Fleischmann and Pons had been altering data without satisfactory scientific explanations, often a hallmark of scientific misconduct.

Close Picks Up the Trail

Sometime in the fall of 1989, Ron Bullough, the chief scientist at Harwell, gave Frank Close a copy of a two-page document summarizing the meeting that he and his colleagues had with Fleischmann at Harwell on March 28. Close had not known about the 2.5 MeV version of the peak, as he explained to me. The 2.5 MeV graph was not shown in the Harwell document, but the text revealed that Fleischmann had indeed displayed a 2.5 MeV peak during his visit at Harwell. When Close read this document, he realized that the value of the gamma-ray peak originally submitted was indeed 2.5 MeV. Until this time, he was one of many people who assumed that equation vii in Fleischmann-Pons' preliminary note that described a 2.5 MeV peak was a typographical error. When Close read the Harwell document, he realized that the 2.5 MeV value in equation vii had not been a mere typo.

Close was perplexed. Why had Fleischmann and Pons initially thought the peak should have been at 2.5 MeV? What led them to submit a corrected 2.2 MeV value in the published version? What was the scientific justification for changing it to 2.2 MeV? On Jan. 12, 1990, Close wrote to Fleischmann and requested an interview. They met at Fleischmann's home in Tisbury in February 1990, as Close wrote to me:

> He sounded flustered and said he didn't want to talk about the gamma-ray graph. This struck me as very strange. So I asked directly, "How did the 2.5 MeV peak get moved to 2.2 MeV?" Fleischmann said that I would have to ask Pons; he said he didn't have the answers. I didn't know how they had made the measurement, so I asked Fleischmann for more information. I didn't know at that time that Fleischmann and Pons hadn't done this measurement themselves but had relied on Hoffman and Hawkins.
>
> I think that Fleischmann's embarrassment with me was that I had inadvertently stumbled on his greatest error — to have relied on such flaky collaborators for such a crucial datum, of which he knew nothing, but had accepted the data as he had been given, and suddenly discovering that it was wrong. The fact that I was asking questions about it made him realize that the issue wasn't going to go away.
>
> But I didn't know how little he knew about it at that time. In hindsight, Fleischmann could not let me know that he was totally at sea, because for a scientist to admit that he is taking credit for something which he has not actually done

is to lose face big-time. Consequently, he put up a smokescreen of what "scientifically" had been done. He said, without giving a name, that someone had changed a linear interpolation to a quadratic interpolation. This was a nonsensical response: You can't move a peak's position by changing a scale inside some range. I tried to discuss this further with him, but he didn't want to talk about it, and he insisted on moving on. So we did, but I smelt a rat.

At some point during our meeting, Fleischmann opened up to me more, and it became clear that Marvin Hawkins had taken at least some of the gamma-ray data. I noted that the explanation might lie with Hawkins, not Pons.

Later, when the full story of the change from 2.5 to 2.2 and its presentation as evidence for fusion became clear, it was Fleischmann's lack of knowledge, as revealed in his flustered responses to me that day, and his insistence that I had to ask Stan, that convinced me that Pons was responsible for perpetrating the fraud. Pons perpetrated it first on his partner, Fleischmann, and then on the world.

The discussion of fraud in Close's letter represents Close's opinion, not mine. My research did not indicate fraud by Fleischmann or Pons.

The Trail to Utah

After Close met with Fleischmann, Close contacted Marvin Hawkins and asked to meet with him in Utah. Hawkins proposed to meet at an off-campus laboratory where he was working on "cold fusion" experiments. When Close arrived on the appointed day in late February or March, Hawkins had suddenly been called over to the university. Close didn't get to meet Hawkins, but he managed to speak with him on the phone. He wasn't able to learn anything new. When Fleischmann learned about Close's Utah visit, he wrote back:

I really must protest directly to you about your behavior during the last few weeks. I had thought clearly there is little point in doing so. There are two reasons for my having to tell you this. The first is that it transpires that you have been associated with a scurrilous TV program on the subject produced by Yorkshire television; the second is that you were in Salt Lake City last week and did not have the courtesy to contact either Stan Pons or myself. We could, at that time, have shown you the relevant records and letters which would have answered the questions you asked me. Instead, you interviewed one of Stan Pons' graduate students, who, to put it politely, has a reputation here for being an unreliable witness. I believe that you also gained illegal access to the National Cold Fusion Institute and, what could be much more damaging to us, to the closed laboratories in the Chemistry Department.

Soon after, in April, Maddox invited Close to stop by his office. Maddox had correctly guessed that Close suspected that Fleischmann and Pons had switched the graphs without providing a legitimate scientific explanation. Close had still not seen a copy of the graph with the 2.5 MeV peak. At the meeting, Maddox confirmed Close's suspicions but did not show any images of the original peak to Close.

Then, a few days after the meeting, Maddox sent Close a copy of the fax that Pons had sent to *Nature* shortly after March 28. The fax shows both the original and replacement graphs for the purported gamma peaks. This was the first time that Close saw and obtained a copy of the original graph depicting a 2.5 MeV peak.

That's when he began to suspect that Fleischmann and Pons had committed fraud and that they had changed the value from 2.5 MeV to 2.2 MeV after Fleischmann's March 28 visit to Harwell.

The Revelation

In January 1991, Close published the first edition of his book *Too Hot to Handle: The Race for Cold Fusion.* Concurrently, he published a feature article in *New Scientist.* He did not directly accuse Fleischmann and Pons of fraud, though he came as close to it as he could: He wrote that "scientific data were altered." In my discussions with him, Close explained the strict U.K. libel laws to me (they have since softened) and told me that his attorneys had placed severe constraints on his choice of language.

On March 17, Bill Broad of the *New York Times* reviewed Close's book and summarized the main point: "The startling assertion by two chemists that they had achieved nuclear fusion in a test tube was based on 'invented' data whose publication involved a serious breach of ethics and a violation of scientific protocol, prominent scientists have concluded." This was misleading because Fleischmann and Pons' claims were based on excess heat.

On Nov. 25 1991, when Close gave his lecture at the National Institute of Standards and Technology, he explained that, of the four parties involved, Fleischmann, Hoffman, and Hawkins didn't seem to know how the peak was moved:

> So Hoffman didn't know how the thing had moved. I then spoke to Marvin Hawkins, the graduate student, and he didn't know, either. I said to him, "Let's get this straight: It's clear that it was measured at 2.5. There are four people involved in this experiment. Fleischmann doesn't know how it's moved. Hoffman doesn't know how it's moved. And now you're telling me that you don't know how it's moved." He said, "I think you got the picture now."

Close's delivery was perfect. The NIST audience erupted in laughter — at Pons' expense.

The Fallout

Any lingering good relations between Fleischmann and Close vaporized the moment Close's book published. Fleischmann felt bitter, as he told Gene Mallove in a 1991 interview:

> When Close came to my house, he didn't have the slightest intention of talking about calorimetry. He wanted to talk about the gamma-ray spectra. So I said, "I can't talk to you about this because all the information on this is in Salt Lake City, so maybe I can dig up some information for you from there." And off he went. Then he produced this dreadful TV program for Yorkshire TV, so I wrote to him and said, "You came to see me on false pretenses. If you told me that you are really a TV journalist, I would never have talked to you."

In fact, calorimetry was one of the topics they discussed the day Close visited Fleischmann. At the time, Close was consulting with a television crew that was producing a documentary on "cold fusion."

In the next few years, the two men exchanged letters. However, Close also received a threatening letter from Gary Triggs, an attorney who was representing Fleischmann and Pons, as Close wrote to me.

"I received a letter," Close wrote, "from their attorney Gary Triggs on Nov. 5, 1992, in which he alludes to stolen documents, and he tries to scare me by suggesting that information I have is either in breach of patents or stolen."

Among other things, Triggs was demanding to know where Close had obtained the graph with the 2.5 MeV spectrum. Triggs was also threatening to sue a publisher of Close's book, according to an e-mail Close sent to Richard Garwin in 1991:

> Pons' attorney is threatening Princeton University Press with a libel action, probably because the book reveals more of their real data than they have. Perhaps the Department of

Energy should bring a counter-action for the waste of public money.

If these people proceed with action against *Too Hot to Handle*, then it is important for the scientific community (and my mental health!) that THEY end up on trial, not me. And to facilitate this, it is necessary that every little piece of documentation can be gathered together. You may know of many people with deep information who might now be encouraged to bring it out.

Triggs never filed the suit. I asked Close whether he ever met Fleischmann again in person. He did, once, by chance, at an event where Fleischmann was speaking:

Martin, in a slight panic I thought, became ever so jolly: "Ah, Frank, you're here. How nice to see you, blah blah. I declined to answer any questions from reporters who approached me because it was Martin's occasion and I hoped to make peace with him. We spoke as he left, and I made a comment that I was sorry misunderstandings had arisen, and he brushed it off. But we had no more contact, as far as I recall. He seemed to oscillate between, on the one hand, seeing me as someone who could help him resolve a major problem and, at the other extreme, as someone who was responsible for all the bad things that had ensued and that I had deceived him. Very sad.

Reflection

When Gary Taubes published his book on "cold fusion," he provided an accurate summary of the gamma incident. Taubes, who once described himself to me as "unforgiving," was compassionate:

It did seem obvious that, under the pressure of going public, Pons and Fleischmann had tried to do what they were hopelessly ill-equipped to do. A person could imagine the fusion pioneers in the frenetic rush, hoping for a peak, seeing one in Hoffman's data, and not asking questions, simply assigning it the proper energy and moving on to the next order of business. It was not good science, but it was certainly human enough. "When the gamma ray peak business came up," said Laura Garwin of *Nature*, "I was staggered by the incompetence of it. Fraud crossed my mind, as well. A good scientist doesn't just say the peak is at one place and then, upon finding out it is supposed to be someplace else, say, 'Oh, it is someplace else.' It is equally explained by stupidity." (Taubes, 313)

Before I spoke with Close, I had independently obtained a copy of the March 11, 1989, version of the Fleischmann-Pons manuscript with the graph showing a 2.5 MeV peak from Bruce Lewenstein's Cornell University Cold Fusion Archive.

But not until I was well into my discussions with Close did I realize the significance of that version. Close was aware of not only the legal but also the moral weight of what he had published. On June 23, 1992, Lewenstein interviewed Close and asked him about his insinuations of fraud. "Was there anything I could have possibly overlooked?" Close said. "I was terrified that there could be something I had missed. You just don't go out and implicitly accuse two scientists of fraud."

In my 2009 e-mail conversations with Close, I asked him, "Why would two men with long-term experience and good reputations in science risk everything by falsifying data in a situation where they knew they would be scrutinized from head to toe?" Close wrote back that he had heard rumors about the quality of some of Pons' earlier work.

Close had hoped that I could put him in touch with Pons so that he could resolve the mystery, but Pons had made it clear to me that he wanted no more business with the past. I never asked Fleischmann about the gamma-ray situation. However, I didn't understand it well enough to have asked a useful question until after Fleischmann had died. Regardless, I doubt that Fleischmann would have told me much more than he had said to other people.

The record is clear. Fleischmann knew it was a big mistake, and he felt embarrassed about it. History will judge whether Fleischmann and Pons committed fraud or whether their manipulation falls more accurately under the category of research misconduct. Fraud requires intent to deceive; research misconduct is simply doing things you're not supposed to do as a scientist. The evidence shows that what occurred fell into the latter category.

Fleischmann's wife, Sheila, shared her thoughts about it with journalist Susan Seddon, writing for *Infinite Energy* magazine, in 1997:

> "Why would he do it? Why would a respected Fellow of the Royal Society try to pull the bogus stunt the media and others accused him of, especially after such an erstwhile eminent scientific career? As if he could possibly have got away with it — it's just nonsensical."

Martin Fleischmann (2004) Photo: S.B. Krivit

CHAPTER 18

Backlash in Baltimore

Spotlight on American Physical Society

B eginning the week of May 1, 1989, physicists finally had their chance to corral the media spotlight that had favored the chemists. The Baltimore Convention Center hosted the annual American Physical Society meeting. A few physicists, speaking for their peers, and one lone chemist hit Fleischmann and Pons hard. Their presentations, as was customary for APS meetings, were neither published by the APS nor peer-reviewed. The parent organization of APS, the American Institute of Physics, did, however, videotape the key presentations.

The Lead-Up to Baltimore

On April 21, Caltech chemist Nathan Lewis (b. 1955) lectured on campus and announced that his team had found nothing in its attempts to replicate Fleischmann and Pons. On April 23, reporter Lee Dye, at the *Los Angeles Times*, wrote that there had been scattered confirmations of the Fleischmann-Pons experiment, but reported that the frustrated Lewis had failed. "Those claims of success," Dye wrote, "only add to the frustration of other scientists who have been skunked in the race to repeat the experiment. Nowhere is that more apparent than at Caltech, where a multifaceted team of experts has labored in vain to come up with something that would either prove or disprove the Utah claim."

Five days later, on Friday, April 28, all within a 24-hour period, researchers at MIT, Caltech, Yale and Princeton went on the offensive. Members of the MIT plasma fusion center planted a story with the *Boston Herald* accusing Fleischmann and Pons of fraud. Caltech distributed three press releases stating that despite extensive efforts, they failed to confirm the Fleischmann-Pons claims.

Moshe Gai, a physicist at Yale, gave a lecture on campus titled "Does Cold Fusion Exist?" and concluded that, based on his failed attempt, it didn't. Gai reported his results to the news media the following day.

Princeton University, the site of another of the country's most well-funded thermonuclear fusion research centers, issued a press release and told the Associated Press that it, too, failed to reproduce the Utah experiments.

Everyone was using the news media to communicate and debate science. Ironically, Fleischmann and Pons were perhaps the only researchers who had submitted — let alone received acceptance of — their scientific manuscript before reporting their work to the news media.

Caltech Press Releases

The Caltech press releases quoted Steven Koonin, who came very close to stating that the Fleischmann-Pons results were theoretically impossible. Also quoted was Lewis, who claimed that he had correctly repeated the Fleischmann-Pons experiment. He provided a list of analyses that he and his researchers had performed to search for confirmatory data, but they came up empty-handed.

Lewis implied that Fleischmann and Pons had fooled themselves. A Caltech press release even stated that researchers on Lewis' team "were able to document no excess heat and no neutron or gamma-ray deviation." It was a profound suggestion, that they were able to document the existence of something that didn't exist, in essence proving a negative.

Lewis was very proud of his team members for their hard work. "I can't overemphasize how very much we owe to the dedication of the team members," Lewis said. "Beginning the day of the initial announcement, they have been working around the clock, and it is that hard work that has allowed us to obtain these results."

It was a rare moment in science history; Lewis had publicly honored his team for attaining nothing.

The Agenda

The official agenda for the May 1989 APS meeting had been determined months in advance. Nevertheless, under the circumstances, a week after the University of Utah announcement, APS squeezed "cold fusion" into the schedule with an unprecedented set of three press conferences and two special evening sessions on May 1 and May 2.

The general APS meeting took place during daylight hours throughout most of that week. News media were interested only in the cold fusion events. From the public's perspective, it would have appeared that nothing else of importance took place at the APS meeting.

On Sept. 29, 2014, I interviewed Edward Frederic "Joe" Redish, 72, a professor of physics at the University of Maryland who organized the "cold fusion" scientific sessions at the 1989 APS meeting. He earned his Ph.D. in theoretical nuclear physics from MIT in 1968.

During my conversation with him, he illuminated several surprising facts about the "cold fusion" sessions at this meeting. The complete interview transcript is available on the *New Energy Times* Web site.

The 1989 Baltimore APS meeting is the turning point in the history of "cold fusion": a time when physicists and one chemist ganged up on Fleischmann and Pons in what *Nature* editor David Lindley described as a "hanging party lacking only its intended victims."

The Planning

Redish was the vice-chair and incoming chairman of the APS Topical Group in Few-Body Systems and volunteered to arrange the special sessions. His thinking, like that of other scientists, was that "cold fusion" reactions, if they existed, involved pairs of deuterons, the nuclei of deuterium atoms, and that therefore his topical group was suitable.

Like Valerie Kuck, the organizer of the American Chemical Society special session on "cold fusion," Redish also immediately envisioned a "Woodstock-type" special session on "cold fusion" for the APS. About a week after the University of Utah fusion press conference, Redish called Bill Havens, the executive secretary of the American Physical Society:

> "Hey, Bill," I said, "with all this exciting stuff about cold fusion, are you guys doing a Woodstock session at the May meeting?"
>
> "Well," Haven said, "I thought we would, but, you know, I talked to the divisions. The people in the nuclear division won't touch it with a ten-foot pole. The plasma physicists aren't interested and don't want anything to do it"
>
> "Fine, I'll do it," I said.

In my 2014 interview with Redish, he said that Steven Koonin, of Caltech, contacted him and asked to be on the program:

> Steve was a friend, somebody I'd known for many years because I'd known him from graduate school. And we had served on committees together, we had been on the Indiana University's cyclotron program committee together, we had worked together on the Nuclear Regulatory Commission report for nuclear physics in the mid-'80s.
>
> Koonin got in touch with me, and said, "We're working on this. I have ideas on this. I'd like to speak on it." And I said, "Great, you're my number one theory speaker. You got an experimentalist?" He said, "Nate Lewis is working on it."

But Redish's memory of the circumstance conflicts with an interview Koonin gave on May 8, 1989, as well as with the call for papers, which already had Koonin's name on it as an invited speaker.

> Joe Redish called me up soon after I got back from Italy and told me that the APS wants to hold a session in

Baltimore, and asked if I would be one of the speakers. He had gotten a preprint of my paper. So I said fine.

Koonin, remember, had tentatively accepted the Fleischmann-Pons and Jones claims, and he was enthusiastic and optimistic at Erice. The first APS "cold fusion" abstracts arrived on April 17. No prominent physicists had publicly attacked Fleischmann and Pons yet.

Redish said that Koonin told him that Lewis was a great experimentalist so he put Lewis second on the program. Redish was soon swamped by at least 50 people who wanted to speak at the session. Originally, he planned only one session but quickly realized he required two. Redish invited Fleischmann, but he declined. Redish invited Steven Jones and his associate Johann Rafelski, and they accepted.

Press Conference #1

The first two press conferences appeared to be scheduled at the last minute, but the third had been announced to reporters in advance. The organizer of the May 1, 4:30 p.m. press conferences was Phil Schewe, a physicist, writer and playwright, who worked for the American Institute of Physics, the umbrella organization of which the American Physical Society is a member society. Schewe was the *de facto* head of media relations for APS. Schewe introduced the first speaker:

> I asked Dr. Robert Perry, an assistant professor of physics at Ohio State. This is an informal briefing. I didn't want to make this seem like a major event all by itself. Dr. Perry is not one of the speakers tonight I asked him to come and give a briefing about the topic of "cold fusion" and maybe give an inkling, not about the specific papers that will be given tonight, but in general, what kind of results you'll hear about at a session like this.

Perry's expertise was in theoretical few-body nuclear physics. His thesis advisor had been Redish, who, as Perry told me, had suggested his name to the AIP public relations staff. He didn't have a prepared talk but instead responded to questions from the reporters. He gave a very brief introduction that revealed the hostile climate. "I'm an innocent bystander," Perry said. "I don't have any axes to grind in all this."

He may not have had an ax to grind, but his views may have been evolving. By April 18, he shared his strong opinion with Steven Koonin in an electronic message. A day later, Koonin sent Perry's message to several people, including Richard Garwin. The subject line of his message was "The Emperor's Clothing":

> I know of no reliable confirmation of any aspect of the Pons-Fleischmann work that would indicate fusion. No one else in the U.S. has detected neutrons. ... Many laboratories have worked night and day to reproduce the Utah results, and no one has seen anything but background. Excess heat has been seen at Texas A&M, but others seeing heat ascribe it to chemical reactions. In particular, reports from Caltech and Los Alamos indicate that large efforts have led to many physicists and chemists studying background neutrons and gamma-rays and seeing nothing of interest.
>
> In general, scientists seem to be remarkably reluctant to criticize the Utah work. Claims that physicists are simply allowing jealousy to cloud their judgment, not wishing to admit that chemists can do what they have failed to do, are absurd. Several reports indicate that Pons has admitted to seeing excess heat in one cell in which normal water was used instead of heavy water. Pons reportedly admitted this again today at Los Alamos. Draw your own conclusions.

When Perry began speaking at the press conference, he told reporters that he was extremely doubtful about the Fleischmann-Pons claims, and he gave them his perspective on the papers they would be hearing later that evening. A reporter asked him to "characterize the reaction of the physicists at the meeting to 'cold fusion' announcement." "I don't know what anybody else's reaction was," Perry said, "but my reaction on March 24, when I first read about it was that it was interesting but I didn't believe it so I set it aside and went on to read another news story."

Among other questions, reporters asked him about how the claims could be understood theoretically, about the different methods by which palladium was prepared, about measuring helium-4. Perry said that the experiments defied explanation and that they "required leaps of faith" to accept, based on the well-understood theory of nuclear fusion. He told them that the experiments were very difficult to repeat and that the levels of reported neutrons were very low.

Press Conference #2

The Nathan Lewis "cold fusion" press conference took place a half hour later, at 5 p.m. Schewe, as he told me in an e-mail, had decided that the news media would need to hear Lewis, so Schewe made special arrangements for him:

> Lewis spoke at the main cold fusion session on Monday night but couldn't attend the scheduled press conference on Tuesday morning, hence the extra one-man press conference for him on Monday. I remember him being in an angry mood. He had dropped his work at Caltech, he said, and put in a lot of time on work that came to nothing.

Lewis, of course, had gone on record on April 21, telling the *Los Angeles Times* what a thorough job he and his team had done, that they had found nothing, and that they "were able to document" that there was nothing there. Three days earlier, Caltech had issued its press releases, which reiterated the same finding of nothing. People in the know knew exactly what to expect from Lewis in Baltimore

"It's my understanding," Schewe said, "that the Caltech experiment does it all; calorimetry, neutrons, their experiment does all of these tests and so they will be the ones to watch. This is Dr. Nathan Lewis, of Caltech, a chemist not a physicist." Lewis addressed the reporters:

As most of you know, we've been working on this since day one, since the Fleischmann-Pons announcement. We were very fortunate to have great students with a lot of enthusiasm. After this was announced, they all ran into the lab and started doing experiments that night. And we've been working pretty much at a breakneck pace ever since then.

We have very open minds, and Caltech is a very good place as far as interactions. The nuclear physicists stopped what they were doing, ripped apart their instruments and said "Whatever you guys need, we will put it in there to do the experiments necessary to see the results." We have a neutron detector that is approximately 10^7 times more sensitive than we needed to measure the rates reported by Pons and Fleischmann. We've used cast electrodes, cold-worked electrodes, we've activated them, not activated them. We have followed the specific directions, we've tried some variations on those directions and we've never seen any neutrons.

We looked for gamma rays from 20 keV to 30 MeV; that's a very wide range. All the known deuterium-deuterium reactions and all the known proton-deuterium reactions give off photons somewhere in that energy range. Our sensitivity, per fusion event, would be on the order of, conservatively, 1,000 fusion events per hour and we see no evidence above the background for any fusion events at that level with those electrodes on all these samples which have been supplied by different vendors.

We have waited far longer than the waiting times that Pons and Fleischmann quote. We've also looked for tritium; we've uncovered artifacts, chemical interferences that will show you a tritium signal when in fact there is none.

We looked for helium. The helium we see in our cells is due to the ambient room air.

We've done calorimetry. We've uncovered a lot of methods that do not work; for instance, not stirring your solutions. You have temperature gradients. In one electrode you inherently generate more heat than the other. The electrodes, being big pieces of wire, are also cooling fins and they are efficient at removing heat from the system. If you do not agitate the system and stir it, the temperature you measure will depend on where you put the thermometer and you can get a very large range of errors this way and those errors placed serious doubt on the accuracy of the numbers that were measured by Pons and Fleischmann. We see no evidence of any excess heat. I'll be happy to answer questions.

Despite the importance Lewis attributed to his group's attempts to repeat the Fleischmann-Pons experiment, it was irrelevant. Failures to replicate are generally meaningless and useless. There are two exceptions.

First, if Fleischmann and Pons, for example, were to have confirmed that Lewis' attempted replication was done correctly, then his attempts would carry some weight.

The second exception is when a failure to replicate is instructive — that is, when a specific and obvious error in the replication attempt can be identified.

Given the well-known poor reproducibility of "cold fusion" experiments, it's difficult to know all of the reasons for failure at Caltech. But in Lewis' case, an instructive fact is available: He did not load the minimum amount of deuterium into his palladium. Lewis had not, as he had claimed, "followed the specific directions." As explained in Chapter 12, Lewis' team failed to load the minimum amount of deuterium into its palladium.

Although Fleischmann and Pons failed to specify this parameter in their preliminary note, Gary Holland, from the University of Utah Chemistry Department, told the Caltech team on April 1 that before anything happens in the cells, they needed to get near 1:1 in their palladium/deuterium loadings. Without such loading, the odds of getting any excess heat were extremely low, although Melvin Miles, a retired Navy electrochemist, believes that in one of the Caltech cells, 76 milliwatts of excess heat was produced.

Members of the news media asked Lewis a few questions about his own work but they asked a lot more questions about the Fleischmann-Pons experiment. Lewis had answers for everything. They weren't necessarily the correct answers, but unless the reporters were experts in electrochemistry, they were at Lewis' mercy.

In response to reporters' questions, Lewis said that his team "absolutely followed the directions" of Fleischmann and Pons. Yet almost in the same breath, Lewis said "Pons and Fleischmann would not talk to me or any scientist I know." He spoke with conviction, and no reporters seemed to notice the contradictions. Reporters did not know that Lewis had taken his cell dimensions from photos he had seen in the newspapers.

A reporter told Lewis that he heard Fleischmann and Pons had told Congress they had put in 1 watt of electrical power and got 4 watts of heat output. Lewis told the reporter that, instead, Fleischmann and Pons had put in 1 watt and got out only a quarter of a watt. The reporter said that Fleischmann and Pons may have deceived Congress. Lewis concurred: "That's what people would tend to think."

There is no evidence of intentional deception. Fleischmann and Pons had errors in the data listed for the 2 mm and 4 mm rods run at 512 ma/cm^2, from which they made their 1:4 claim. After correction, the values for those rods still showed excess heat. Beyond that, the 1989 preliminary note showed nine other runs that produced excess heat.

In their 1990 paper, Fleischmann and Pons published an updated table of the 1989 data. The correct values for those runs were 3.35 W and 0.38 W. Although that may not seem like a large amount of excess heat, the data are significant; their measurements were accurate to 0.001 W. The corrected value for their best experiment reported in their preliminary note was 5.69 watts in and 9.04 watts out.

A reporter asked Lewis whether he could explain the Fleischmann-Pons experiment that vaporized most of a 1 cm palladium cube. Lewis said he could explain that easily as an ordinary hydrogen-oxygen chemical explosion. Lewis was mistaken again. A chemical explosion strong enough to vaporize most of that cube would probably have destroyed most, if not all, of the laboratory.

A reporter asked Lewis to comment on the section of Fleischmann and Pons' preliminary note where they wrote that the "bulk of the energy release is due to an hitherto unknown nuclear process or processes," rather than a fusion process. Lewis spoke not only for himself but also for Fleischmann and Pons. "There is no evidence for any unknown nuclear process," Lewis said. "We find no evidence for anything other than conventional chemistry in these cells."

The 33-year-old Lewis, who had a doctorate in inorganic chemistry, and his graduate students had been working on the Fleischmann-Pons electrochemistry experiment for less than five weeks, an experiment the two electrochemists had worked on over the course of five years.

Consensus of Failure

The first "cold fusion" scientific session at the APS meeting began at 7:30 in the evening with the four invited speakers: Steven Jones, Johann Rafelski, Steven Koonin and Nathan Lewis. Most of the remaining presentations fell into two categories: a) proposed theoretical explanations and b) unsuccessful searches for gamma and neutron emission.

Eighteen hundred people sat and listened as the physicists reported the mostly null results of their searches for gamma and neutron emissions. However, Fleischmann and Pons had not observed any gamma rays emitted directly from the cell; their neutron flux was only slightly above background. Such searches were largely irrelevant. Four speakers at the American Physical Society meeting reported possible partial confirmations (Chapter 21), but no one in the news media reported this. The failures to replicate, however, made the headlines.

A Few Words From the President

James Krumhansl, the president of the American Physical Society, opened the session with a brief statement:

I've been instructed to keep my statement brief, and I will. I would like to welcome *all* of the audience, both members and visitors to our scientific meeting. The American Physical Society has, as its purpose, the advancement and diffusion of the knowledge of physics, and as a part of science, in the overall sense, we do our part through open meetings and *refereed* scientific publications. And it is in that spirit that we want to provide this forum for the discussion of the subject tonight, that is, "cold fusion," its possibilities and implications on a scientific basis, thank you.

When Krumhansl said "refereed scientific publications," a good portion of the audience broke out in laughter while he was midsentence. No evidence from the video indicates he was trying to be funny. But once the audience began laughing, he couldn't hold back his own smile. Everyone understood that he was referring to Fleischmann and Pons' preliminary note.

Redish's Moment in the Spotlight

As he had planned, Redish took the opportunity to speak about the scientific process:

I would like to address some remarks to the ladies and gentlemen of the press. You are serving for us as interpreters to the general public. And it's important to understand how certain aspects of the current debate illustrate the normal processes of science. We are searching here for real answers, for the truth of how the world works. We don't accept a new idea easily because it's so easy to get it wrong.

We challenge every new idea very intensely in order to make sure that it is real before we incorporate it into our scientific map of the world. We have to do this. Given human nature, it's very easy for us to replace truth by wishful thinking. It's the process of challenge, it's our demand for repeatability and our insistence on a detailed consistency that makes science work and gives it its power. Usually, preliminary results are challenged internally by the working group, then by peer review and, often, again in the open scientific literature by other groups before the science goes public.

What you are seeing here is a part of the way that we work. I see it as a plus that the general community is seeing how science functions, but it's important that they not misinterpret what's going on. There are always a range of results in the scientific community. Some are well understood and firmly believed. Others are in process and taken as working hypotheses but still under challenge.

The most intense challenge happens when a new result appears at first to contradict our expectations which are built up from what we know. Tonight we will discuss such a case. Whether that deuterium-metal interaction can lead to nuclear fusion or not, I don't know yet, but I can promise you that the standard working-through of science will find the answer before long.

Now, since I'm an educator as well a scientist, I have to mention a second point. You are also one of our links to the children, our future scientists. To them, you must make clear that science is often like this, even if it's not so publicly visible. It's hard work, it's very important and it's very exciting, and a great deal of fun.

Tonight you will have four kinds of talks. There will be experimental talks describing the results of measurements. There will be analysis talks trying to interpret what the raw data means. There will be conservative theoretical talks trying to see what's possible in the standard framework. And finally, they'll be some speculative theoretical talks, considering whether we've missed something, whether there is some mechanism that we've overlooked that makes it all possible.

Redish may have been overly idealistic about the behavior of some of his fellow physicists who followed him that evening. Koonin was first.

His depiction of how new ideas should be challenged in science is vulnerable to critique: If science is determined by committees or consensus, then science risks being dogmatic.

Cartoonist Sidney Harris' 1989 depiction of the "cold fusion" conflict

CHAPTER 19

"Incompetence and Delusions"

Koonin, Lewis Make Their Mark

Edward "Joe" Redish finished his opening remarks at the 1989 Baltimore American Physical Society meeting and turned the microphone over to Caltech theoretical physicist Steven Koonin:

> It's with some trepidation, that I open this theoretical talk with a discussion of experiments, and all the more so because I want to talk about the Fleischmann-Pons-Hawkins experiments. ... [They claimed their production of heat was] greater than 10 watts per cubic centimeter beyond the input for more than 120 hours. That corresponds to 4 MJ per cubic centimeter of palladium. That's an enormous number, 400 electron-volts per atom, and if you believe it, it's logical to conclude that it can only be nuclear physics. You can't get chemistry to give you that kind of energy.

Koonin began by taking the opportunity to perform a mini peer-review of the Fleischmann-Pons neutron claims and gamma-ray graph. According to his analysis, Fleischmann and Pons had measured gamma-ray energy at 2.200 MeV. "What makes 2.200 MeV gamma rays? Koonin asked. "Well, it turns out that there is a decay of radon-222."

It was witty, and Koonin sounded smart. The audience loved it and laughed. But there is no such 2.2 MeV gamma line emitted during the decay of radon-222. Koonin's implication was that Fleischmann and

Pons had made an enormous faux pas and mistaken an emission from an ordinary radioactive gas (often found in the basements of buildings) for evidence of fusion. It brought some comic relief (at Fleischmann and Pons' expense) to the uncomfortable question of "cold fusion."

"I don't know how much radon there is in their laboratory," Koonin said, "but I do know that they mine uranium in Utah." The laughter grew even louder, and this time it was accompanied by applause.

But Koonin's unscientific and untested speculation was irrelevant. Koonin had not seen Petrasso's presentation, scheduled for Tuesday night. Petrasso's investigation confirmed that a) the peak — caused by who-knows-what — was actually measured at 2.5 MeV and b) Petrasso knew of no possible source for a 2.5 MeV gamma-ray.

Koonin quoted on TV at APS Meeting: "It's all very well to theorize about how cold fusion in a palladium cathode might take place. One could also theorize about how pigs would behave if they had wings. But pigs don't have wings."

Koonin explained why, according to well-understood theory, getting deuterons to overcome the Coulomb barrier at room temperature was a problem, and he talked about those issues for another five minutes.

"Let me now turn to a speculation on this first problem," Koonin said. "Theorists are allowed to float trial balloons. That's well accepted in the culture. Experimentalists are not allowed to float trial balloons." The response from that remark was nearly half a minute of laughter directed at Fleischmann and Pons.

Koonin spoke about a variety of attempts he made to examine possible routes for high reaction rates of deuterium-deuterium fusion at room temperature. To nobody's surprise, none of the routes was feasible. He reached his conclusion, which included the most hostile words of the "cold fusion" conflict:

> Let me turn to the University of Utah experiments. The crucial question seems to be, "Is the heat due to fusion?" In my opinion, the spectacular effect they claim has not been proven by the usual standards we expect in scientific discourse. Secondly, we have no reliable report of reproduction of these experiments despite strenuous efforts by many groups around the world who bring to bear resources and expertise far greater than Pons and Fleischmann had to do this. Moreover, even the people at Harwell, who have professor Fleischmann as a consultant, have been unable to reproduce the effect.

> Finally, one cannot accommodate the lack of radiation by any accepted theory. My conclusion, based on my experience, my knowledge of nuclear physics, and my intuition, is that the experiments are just wrong and that we are suffering from the incompetence and perhaps delusion of Drs. Pons and Fleischmann. *[Cheers and applause]*

> Let me conclude by remarking that the phenomenon of "cold fusion" was not unknown to the ancients! This is an Aesop's fable, which I would like to read to you, it will just take a minute. It's called "The Great Leap at Rhodes," and perhaps we might want to relocate it to Salt Lake City.

> A certain man who had visited foreign lands could talk of little when he returned home except the wonderful adventures he had met with and the great deeds he had done abroad. One of the feats he told about was a leap he had made in a city called Rhodes. That leap was so great, he said,

that no other man could leap anywhere near the distance. A great many persons in Rhodes, he said, had seen him do it and could prove what he said was true. "No need of witnesses," said one of the hearers. "Suppose this city is Rhodes. Now show us how far you can jump." *[Laughter and applause]*

Koonin: "This Is Fraud"

In my conversation with Redish, he filled me in on some of Koonin's backstory. He told me what happened 15 minutes before the session started:

> The one thing that I do remember right before the session is that we came into the room and there was all kinds of stir. There were cameras and people around. We went into some room where the first few speakers were getting ready, and Steve Koonin was there. He said, "Joe, this is fraud. They faked this." He was convinced that it was faked.
>
> And I panicked. [Gasps] I said, "Oh, God. Steve, don't say that. Please. Please. Do not." I said to him something to the effect, "Don't say anything that can be considered libelous, that could get us into a suit."
>
> I didn't want this to be where the thing went. I didn't want it to be about accusations, I wanted it to be about the science.
>
> He kind of grumbled, and he said, "Well, OK, maybe." He swallows his words when he doesn't want to talk to you. So he went out. I didn't know what he was going to do.

Redish, as he told me, was relieved that Koonin didn't accuse Fleischmann and Pons of fraud. The second-to-last sentence in Koonin's last slide was crossed out. Redish thought that "fraud" was what the sentence had said and what Koonin had planned to say. I asked Redish why he anticipated that Koonin might have accused Fleischmann and Pons of fraud:

Well, two reasons. First, I saw he was really mad. He was very angry. He was not being calm and cool and collected. He was upset about it, and he thought it was fraud. I did not want him to say that because I felt it would have distorted the discussion away from the science and that there would have been a big argument, even whether he had been libelous. I just wanted to keep the focus on the science and not on the personalities.

The applause after Koonin's talk seemed to indicate concurrence by the audience. The videotape shows Redish with a big smile on his face after Koonin's talk.

As it turns out, the expression on Redish's face was relief that Koonin didn't use the word "fraud." I asked Redish why "cold fusion" was so anger-provoking for theoretical nuclear physicists?

It wasn't anger-provoking for most of us. Many of us were excited. The anger wasn't shared until the data started coming in. At the beginning, it's exciting. It's only when it starts to fall apart, you say, "This guy is wasting my time."

For all of us, we're doing exactly what we want to do, we're excited about science, we want to do more science, and if somebody misleads you and causes you to waste six months of your research career on something because they've faked it, then you get angry.

If somebody got a hint of something, and it says, "Well, maybe they've got something here, but they don't know what they've got and they don't understand it. Maybe I could understand it." Man, that's the coolest thing. You hope for that. That's what you want. So it wasn't uniformly negative. Some people are starting to get irritated.

I asked Redish whether he thought there was any possible constructive purpose for Koonin's personal attack on Fleischmann and Pons. Redish didn't think so:

> I don't think it's a good way for science to function. I would say "No." I would say we shouldn't be personal about the arguments. The arguments should be about the science. OK, [Fleischmann and Pons] got it wrong. We [all] get it wrong; we get it wrong lots. We try stuff, and it doesn't work, try stuff and it kind of looks like it works, and we get really excited and follow it down and, sometimes, we win the Nobel prize with it.

Mysteries

Years ago, I exchanged a few e-mails with Koonin to see whether any of the more-recent research had changed his opinion. It hadn't. Sometime later, I met him in person, at an APS meeting in Baltimore.

I introduced myself, and he was cordial but clearly no longer interested in the topic. I did ask him why he made such a hostile, personal attack on Fleischmann and Pons. "I call things as I see them," Koonin said.

It wasn't hard for me to see things from Koonin's point of view, but there were two mysteries. The first was his assertion in his talk that there were "no reliable reports of reproduction of these experiments despite strenuous efforts by many groups around the world." At least 16 groups had reproduced some aspect of the Fleischmann-Pons claims by then.

The second mystery is how and why Koonin made such a dramatic shift in his demeanor. At Erice, he was eagerly exploring possible theoretical avenues to explain the Fleischmann-Pons results. At Erice, he was joking and laughing with Fleischmann. Now, he was making jokes about, and causing people to laugh at, Fleischmann.

I have two guesses about his change of mind. The first is that Koonin learned from Garwin that Fleischmann and Pons had switched the values in their gamma-ray graph. The second is that Fleischmann and Pons participated in the Utah lobbying effort to get money directly from Congress.

Nathan Lewis, "Objective Scientist"

After Koonin, Caltech chemist Nathan Lewis stepped to the podium. "There are two pieces of news," Lewis said. "One, I'm a chemist, and two, I'm an objective scientist, so hopefully that will please you."

It sure did. The audience cheered and applauded. Lewis began by repeating more or less the same things he told reporters during his press conference earlier that evening. Lewis' group, he said, saw nothing at all, despite the fact that they had all tried so hard. He said he learned about the Fleischmann-Pons experiment from their preliminary note, the newspapers, and the rumor mill.

Lewis said that the Caltech instruments were so much better, his team was bigger, and the error limits were smaller. Lewis knew exactly what Fleischmann and Pons had done, he said. Caltech had made an exact replica of the Utah experiment, and Lewis knew exactly what was possible and what was not possible. So said Lewis.

As a result, Lewis said, he was able to "place strict upper limits" on the products of the Fleischmann-Pons' experiment. "Strict upper limits" meant that no positive data existed for "cold fusion," not in his lab, in Fleischmann and Pons' lab or in any other lab in the world. Lewis spent the first 10 minutes talking about his group's work; he devoted the remaining 20 minutes to critiquing Fleischmann and Pons' preliminary note.

"Let's turn to heat," Lewis said. "Heat is confusing to almost everyone, including almost all of my electrochemical friends. I will try to walk you through this, and you will be shocked. Guaranteed! Now, let's see what really happened."

Lewis spoke rapidly, before scientists who were not chemists, let alone electrochemists, and who had no way to carefully analyze his presentation in the moment. He grossly misled the audience of physicists.

He wasn't so cocky a few months later. For three days in an October 1989 workshop, Lewis sat with Fleischmann and Pons while they explained the results of, by then, two dozen successful excess-heat experiments. With 50 other scientists in the room, many of them chemists, the transcript of the meeting shows that Lewis offered virtually no criticism about Fleischmann and Pons' work.

Also, in March 1990, the Fleischmann-Pons 55-page full paper was published in the *Journal of Electroanalytical Chemistry*. It included an appendix that reviewed calculations from their preliminary note. As mentioned in Chapter 18, they corrected the value for the largest run, and it still showed more heat energy out than electrical energy in. The 1990 paper confirmed there had been 10 other runs with excess heat in their 1989 preliminary note.

Creating a Stir

For a brief moment, Lewis spoke about his work and how his group calibrated the calorimetry for its cell. He quickly returned the subject to Fleischmann and Pons and his guesses about how the two electrochemists had calibrated their cell. "We think they flipped the switch on first with the resistor," Lewis said, "and then flipped it off and then flipped on the cell. The cell evolves bubbles, and so we don't know for sure which way they did their calibration. We know which way we are doing it. We know they did not stir, and I'll show you how important that is later."

But that was mostly untrue. Lewis implied to the physicists that Fleischmann and Pons did not take steps to establish a uniform distribution of temperature within the cell, necessary to ensure accurate thermal measurements. Fleischmann and Pons were not experts in nuclear measurements, but when it came to electrodes, electrolytes and electrochemistry, they were experts.

Fleischmann and Pons deliberately designed the cell narrow enough so the bubbles in the electrolyte would automatically stir the electrolyte and so they could avoid the use of unwanted mechanical stirring devices. The technical term for the bubbling is sparging, as they explained in their 1989 paper. "Stirring in these experiments was achieved," Fleischmann and Pons wrote, "where necessary, by gas sparging using electrolytically generated deuterium gas." Their design provided a benefit besides simplicity. A mechanical stirrer would have contributed heat to the system and required added corrections to the calorimetry.

Lewis told the physicists what would have happened if the electrolyte in the cell was not stirred. But Lewis, as he explained during his presentation, knew that Fleischmann and Pons had used sparging to accomplish the stirring because he read their preliminary note.

Without having tested, he said that sparging didn't work, leaving the physicists with a potentially misleading impression. He couldn't have tested it even if he wanted to because he didn't know the exact cell dimensions.

"We built an exact cell," Lewis said, "as best we could, from all the press photos that we had of the Pons-Fleischmann cell. [roaring laughter from the audience] We measured the ratio of Pons' wrist to his arm and got a good scale-length — [more laughter from the audience] — and we built the cell! And in the cell, we drilled three holes ... ."

The physicists were thrilled with Lewis, thoroughly enjoying his show and his destruction of Fleischmann and Pons' reputation. Nevertheless, his "evidence" of Fleischmann and Pons' big error was irrelevant and wrong. Fleischmann told me why in a 2003 e-mail.

"*Fortune* magazine said that the cells we had used were not photogenic, and they wanted to photograph a really big cell," Fleischmann wrote. "Thus is born a really big confusion. The group at Caltech used this photograph to scale their apparatus, and the use of small electrodes in such cells would have led to miserable and inexplicable results."

Some people may not have agreed with Fleischmann's label of "confusion." Given the threadbare technical details, his and Pons' public display of the wrong cell could be interpreted as misleading.

The skinny cell, by Fleischmann and Pons' design, allowed the bubbling in the cell to stir the electrolyte and eliminate temperature gradients. Lewis' fat cell was not, as he confidently claimed, an exact replica.

Pons posing for the camera, holding fat cell; Fleischmann holding actual cell
Images: University of Utah

A week earlier, Pons and Fleischmann had been in Washington. They had testified to Congress on behalf of the University of Utah about their research. In a follow-up letter to congressional staff member Kathryn R. Holmes, Fleischmann responded to the Lewis matter:

> At least one vocal commentator on our work copied the cell design which we demonstrated at the committee meeting. We have not, as yet, used cells of this size, which have been designed for work with very large electrodes. As they are the largest cells we have made, we thought it most convenient to use these as a visual aid. However, it turns out that other [scientists who have attempted to replicate our work] have used electrodes smaller than our smallest electrodes in cells of such enormous size ... at currents which are lower than our lowest currents. It is totally unsurprising, therefore, that these workers have observed temperature gradients in their cells."

Fleischmann's message was that the cells needed to be small, the electrodes needed to be large (to get large effects) and the current densities needed to be high. From his perspective, most of the people failing to get positive results were doing all the wrong things.

Peer-Review Through Press Conferences

Lewis had reached the end of his press conference:

> Finally, if we're going to have peer-review through press, if we're going to have publication through press conferences, we should have peer-review through press conferences, too. If we ever have another press conference on the subject, I want the reporters to ask all of these questions *[laughter from the audience]*.

Lewis went through his list. Three weeks after this press conference, Lewis submitted his manuscript to a journal. Two-thirds of Lewis' presentation had been a critique of Fleischmann and Pons' work. Lewis' choice to use the APS platform in this manner was a risky wager. It was certainly not demonstrative of the customary ethical practice of responding critically to another scientist's published paper by submitting a comment to the journal that had published the paper.

Lewis never published his critique of Fleischmann-Pons' work in any peer-reviewed journal. He published a paper in August 1989 about his own failure at "cold fusion." But that paper was only about the Caltech work; it did not critique the Fleischmann-Pons work.

Most of Lewis' statements about Fleischmann and Pons in his APS presentation were based on misunderstandings, hearsay, guesses and interpolation. Few people in the audience would have noticed the quiet hiccup in his voice and the twitches in his shoulders, all of which are evident from the video recording and reveal his nervousness. His beaming self-confidence and certainty carried the audience and the attending news media to the inevitable conclusion.

Lewis' closing comment was ill-considered. The list of questions he offered — proposed as peer-review through press conferences — was not for peers. The questions, as he said, were for the reporters.

One final technical point: Lewis also told the audience that, in three of the experiments, Fleischmann and Pons had misrepresented the percent of power break-even by mathematically rescaling the length of those electrodes to the six other rod-type electrodes.

However, Fleischmann and Pons had disclosed in their preliminary note that they had rescaled the length. The reason is that they were always interested in the amount of energy per unit of reactant material. In this set of experiments, most of their samples were 10 cm long. Therefore, for their other-sized samples, they had to normalize the size so they could present a meaningful heating rate across all samples.

Additionally, Fleischmann and Pons had assumed the heating effect was based on volume, rather than surface area, so they normalized based on volume. As it turns out, the heating effect is based on surface area: The reactions take place within the first micron of the surface. Therefore, Fleischmann and Pons had underestimated their projections by *two orders of magnitude* for a 1mm wire. For thicker wires, it is more.

When Lewis walked out of the meeting room, one of the television crews in the hallway asked him to repeat a comment he had said during his scientific presentation. Lewis, with a big toothy smile, happily obliged. "This experiment hasn't been reproduced," Lewis said, "by any national laboratory or any university yet without a good football team."

In 2004, I asked Lewis in an e-mail whether he had heard about the more-recent research. By that time, Charles Beaudette's book on "cold fusion," which had focused on excess heat, was well-recognized in the field. Lewis responded that he didn't know anything about work performed in the previous decade. Two years later, I met him at a science conference in Los Angeles. He had just given a presentation on global energy and was promoting solar energy research.

After his talk, I introduced myself. I asked him whether he'd be willing to answer a few questions about the "cold fusion" history. He said, "No, I don't talk about that anymore. You can read about it in Gary Taubes' and Frank Close's books." I told him I had read the books and still had questions. He turned and walked away.

CHAPTER 20

Judgment and Verdict

Derision and Denunciation

C altech scientists Steven Koonin and Nathan Lewis, and a few other scientists, gave dramatic presentations at Baltimore. The mockery and derision, however, camouflaged the fears and discomfort prevalent among the attendees.

Although most of the news media focused on the few scientists who used the American Physical Society meeting to attack Fleischmann and Pons, the three dozen other presenters did not speculate on why or how Fleischmann and Pons had made sophomoric mistakes. Instead, the presenters stuck to reporting their own research and results, as is customary at scientific meetings. But those presentations didn't get media attention.

After Lewis spoke, by all accounts "cold fusion" was dead. Half of the audience left the room. By the end of the evening, only a hundred people remained. In a newsgroup, Jon Webb, a science student at Carnegie Mellon, in Pittsburgh, Pennsylvania, wrote an eyewitness account:

> We found the talk by Lewis of Caltech to be devastating in its implications for the correctness of the experimental results reported by Pons and Fleischmann. After Lewis' talk, about half the people walked out. As they were leaving, the session chairman tried to get people to quiet down, saying "We have to get back to work." Someone yelled from the back of the auditorium, "We are!"

Of the remaining speakers, two more physicists — Walter Meyerhof and Douglas Morrison — had a significant impact on their fellow scientists as well as the news media.

Meyerhof

Walter Meyerhof was born to Jewish parents and, with luck, determination and assistance, escaped Nazi Germany. Eventually, nuclear physics became a major focus of his work. He was recognized for establishing nuclear physics research at Stanford University in California, and he chaired the Physics Department from 1970 to 1977. His thick German accent stayed with him. He spoke methodically, but he could not hide his disbelief and cynicism.

According to a news article from Stanford University, Nobel laureate Steve Chu, a former chair of Stanford's Physics Department, wrote that "Walter was a real gentleman and scholar."

If Meyerhof, 67 at the time, had performed any of his own experimental research, he didn't present it at the Baltimore APS "cold fusion" session. Nor did he present a theory that might help explain the massive amounts of heat that Fleischmann and Pons had reported.

Instead, he used his speaking opportunity at the meeting to assume bad faith on the part of Fleischmann and Pons and to attack the credibility of their work. He assumed that they lacked the competence to measure heat accurately, and he offered his theoretical explanation of how they went astray. He also critiqued the excess-heat experiment reported by his colleague Robert Huggins, a professor in the Materials Science and Engineering Department in the Stanford School of Engineering. Huggins had started the Stanford Center for Materials Research and was its director for 17 years. Meyerhof began with a play on words:

> A short title for my talk might be called "de-fusion." The problem, as it has been pointed out before, is that Fleischmann and Pons report 4 MJ per cubic centimeter of palladium. ... There is no known chemical or physical

process that can explain this much so-called excess energy. What is the solution to this problem? One solution proposed was "hitherto-unknown nuclear processes." Another solution, as I propose to do, is to examine the excess-energy determination by a method called steady-state calorimetry.

Meyerhof explained his theoretical model in the context of the Huggins experiment. Meyerhof didn't appear to have any experience in electrochemistry, but this didn't stop him from speaking with conviction.

Walter Meyerhof speaking at the May 2, 1989, Baltimore APS press conference

"The setup of Fleischmann and Pons is more complicated," Meyerhof said. "I got this [measurement] from a picture in *Science* magazine. And, just like Dr. Lewis, I calibrated the size of the cell by the width of Dr. Pons' hand." No further analysis of Meyerhof's theoretical model is needed.

Here are Meyerhof's conclusions:

1. The position of the thermometer in the steady-state calorimeter can simulate either excess or deficient power. It can go either way, depending on where you put your thermometer.
2. There are no unknown nuclear processes that are needed to invoke the findings of Fleischmann and Pons. Finally, I couldn't help myself to make a social comment and conclusion, and that is the following:

> Tens of millions of dollars are at stake,
> Dear sister and brother,
> Because scientists put a thermometer
> At one place and not another.

[Laughter, cheers, applause]

Morrison

Scottish particle physicist Douglas Morrison, who had worked for many decades at CERN also took the opportunity at the APS meeting to present an un-peer-reviewed oral critique of the work of Fleischmann and Pons and, to a lesser degree, the work of Steven Jones.

Morrison's presentation included his personal survey of "cold fusion" results that he had heard about in the previous five weeks. He began with a one-sentence dismissal of all the transmutation work done between 1912 and 1927. Then he gave a brief history of the work of Paneth and Peters and how, in 1926, they had initially thought they had produced helium from a "cold fusion" type of experiment.

"Starting with the history," Morrison said, "the earlier reports of the conversion of hydrogen to helium in palladium go back to the early 1920s. People were interested in how the sun gets its heat, and these experiments were all very quickly shown to be false."

Morrison explained to the APS audience that, a year later, in 1927, Paneth and Peters retracted their results and said that the helium they measured was from leaks from the air. This is only half-true, and Paneth and Peters did not completely retract their results, as is described in Volume 3 of this series: *Lost History*. Morrison's assertion that all the low-energy experiments showing transmutations in the 1920s were failures is also false.

Douglas Morrison speaking at May 2, 1989, Baltimore APS press conference

But Morrison apparently did not do a thorough investigation of the Paneth and Peters' history, and he lavished praise on Paneth, the lead author, for retracting the claims, for acknowledging their "mistake":

> And the point is Paneth is a great scientist. I'm interested in the history of wrong results in science. Most people don't withdraw; they continue, like the famous question of Blondlot's N-rays. He never withdrew, and in fact, a year or two after he was exposed, he published a book claiming that he was right, even though no one believed it. But Paneth was exceptional; he was a great scientist.

Morrison was using the sophisticated rhetorical technique of comparison. (Pinch, 1995) By establishing Paneth as a "great scientist" for withdrawing his claim, he was preparing the audience to perceive Fleischmann and Pons as "bad scientists."

Morrison then discussed the ideas of Irving Langmuir, a physical chemist who had studied several examples of claimed scientific phenomena that were illusory. From his studies, he derived seven criteria for "Symptoms of Pathological Science." Of the seven symptoms that he listed, three are useless as criteria.

His criterion "there are claims of great accuracy" carries no distinction from valid scientific claims. His criterion "fantastic theories contrary to experience are suggested" fails to identify why a new theory is intrinsically wrong and merely suggests why the theory is unlikely. His final criterion, about the ratio of supporters to critics, suggests that scientific validity is based on a popularity contest. (Langmuir, 1989) As a result, Langmuir's criteria has limited value and leaves room for a better model of pathological science.

Continuing with his use of the rhetorical device of comparison, he praised Lewis and Meyerhof, injecting some humor along the way:

> Last night, the Utah excess heat was explained by the brilliant detective work of Nathan Lewis of Caltech, by the theoretical work of professor Meyerhof. This was like the detective show on TV, "Columbo," when you start off and you know what the crime is but you don't know how it was done. *[Audience laughs]*
>
> The gammas — this was the graph that convinced many people that what Martin Fleischmann was saying must be right — made really a beautiful graph, really superb. The peak was about 2.21. The first version had 2.5, so they said they had a little bit of trouble with the energy calibration, and then it came spot-on.

Morrison was taking more than poetic license. Fleischmann and Pons gave no explanation for switching the peaks, let alone one about "energy calibration." The peak did not come "spot on." He knew it had been switched. He could not have been less sincere.

Morrison spoke about neutrons and the challenges that many people faced in detecting them in "cold fusion" experiments. He made another remark that seemed designed to insult Fleischmann and Pons' competence as electrochemists: "There's an enormous amount of literature about palladium and palladium hydrides. My desk is stacked with all of these things. People who have supposedly been working on it for years just haven't read the literature. My conclusion from all this — the only reasonable conclusion — is that the experiments are mistaken."

For anyone who knew Fleischmann, Morrison's comment was ridiculous, verging on slander. Fleischmann was one of the world's experts in palladium hydrides and had studied them since his youth.

Like Lewis, Koonin, and Meyerhof, Morrison exuded certainty and conviction, not to mention smugness. Among the people who used the rhetorical devices of comparison, ridicule and humor, Morrison was the most sophisticated. The only things missing from Morrison's talk were facts that showed why the Fleischmann-Pons experiments were wrong.

In a conversation with me, author Charles Beaudette once summarized the obligations for scientific critique of experimental claims: The critic must identify a specific error of protocol, a mistake in the data analysis, or an unstated assumption. Morrison had revealed none. Trevor J. Pinch explained these rhetorical techniques well in the book *Science, Reason, and Rhetoric*:

> The laughter generated builds rapport and solidarity against its luckless victim. More subtly, it allows the critic to appear to be "more scientific than thou" while engaging in the dirt and real politics of the controversy. In short, humor allows scientists to have their cake and eat it. They can accuse their opponents of breaking the scientific rules and at the same time break the rules themselves. (Pinch, 1995, 173)

Morrison, like a sportscaster, presented a tabular list of "cold fusion" experimental reports from around the world. He separated them into five regions and displayed columns for characteristics of the individual reports. Morrison gave the tally for the first column, indicating whether the researchers saw any effect:

Italy: (Yes-4, No-0)
Rest of Western Europe: (Yes-0, No-10)
Eastern Europe: (Yes-6, No-0)
USA: (Yes-6, No-7)
Rest of the World: (Yes-7, No-0)

> Experimental quality can be graded 1, 2, or 3, but 1 means first-class. For the rest of Western Europe, instead of having all "Yes," as it was in Italy, it's all "No." For Eastern Europe, it's all "Yes." *[Audience laughter]* And the quality of the experiments — well, it's tricky.

Morrison did not need to say that the experiments in Eastern Europe, in his judgment, were deficient. He had accomplished his objective by saying that the quality was "tricky" rather than "first-class." Morrison's point was not subtle. Western European researchers who had reported null results, according to Morrison, were more skilled than the Italian and Eastern European researchers who had reported positive results. Morrison continued with his regionalization thesis, using humor to engage and defuse the defenses of the audience while he made subtle *ad hominem* attacks against the researchers who claimed positive results:

> So this is the problem. There is a regionalization of results. *[Audience laughter]* And then you've got the "No" and the "Yes" groups and the USA, which is a split personality. Georgia Tech has switched already. Texas A&M — I thought it had switched officially — but if it hasn't, it should from what I know of the results. *[Audience laughter]*

In his May 7, 1989, newsletter, Morrison gave more details:

> This survey gave disturbing statistical evidence that the results correlated significantly with the region of the world.

Further, this effect seemed to be related to the [psychological] climate of information available. Thus, if there was a general expectation [in that region] that the initial reports on cold fusion were true, then all the results reported [in that region] were confirmations.

On the other hand, in those regions of the world where concerned scientists and responsible science reporters had been in close contact, then all the results reported were negative.

The APS audience found his presentation entertaining, and they enjoyed many laughs at the expense of science. Morrison gave his concluding comments:

Utah is very special. As you know, it has a serious economic problem, and they are delighted by fusion. So in the same way as California has its Silicon Valley, in Utah they want to have the Fusion Valley and solve their problems.

There is a social responsibility of scientists, because we are not alone in this business. There's a wonderful dream of this energy that we can get without pollution, but what are the facts? The chemists think it is not chemistry. The physicists think it's not physics or fusion. And then there was this 18th century Scottish philosopher, David Hume, who summed it up: "Would you rather believe that all the laws of nature are wrong or that a man makes a mistake?" *[Audience laughter and applause]*

To be more serious for a moment, this is a time for physicists and chemists to continue working together and not to criticize one another. It's also time for humility for all of us and to have sympathy for everyone.

It was a sophisticated technique with which to close his talk. He left the audience with the final but false impression that he was sincere, humble, sympathetic and non-judgmental.

Trevor Pinch, though he did not analyze Morrison's talk, made the astute observation that both Lewis and Koonin (and, as we see, Morrison and Meyerhof) used the rhetorical strategy of postscripts. Koonin used the "Leap at Rhodes," Lewis used the "Peer-Review by Press Conferences," Meyerhof used his "Thermometer poem," and Morrison used Hume's allegory:

> [Koonin's] rather unorthodox end to his talk works similarly to the humorous interludes throughout [his talk. His postscript] section is clearly humorous. Yet its message is again serious: Fleischmann and Pons are fools making grandiose claims which they cannot substantiate and which any smart person can expose. ... The postscript allows Lewis to make particularistic, *ad hominem* claims with no evidence and little argument. (Pinch, 1995, 172)

Press Conference #3

For reporters covering the APS meeting, the Tuesday, May 2, 1989, 10 a.m. press conference was the final "cold fusion" event. Nine physicists sat at the head table: Moshe Gai (Yale University), Johann Rafelski (Arizona State University), Steven Earl Jones (Brigham Young University), Steven Koonin (Caltech and University of California, Santa Barbara), Richard N. Boyd (Ohio State), Douglas Morrison (CERN), Walter E. Meyerhof (Stanford University), Edward F. "Joe" Redish (University of Maryland), and James Krumhansl (president of the American Physical Society, Cornell University). Phil Schewe gave the introduction. The purpose of the society, he said, was to educate its members. He portrayed the previous evening's events in a positive light:

> I would hope that some of you who were there last night recognize how constructive the interplay between the chemists at Caltech and the physicists and the other participants at the various individual papers was in promoting this objective which we have.

Redish and Krumhansl, sitting at the head table, were no longer smiling. Schewe asked each of the other physicists to give a short synopsis of his paper. Meyerhof was the first, followed by Morrison. They essentially repeated what they said in their talks. The third panelist was Richard N. Boyd, a professor of physics and a professor of astronomy at Ohio State University. Here are his opening comments:

> We set out to do an experiment, primarily to test the Fleischmann-Pons result at the University of Utah. The reason for that was primarily that it looked like it would be easy to confirm or reject that result without a great deal of effort. The chemical preparation of the electrodes seemed to be very important in the paper that they wrote, so we spent a great deal of time worrying about that.
>
> The question here is whether you're loading up the palladium with as much hydrogen as you need in order to bring about the effects that Fleischmann and Pons claimed. We did a careful job of that, and we even had a confirmation of that because, with a slight provocation after we had finished one of our experiments, one of our electrodes lit up like a light bulb.
>
> What we set out to do was test the neutron fluxes which were claimed by the Pons-Fleischmann group and which, in another sense, would have been required if what they were really observing and the way of heat output as explained by cold fusion. It was quite easy to disprove the second part.

Although it was 25 years later, I was able to contact Boyd by phone in 2015 and discuss this.

KRIVIT: You said, in 1989, that you had a confirmation.
BOYD: I doubt it, or at least not for very long.
KRIVIT: Well, that's what you said.
BOYD: We certainly didn't think that we had a confirmation of "cold fusion" for very long.

During our conversation, I asked Boyd about the electrode that lit up like a light bulb. He described it instead as a bright red glow. I do not know the explanation for the inconsistency between the 1989 "lit up like a light bulb" and the 2015 "bright red glowing thing."

The Jones Vote

Steven Koonin was next at the APS press conference. He summarized his presentation from the prior evening. Koonin kept to the science and was serious: no fables, no insults, and no flying pigs.

Steven Koonin at Baltimore APS press conference, the day after he accused Fleischmann and Pons of incompetence and delusion

He dismissed the Fleischmann-Pons claims as wrong, not because of an error with their work but because of circumstantial and theoretical reasons: "Nobody's been able to reproduce it. That, together with a lack of theoretical basis — namely I can't find any way to hide the radiation — suggests to me that the experiments are wrong."

Not only had Koonin been unable to find a theoretical way to hide the radiation, but he had been unable to find a way to get the reaction rates high enough to account for Fleischmann and Pons' heat.

Steven Earl Jones speaking at May 2, 1989, Baltimore APS press conference

Steven Jones was next, and he repeated, more or less, the same things he had said in Erice and before the congressional committee. He also gave another one of his kindergarten-level analogies, comparing his own results to a penny, and the heat results of Fleischmann and Pons to a trillion pennies. He also brought along a prop similar to the one he had used in Congress to make another analogy:

> The comparison I used for Congress last week, which I'll use again today, is this little plant, little sprout. It's no bigger than the one I had before Congress. This represents cold nuclear fusion, this little sprout. Now it's just sprung out of the ground, really. Even though we've been working on it for four years, it's still just come before the scrutiny of the scientific community, and it's being scrutinized very hard. Now some people, the University of Utah people in particular, have a claim that this is a tree. They're sure it's a tree, or quite sure. It's going to grow up very quickly and give us enough wood to provide all our energy needs for generations. I don't think it's a tree; I've said this repeatedly and consistently.

Jones continued speaking to the reporters and got more mileage out of his fertilizer joke:

> Don't sell your oil well. Don't even invest in palladium, and so on. I mentioned to Congress that I don't think this tree needs a lot of fertilizer right now. I realized that has a little double entendre, but what I meant was that there's a lot of fertilizer out there, but it doesn't need a whole lot of money right now. It just needs to grow, and we need the standard level of grants to check this out and see what happens, clarifying that point. I do think that the little sprout is living based on the work that I've done. It's a very tiny sprout. I don't think it will grow into a tree, but I do think it will grow into a pleasant flower, a new addition to the garden of physics which we can delight in because of its beauty, not necessarily because it provides wood but because it gives us a picture of nature which is unique.

A common rhetorical strategy used by Jones was his injection of lightheartedness and joviality during a serious and ugly public conflict. Then Jones got creative with the panelists. He asked them to vote on whether they agreed that the Fleischmann-Pons experiment was now considered dead.

"How many think we can rule that out at this time?" Jones asked. A big smile swept across his face. His collaborator, Johann Rafelski, began chuckling. Jones' poll was a superb strategy to transcend the cumbersome and technical language of physicists. Instead, he provided the news media with a concise and easy-to-understand factoid that was perfect for reporters and their audiences.

Nobody sitting at that table had any doubts about the consensus among physicists. Jones shrewdly used peer pressure to his advantage. His voting stunt was a masterful way to further tear down Fleischmann and Pons. A frame-by-frame analysis of the videotape shows that, before Jones had finished asking his question, he began to raise his hand as he telegraphed his vote to the other physicists.

With his hand raised high, Jones leaned forward to look at the other physicists to see how they would vote. For a moment, the somber-faced judges of science silently pondered their forthcoming verdict. Koonin needed no time to decide and was the first to raise his hand. The others needed a moment to think and perhaps to watch their peers.

Moshe Gai cautiously raised his pen, signaling his vote. Meyerhof, who had been watching Jones the whole time, was next to raise his hand. Morrison had also been watching Jones. Morrison decided, then began to look away from Jones as he raised his hand. Boyd cautiously raised his hand next. All but Rafelski raised their hands. Krumhansl and Redish were out of view, but other video clips show their raised hands. The result of Jones' vote was carried by all major national and, I presume, international media outlets. Here are a few examples:

> *New York Times*: "vote by eight of nine physicists calling the Utah experiment dead."
>
> Associated Press: "A panel of nine scientists on Tuesday disparaged Utah researchers' claim of achieving fusion in a jar, suggesting they were fooled by faulty measurements."
>
> ABC-TV: "At this point, the bulk of scientific opinion is [doubtful] that cold fusion exists as Pons and Fleischmann reported it."

Immediately following the vote against Fleischmann and Pons, Jones called for a second vote and asked the panel whether its members thought his results were real.

"However," the AP wrote, "the panel also voted 6-3 against ruling out that experiments in Utah and elsewhere had produced neutrons."

Thomas H. Maugh II, writing for the *Los Angeles Times* on May 30, 1989, was sympathetic toward Jones, oblivious to Jones' secret maneuverings:

> Most of this year has been rather painful for Jones. The low-keyed 40-year-old physicist has been falsely [Ed: Perhaps not. See Chapter 2] accused of stealing ideas and of sloppy experimental work. In the rush for scientific acclaim, his once-friendly relationship with his two competitors has

degenerated into a series of misunderstandings and angry phone calls.

He has traveled the world defending his research and his claims of priority, and at nearly every stop, he has been treated as an intruder trying to ride the coattails of Pons and Fleischmann. Jones says the pace of the last few months has cost him 13 pounds [in weight].

Through it all, he has kept his composure, distanced himself from the wilder claims of his colleagues, and generally impressed other researchers with his levelheadedness and persistence. Whether cold fusion is eventually proved or disproved, some people believe that he is the only Utah researcher who will emerge with his reputation intact.

Jones cautions that, based on his results, 'Cold fusion is hopeless as an energy source.' But he believes that his discovery can solve many scientific mysteries.

It was over. Before Baltimore, Fleischmann and Pons had been treated like honored guests. Not only were they showered with praise by members of Congress, but they also had been invited to the White House.

Their appointment with John Sununu, White House chief of staff for President George H.W. Bush, had been scheduled for May 4. One day after Baltimore, and one day before the White House meeting was supposed to happen, Sununu cancelled because of a "scheduling conflict." Politically, Fleischmann and Pons had been destroyed.

CHAPTER 21

Death, Aftermath and Reflection

Critics Feel Better

Despite the well-publicized bad news from the Baltimore American Physical Society meeting, a few scientists did report supporting evidence for some of the "cold fusion" phenomena. Details of their presentations are hard to come by because they were not video-recorded or, if they were, not retained by the APS, and no journalists reported these outliers.

Many years later, Martin Perl, a 1995 Nobel Prize winner in physics, gave the 2005 keynote speech at the International Congress on Nanotechnology in San Francisco, offering a useful lesson for scientists, and it directly applies to these outliers in 1989.

Perl told the science students and scientists in the audience that they must be vigilant about their biases. He told them to be cautious about discarding data that didn't fit with their expectations, because novel discoveries often are found in the unexpected. He warned them to be careful about their assumptions, particularly when they find "noise" in their data. Perhaps the "noise" might be a faint signal of something on the verge of being discovered, Perl said.

He also told the audience to select colleagues carefully. For example, he said, if you tell someone about a new idea, and the person immediately dismisses it, saying, "That's impossible because it violates some law," then perhaps it's best to find a new colleague.

The Outliers

In his May 7, 1989, newsletter, Douglas Morrison briefly mentioned four people at the American Physical Society meeting who reported possible confirmatory data.

Dieter Seeliger from the Technical University of Dresden reported a small flux of excess neutrons from his group's "cold fusion" attempt.

J.R. Granada of the Centro Atomico Bariloche and Instituto Balseiro in Rio Negro, Argentina also saw a slight indication of excess neutrons.

At Miami University in Oxford, Ohio, Joseph Cantrell, in the Department of Chemistry, and William E. Wells, in the Department of Physics, reported excess heat in their "cold fusion" experiment. Morrison, however, missed their talk because, as he wrote in his newsletter, he was having an "important conversation" with Steven Jones.

These outliers, these exceptions to the consensus, were written off as "noise." For the moment, "cold fusion" was declared dead.

It's Dead

History at that moment was written based on the negative reports given at the American Physical Society meeting. Koonin's "incompetence and perhaps delusion" insult was widely quoted. Paul Recer, at the Associated Press, quoted Nathan Lewis, who said there was "no unknown nuclear process, and nothing but conventional chemistry." Malcolm W. Browne, at the *New York Times,* reported that "experimenters appeared to be unanimous" in discounting Fleischmann and Pons' claim. Jerry Bishop, at the *Wall Street Journal*, reported that "frustrated and angry physicists cited a half-dozen possible errors" made by the two chemists. Thomas H. Maugh II, at the *Los Angeles Times,* quoted Nobel Prize winner Leon Lederman saying the head of the University of Utah "ought to be fired."

"The consensus," Maugh wrote, "was that, unless Pons and Fleischmann can produce some dramatic new evidence to support their contentions, their claims are likely to fade into obscurity along with polywater, N-rays and other highly publicized scientific 'breakthroughs' that were subsequently discredited."

Cartoon published in the Boston Herald. *Artist P.J. Wallace*

The serious news also was supplemented with satire in print as well as in novelty merchandise. Somebody made a "Department of Fusion Confusion" cold fusion coffee mug. It included a miniature test tube, a swizzle stick, a clothespin and a cork. It was "guaranteed to work unless you are from Boston or have any affiliation with Boston physicists." Its patent was pending, as of May 1, 1989.

Somebody at Caltech made T-shirts that depicted the Caltech cold fusion group pouring cold water over the University of Utah. The Cornell Cold Fusion Archive contains half a dozen such novelties.

Some of them, for example the Fusion Fun Kit, are creative and humorous. The kit was accompanied by two pages of instructions and, of course, a template for a press release. The contact person on the template was Dr. Swami Swaminathan, Ph.D.

The $4.95 Fusion-in-a-Bottle came in a "neutron-free version" do-it-yourself kit. Of course, it too came with a template for a press release.

As Redish Remembers

Joe Redish remembers the APS "cold fusion" session as a highlight of his scientific career. He explained why in a 2014 interview with me:

> It was something that drove the whole community. There was history being made, positive or negative, whichever way, and I was a part of it. Everybody I knew was interested in the question of "cold fusion," every nuclear scientist. I mean, the community was fascinated by this. They wanted to know, Is this real, is it not real, is it possible? It was absolutely where everybody was at.

There is a disparity between Redish's recollection and the many physicists who joined Koonin and Lewis in laughter, derision and applause at Fleischmann and Pons' expense.

Koonin, the Following Week

On May 5, 1989, Koonin wrote an e-mail to Richard Garwin:

> On the political front, I will be at the Electrochemical Society meeting on Monday where Pons, Fleischmann, Jones, Lewis, Huggins, and people from Texas A&M and Case Western are supposed to talk. Lewis and I will be wearing body armor, but it should be fun. Unless they concede completely, the final round will likely happen in Santa Fe.
>
> Have you seen the MIT preprint by Petrasso et al. on the analysis of the gamma spectrum? They took the Fleischmann-Pons spectrum as shown on TV and conclude that the "peak" shown is actually a 2.5 MeV artifact somehow shifted down to 2.2 MeV. The whole thing really sounds like fraud now, although I can't say that in public.

On Monday, May 8, 1989, Douglas Smith, a science writer in the news office of Caltech, interviewed Koonin a few minutes before the start of the Electrochemical Society meeting at the Bonaventure Hotel in Los Angeles. Koonin was on leave from Caltech and working at the Institute for Theoretical Physics, at the University of California, Santa Barbara. Smith asked him about the Erice Fusion Forum workshop and then about Baltimore:

> A lot of press was at Erice. It's weird talking to reporters. I never had that experience before. They clearly know what they want to get out of you. And it's not necessarily what you want to say. It was fun. It's, of course, a kick to wake up in the morning, twice in a row, and see yourself on TV and read about yourself in the papers. It's fun.
>
> I was very careful with the words I used. I thought a lot about that when I came out with the "incompetence" and "delusion" stuff. I talked to a lot of people before I settled on those words. I spoke to a lot of scientists who had been in the public arena.
>
> I would say it was probably a week before Baltimore, sometime around April 21, when serious doubts started to emerge.

No scientific event or information emerged publicly around that time to suggest scientific reasons to doubt the Fleischmann-Pons claims. To the contrary, there were reports confirmations of neutrons from room-temperature nuclear experiments from reputable government laboratories in Italy and India. There were also reported confirmations from Czechoslovakia, Mexico, California and Brazil — all within a 48-hour span, between April 17 and April 19.

On April 19, 1989, the U.S. Department of Energy began to take the possibility of "cold fusion" seriously and called an emergency meeting in Washington, D.C. to discuss the research with the heads of national laboratories.

On April 20, Garwin's "bet against cold fusion" published in *Nature*, as did an editorial in the same issue saying that Fleischmann and Pons had withdrawn their manuscript.

On April 21, Lewis gave a seminar at Caltech on "cold fusion" and announced that he failed to confirm the Fleischmann-Pons experiment.

Two things happened in Washington on April 21: U.S. Energy Secretary Admiral James Watkins issued a press release announcing three initiatives to pursue and encourage research into the Fleischmann-Pons phenomena; Robert A. Roe, (D-NJ), issued a press release and announced the congressional hearing on "cold fusion." Koonin continued to explain to Smith his view of the events:

> So I would say that, by about the 17th or 18th, we were starting to have real problems, Lewis is turning up all these errors in the calorimetry, the radon business is appearing on BITNET *[Ed: an early U.S. and university computer network used for e-mail communications]*, nobody could think of how cold fusion could work theoretically. I decided finally that I was going to hit Fleischmann and Pons really hard in Baltimore. I dry-ran the talk before a group of physicists, asking if the tone was right, and we did it!
>
> I think we've done well, assuming we're right. My rationale is that I believe strongly that we are right, but there's always a chance — one in 1 billion — that we're wrong. And the way I look at it now, we'll all be so happy if we are right. It's really been fun. It's been tremendous fun. My daughter saw me on TV. She keeps asking everybody if they've been on TV. And my mother was impressed.
>
> Another strange direction this has taken is that there's a conference in Santa Fe the third week of May, and I heard that 4,000 people have registered. *[Ed: An internal DOE document dated May 17, 1989, stated that 500 participants had registered.]* The Department of Energy told Los Alamos National Laboratory, "You have to run a conference." I don't think that Los Alamos wanted to do it, and apparently nobody else did, either, because it will be a bust once we get on the program. The Department of Energy put together an

advisory committee with some rather distinguished people on it, and I'm on that. I think basically, in the end, we will tell them it's nothing that you can make energy from, but there may be some interesting physics.

I think that the Institute for Theoretical Physics comes out like a hero, assuming that Fleischmann and Pons don't pull a rabbit out of a hat tonight. We waited, we didn't jump in, and we did a good interdisciplinary job, and we had a lot of fun doing it.

It consumed my life for the past six or seven weeks. I've been thinking about nothing else. My efficiency is slowed down because I get interrupted by reporters, people who want to know what's going on. But I feel like, last week at Baltimore, we got the word out. In some sense, I think it was a public service, to really make it known that there are real doubts about this whole business, especially when something like this makes the cover of *Time, Newsweek,* and *Business Week* in the same week.

I feel like we did a good job, but I don't feel happy about it in the sense that I think we destroyed those two guys and the grad student. Can you imagine what his résumé is going to look like? It was also funny — once I started appearing in the news, all these people out of my past called me up and said, "Hey, I saw you in the *New York Times* or on TV, doing your thing."

There was even this taxi driver in Baltimore when I left, and he recognized me: "Weren't you on that panel last night?" There's no instant celebrity stuff, everybody gets 15 minutes, and I think I've had mine. I've had a few letters from people who say I came out too strongly, mostly not from scientists. Again, if you don't know the details of how they screwed up, it looks a lot harsher than it really is.

At that point in the interview, Smith said to Koonin, "Delusion and incompetence are fairly strong words. They are one level below the F-word." Koonin responded:

Which is where it was meant to be. And I think, as the facts come out, it's going to look more and more like fraud. I feel like an assassin, in a way. But you got to call 'em as you see 'em. And if you look at the response we got at the meeting, it showed that it was on everyone else's mind, too. Everyone sort of sat there for a second, going, "Oh, no it's all wrong. It's all a big mistake." Lewis and I gave back-to-back talks in the session, a one-two punch. And everyone walked out after he finished. There was nothing more to be said.

I was in Baltimore for three days, and I only went to one scientific session, a single 20-minute talk. The rest was interviews, committee meetings, and talking to colleagues in the halls about this cold fusion business. That is not the way you normally do science conferences, I felt like I was constantly on the go.

After the Baltimore meeting, I went to Cornell University and gave a seminar. After the talk, a historian of science came up to me and said, "Save all your transparencies; save your notes. You guys are moving a little too fast for us right now, but we will catch up to you." And that drove home to me the fact that we're probably in the midst of something that's at least socially important, if not scientifically.

That historian was Cornell professor Bruce Lewenstein, who preserved a copy of this interview in the Cornell Cold Fusion Archive.

Two Paths in Failed Scientific Replications

Nathan Lewis' response to his failure to repeat Fleischmann and Pons' results offers a lesson on the topic of scientific replication. A researcher who has failed to repeat the results of another scientist has two paths from which to choose in how to report the failure. The researcher can take a position of certainty or uncertainty about whether he or she has repeated the experiment with sufficient accuracy.

A researcher choosing the path of uncertainty allows for the possibility that the original claim may be valid.

J. Kirk Dickens, a nuclear physicist who worked at Oak Ridge National Laboratory for 46 years, demonstrated this thoughtful approach when he concluded his talk in Baltimore:

> In closing, I would like to show you my favorite comment from the *Chicago Tribune* of April 8. I'm sure every one of you has seen it. For those of you in the back of the room who can't see it, it's an oil-slicked bird asking an oil-slicked sea lion, "Is there any more word on how those fusion experiments are going?"
>
> I feel as Dr. Perry said today at his conference: We shall continue, and we shall continue because we should stay in the scientific arena as long as it is not understood.

A failure to confirm results can occur for numerous reasons, none of which have bearing on the legitimacy of the original experiment. In his 1985 book *Changing Order: Replication and Induction in Scientific Practice*, Harry M. Collins writes about the sometimes-difficult task of knowledge transfer from one scientist to another who is attempting to repeat an experiment. Successfully repeating an experiment, Collins writes, requires attention to innumerable procedural details and subtleties. Not every novel experiment can be immediately and easily reduced to a concise set of instructions that can be published in a journal.

The most elusive kind of replication failure occurs when an original researcher is unaware of a critical factor and therefore fails to communicate that factor to an attempted replicator.

Another kind of failure occurs when an original researcher is so familiar with a critical factor that the original researcher forgets to communicate that factor. An example of this seems to be Fleischmann and Pons' failure to include in their preliminary note the requirement of an unusually high degree of deuterium loading into palladium.

A researcher choosing uncertainty does not, after reporting the failure, go on the offensive and use that same opportunity to explain why the original experiment is a mistake.

An example of this occurred when Matthijs Broer, of AT&T Bell Labs, reported his group's failure at Erice. Broer presented his results straightforwardly without speculating about why Fleischmann and Pons' claims might have been mistaken. During the question-and-answer session with Fleischmann, Broer made his neutrality clear and tried to learn from Fleischmann whether he was missing any crucial parameters.

The path of certainty does not leave open much, if any, room for the possibility that the original claim may be valid. Nathan Lewis demonstrated this well at the APS meeting.

More often than not, in the cases I have examined in this book and in *Lost History: Explorations in Nuclear Research, Vol. 3*, scientists who have chosen the path of certainty have swayed opinions in the scientific community and the public at large, not from the strength of their facts but from their position of authority or perceived authority.

Once-in-a-Lifetime Experience

Frank Close and other people who were critical of Fleischmann and Pons knew how big the conflict had been, as Close wrote:

> As Lewis recalled, "It was a rare time; you lived and breathed it. There were big stakes. It mattered." Mike Sailor, the colleague on those experiments who had rushed across Los Angeles in Lewis' car seeking palladium, encapsulated it thus: "Boy, do I miss those times." And so do I.
>
> As I met again my several friends, many of whom I had not known before 1989 and probably would never have met otherwise — Lewis the chemist, Petrasso in fusion, and many others across the scientific spectrum — it was uniformly clear that we would be unlikely to encounter such times again in our careers. We were returning to our mainstream researches with some sadness that the story had not opened the great breakthrough, that we could not tell our grandchildren that we were in at the start of the new age.

"We've learned just how easy it is to fool oneself into believing that there is an effect."
—Nathan Lewis

Nathan Lewis quoted in Science

We have to adjust to the more normal, hard, dedicated application, the 99% perspiration and 1% inspiration that is the reality of scientific research, changed by what we experienced and going our separate ways. We had had our days in the sun. The central characters may still hope that theirs will return. (Close, 1992, 359-62)

Comfort in Death

David Lindley, the assistant physics editor for *Nature*, in his May 4, 1989, editorial, used morbid metaphors:

> The APS meeting in Baltimore took on the atmosphere of a hanging party lacking only its intended victims — Stanley Pons and Martin Fleischmann. ...
>
> At the end of the session at Baltimore, physicists were left with the comfortable feeling that fusion was dead.

Physicists indeed seemed to be comforted by the Baltimore outcome. But "cold fusion" wasn't completely dead. A humorous quote from the movie *The Princess Bride* illustrates the nevertheless very serious circumstances of "cold fusion."

"There's a big difference between mostly dead and all dead," Miracle Max said. "Mostly dead is slightly alive." These words of wisdom are uttered by Max, the wizard, when he is presented with the princess's recently tortured and near-dead amour, Westley.

The "cold fusion" story after May 2, 1989, was far from over, but to a large degree, it was mostly dead.

Before Baltimore, U.S. Energy Secretary Admiral James Watkins had announced three initiatives to investigate and encourage research into the anomalous phenomena.

One was the meeting at Santa Fe, to be hosted by Los Alamos National Laboratory. Another was the Department of Energy-sponsored review of "cold fusion" by independent experts. The last was his directive to DOE lab directors to intensify their "cold fusion" research efforts and to send him weekly reports of their progress.

Top management at the equivalent energy agency in Italy was equally excited and interested in the field and the success of one of the Italian groups that had confirmed some of the results.

After Baltimore, none of Watkins' initiatives had any hope of supporting and encouraging the research in the United States. The stigma attached to the field in the United States was too great. Any scientists or students who wanted to pursue the research now risked the judgment of their peers, sponsors or supervisors. They would be perceived as engaging in false or fraudulent science. Researchers who nevertheless persevered became targets of scorn and derision from their colleagues.

Public money for the research in the U.S. became virtually nonexistent. Most journal editors in the U.S. and the U.K. would not send submitted manuscripts out for peer review. "Cold fusion" became the poster child for bad science.

At least for the next two decades, it wouldn't matter how good the positive results were. It made no difference that the magnitude of the observed effects was inconsistent with Irving Langmuir's criterion for pathological science; the stigma was just too great to overcome the hostile politics.

International Fallout

The timing of repercussions in other countries from the rejection in the United States varied. Most subsequent international research in "cold fusion" occurred in China, Japan, India, Italy and the former Soviet Union. I've had difficulty getting clear information from the Chinese and Japanese researchers.

It is unclear whether anybody in Germany stayed with "cold fusion" research beyond the first few months. On May 8, 1989, a German group led by Gerhard Kreysa, the head of the Dechema Institute, demanded to speak in Los Angeles at the Electrochemical Society meeting but was rebuffed by Elton Cairns, the organizer. Cairns didn't think Kreysa had anything useful to contribute.

Fleischmann told me in a June 4, 2004, letter that Kreysa's group reported just a single failed experimental run before concluding that "cold fusion" was a failure. Fleischmann wrote that, nevertheless, the Kreysa group did a lot of "damage behind the scenes" politically.

India's researchers got off to a very strong start in early 1989 with their "cold fusion" research, but by early 1990, that program also succumbed to the negative politics and perception of "cold fusion."

Soviet Union

Igor Goryachev, a retired scientist from the Kurchatov Institute who has been involved in "cold fusion" research and conferences since 1989, provided me with the clearest explanation of the early "cold fusion" events in the Soviet Union, which soon became Russia:

> I can tell you that, actually, in 1989, there was no negative attitude toward cold fusion here. On the contrary! The information about the University of Utah press conference was received by the physics community with satisfaction because it correlated with some experimental discoveries made by Russian physicists in the mid-1980s.
>
> Immediately, active discussions about "cold fusion"

began in our scientific community and in our leading research institutes. In April 1991, a conference on the phenomenon was organized by the Joint Institute of Nuclear Research in cooperation with University of Moscow. This took place in the town of Dubna. More than 100 scientists attended the conference. Soon after that event, a federal grant was established to investigate "cold fusion." In 1993, the First Russian National Annual Conference on Cold Fusion took place.

The difficulties began soon after that. The reason for the trouble is that the Russian Academy of Science established a so-called commission on fake science. Some members of that commission began fighting against new discoveries because those findings and inventions "contradicted modern physics." They identified "cold fusion" as fake science. This commission still exists. The official name is Commission on Pseudoscience and Research Fraud of the Russian Academy of Sciences and is nicknamed the Kruglyakov Commission, after the name of the founder and chair, academician Edward Kruglyakov.

Goryachev told me that two groups in the 1980s had reported "cold fusion" results. One was B.V. Derjagin, Andrei Lipson and their colleagues. (Kluev, 1986)

The other was M.A. Yaroslavskiy. The federal "cold fusion" grant, for the equivalent of $250,000, was issued by the State Committee of the USSR on Science and Technology.

Italy

The response to "cold fusion" in Italy was shaped by unique circumstances. As I reported in Chapter 13, top management of the main laboratory of the Energia Nucleare e Energia Alternativa in Frascati enthusiastically welcomed the "cold fusion" confirmation by Francesco Scaramuzzi's group.

Unlike the rest of Western Europe, which had relatively good financial support for research programs in nuclear fission and fusion in 1989, Italians had just rejected conventional nuclear power. So the Frascati lab was in trouble. Antonella De Ninno, who worked in Scaramuzzi's group, explained:

> In November 1987, in Italy, there was a popular referendum against nuclear power plants. It was just after the Chernobyl accident, and Italians decided to quit nuclear power forever. ENEA was at that time, together with ENEL, the main public institution involved in the management of the existing nuclear plants and in the construction of new plants. After the referendum result, the Italian government took some time to rethink the role of ENEA and, in the meanwhile, suspended funds. The budget had been gradually reduced since 1986.
>
> Scaramuzzi's visit to Parliament restarted the flow of money to ENEA, and ENEA management also promised to fund our "cold fusion" research. We received offers of collaboration from many laboratories all over Italy.
>
> During the following months, May, June, and July, we received many visits from scientists from private companies (I specifically remember British Petroleum) and people from public laboratories. ENEA management must have invited them because they didn't contact us directly.

Meanwhile, for the rest of Europe, Euratom, the European Atomic Energy Community, commissioned studies on "cold fusion." Euratom is an international organization founded in 1957 to support and encourage research and technology in nuclear fission. The organization later became involved in supporting thermonuclear fusion research. In a historical retrospective written in 2008, Scaramuzzi reflected on the results of the Euratom-sponsored research and the lack of reported confirmations of "cold fusion" in the rest of Europe:

The most striking feature in this geography is the almost total absence of [cold fusion] research activities in the rest of Europe. After the negative results obtained in the [cold fusion] experiments performed in [Europe in] the spring/summer of 1989, mostly under the request of Euratom, all interest in cold fusion seemed to have disappeared.

Many researchers in Europe who attempted and failed to repeat the "cold fusion" experiments in 1989 became discouraged. The situation was made worse by their growing resentment and anger toward Fleischmann and Pons, as De Ninno explained:

By the autumn of 1989, the mood in Italy had changed drastically. Many colleagues began to be very doubtful and nasty against those of us researching "cold fusion." ENEA management dropped the promises of funding, and our colleagues in ENEA involved in the neutron measurements said that they didn't want to be involved anymore, I suspect, because they had pressure to quit.

Remember, ENEA-Frascati employs the biggest community of researchers who work in thermonuclear fusion in Italy. My colleagues here were very doubtful about our findings. The stigma was in place by the end of 1989, and it was considered very inappropriate — unscientific — to show interest in "cold fusion" by then.

The Scaramuzzi group ran tests from October 1989 through March 1990, verifying to the best of their ability that no artifacts had been responsible for the first set of neutron signals. Even though the group obtained additional confirmations, it was no use. The political climate for "cold fusion" had shifted at ENEA. As attractive as "cold fusion" research seemed to ENEA management when it held the April 18 press conference, the alluring topic had lost its luster.

De Ninno is not sure why ENEA management got cold feet and deferred providing special funding for "cold fusion" research. It may be related to the negative report written and leaked by members of the United States Department of Energy "cold fusion" review panel in the summer of 1989.

Not the End of Science

How is it that such a well-informed person like John Maddox failed to mention the dozen-plus reports of worldwide partial confirmations of Fleischmann and Pons in his dismissive editorial? How is it that respectable scientists assumed bad faith and called other respectable scientists frauds in the absence of any logical intent to deceive?

The answer is not simply jealousy or competition for research money. Nor it is limited to the fact that Fleischmann and Pons' claims challenged prevailing thought imposed by 50 years of nuclear science. It's something bigger.

There was, and still is, a sense, even among scientists, that we have learned everything major there is to know about science, and only minor details remain to be discovered. Such shortsightedness is nothing new in science, which, paradoxically, is a field that places a high value on discovery.

Turn the clock back to 1909, and look what Frederick Soddy, an early pioneer in radioactivity and nuclear transmutation, developer of the concept of isotopes, and a 1921 Nobel Prize winner in chemistry, had to say:

> Radioactivity is a new primary science owing allegiance to neither physics nor chemistry, [and] as these sciences were understood before [radioactivity's] advent, ... the old laws of physics and chemistry, concerned almost wholly with external relationships, do not suffice. ... Is it possible to give, by the help of an analogy to familiar phenomena, any correct idea of the nature of this new phenomenon "radioactivity"?
>
> The answer may surprise those who hold to the adage that there is nothing new under the sun. Frankly, it is not possible because, in these latest developments, science has

broken fundamentally new ground and has delved one distinct step further down into the foundations of knowledge. (Soddy, 1909, 3)

Had anyone twelve years ago ventured to predict [the transmutation of] radium, he would have been told simply that such a thing was not only wildly improbable but actually opposed to all the established principles of the science of matter and energy. So drastic an innovation was, it is true, unanticipated.

Radium, however, is an undisputed fact today, and there is no question, had its existence conflicted with the established principles of science, which would have triumphed in the conflict. Natural conservatism and dislike of innovation appear in the ranks of science more strongly than most people are aware. Indeed, science is no exception. (Soddy, 1909, 4)

CHAPTER 22

The Press Conference From Hell

Seizing the Microphone

A week after scientists at the American Physical Society meeting decreed the new science dead, electrochemists descended in force on Los Angeles, California. It was the 175th meeting of the Electrochemical Society (ECS), a scientific society founded in 1902.

The ECS meeting was the third of three major U.S. scientific societies that, in early April, provided a forum for "cold fusion." Elton Cairns, the senior vice president of the society, was responsible for creating the ECS "cold fusion" session.

Sixteen hundred scientists from around the world came to the Bonaventure Hotel for the meeting, which started on May 8. Many of them attended the "cold fusion" session that Monday evening.

Not surprisingly, Fleischmann and Pons elected to participate; this was a society of their peers. To their dismay, the press conference following the scientific session was perhaps the most tortuous experience of Fleischmann and Pons' public involvement with the controversy.

Elton Cairns

On Sept. 30, 2014, I interviewed Cairns to get his perspective on the "cold fusion" scientific session as well as the press conference. The full interview transcript is available on the *New Energy Times* Web site.

Elton Cairns, senior vice president of the Electrochemical Society, 1989

Cairns told me that, as soon as he put out the announcement for the "cold fusion" session, he was overloaded with abstracts and had to reject most of them. He explained his selection process to me:

> I favored those papers that I thought were based on the best quality of research and were trying to find evidence for something happening. There were lots of papers that said, "I tried it, and it didn't work."
>
> Well, that doesn't shed any light on anything. Many of the offerings were of that sort; they didn't say why it didn't work. Or why it should've worked. They just said, "It doesn't work." That doesn't help. I turned all those away. The ones that I ended up with are those that I thought would be interesting from a scientific and technical point of view.

Even though Cairns is an electrochemist, he told me that he was never convinced that Fleischmann and Pons had truly measured any excess heat. "I mean," Cairns said, "how famous would the field of electrochemistry be if we had cold fusion. Just think of it. *[He laughed]*

Electrochemists would rule the world!"

Cairns' chief complaint about Fleischmann and Pons was that they had not published sufficient information in their preliminary note to show an excess of integrated energy rather than simply an excess of power. (Power is the rate at which energy is produced or emitted.)

I asked Cairns whether he was familiar with the full paper that Fleischmann and Pons published in early 1990. He said no, he had not paid attention to the research after the Los Angeles meeting. It was evident that he had had disappointing experiences with some of the "cold fusion" claimants:

> You can imagine that a lot of the scientists who were trying to reproduce the Fleischmann-Pons results were very interested in getting into the news and being famous, too. There were a lot of hangers-on who wanted to shine in the reflected light of Fleischmann and Pons. I just did not want to have that kind of circus atmosphere in the technical symposium.

In spite of his disappointment with many of the "cold fusion" claimants, he was not as bothered by the poor repeatability as were the physicists. It was easy to understand from an electrochemist's perspective:

> Repeatability was a concern, but it can be understood that, to do an experiment properly, one needs to control all of the variables. If you don't know what all the variables are, there can be some that unwittingly you're not controlling. And if you don't control them, then your results can scatter. I was willing to give the benefit of the doubt on that issue that there was something important that was not under control, so the lack of reproducibility was not the major issue in my mind.

I asked Cairns why physicists were so much more concerned about repeatability than chemists were. Cairns responded:

The partisan answer would be that physics is much more easily controlled than chemistry. Relatively speaking, it's much easier to design a fully controlled physics experiment than it is to design a fully controlled chemistry experiment. From the point of view of a physicist, this looks really bad if the chemist can't control everything. But from the point of view of the chemist, it's perfectly understandable because chemists know how difficult it is to identify and control all of the variables that affect the results.

Cairns commented on the press conference:

I was disappointed because both the press and the scientists did not behave in an orderly, organized, logical way to try to discuss what was really happening. It was more of a sociological experiment on how all these people were interacting with one another and who would get the most publicity and who would make the most newsworthy claim — things like that, rather than trying to do a good job of explaining the science logically and engaging in technically useful questions and discussion. It was more of a free-for-all.

If cold fusion had been shown to be true, there would have been a really, truly earth-shaking event. It would have just completely changed the energy picture altogether. It's because of that issue, what impact it would have had if true, that it sticks in people's memories.

The Scientific Session

There is no good record of the scientific session. Cameras and tape recorders were not allowed in and no recordings seem to exist. The presentations do not appear to have been published in proceedings.

Lee Siegel, writing for the Associated Press, was one of the few reporters whose company paid the $200 conference fee so he could gain entry. Here is part of his report:

[In Baltimore,] some critics said Pons and Fleischmann were deluded into thinking that their device produced excess heat through fusion because they measured temperatures in the jar improperly and were actually measuring hot spots.

However, Fleischmann, appearing before 1,800 scientists here, said that there were no hot spots in his fusion device and that the excess-heat measurements were accurate. He even showed a short movie of gas bubbling through the fusion device to support his claim that heat was distributed evenly throughout the apparatus.

"This argument of ineffective (heat) mixing really doesn't hold water," Fleischmann said.

Pons, using technical language, lashed out at critics who contend that neutrons allegedly generated by fusion in the experiment may really have come from naturally occurring radon gas in the Utah laboratory.

He said that he and Fleischmann measured gamma rays — an indication of neutrons and thus fusion — that came from their experimental device but were not present elsewhere in the lab.

Scientists who have been unable to reproduce the findings of Pons and Fleischmann have failed to use adequately large palladium electrodes, Pons said.

The meeting was the first appearance by Pons and Fleischmann before their fellow electrochemists, although they previously discussed their findings at a meeting of the American Chemical Society.

Before the session started, scientists eager for more information on the scientific controversy were forced to crowd like sardines into a hotel hallway while Electrochemical Society officials prepared the meeting room.

Security guards repeatedly had to ask people to stand back and not rush into the meeting room once the doors were opened.

Also present Monday night was Steven Jones of Brigham Young University, who showed a series of slides of

3-year-old notebook pages to refute accusations by Pons and Fleischmann that he borrowed heavily from their work.

People waiting to get into the scientific session

Siegel was doubtful about "cold fusion," as he told me in an e-mail:

> When my boss, the news editor in Los Angeles, put the "cold fusion" news story in front of me and asked what I thought, I said one word: "Bullshit."
>
> The Electrochemical Society really pissed off the media because the society made reporters pay the meeting registration fee. I still remember the agony of that night, having to run to a pay phone and call in a new lead to my story.
>
> If you want, you can use this for your book, pointing out the irony that I later became a University of Utah spokesman.

By all accounts, aside from what Siegel wrote, nothing else of significance happened during the scientific session, which ran from 5:45 p.m. to 8:05 p.m. From my knowledge of the history, four of the presenters reported confirmations of some aspect of the Fleischmann-Pons experiment. Siegel was also at the hellish press conference.

Near-Bloodshed

Although valuable scientific information was exchanged between scientists and reporters in the press conference, it did not happen without near-bloodshed. The press conference began about 10 p.m. on Monday evening, May 8. Bruce Deal, the president of the Electrochemical Society, presided, breaking up several arguments before angry words turned to flying fists and chaos.

This press conference was described in the 1997 book *Truth and Consequences: How Colleges and Universities Meet Public Crises,* by Jerrold K. Footlick. Author Charles Beaudette excerpted part of Footlick's account and republished it in his book *Excess Heat: Why Cold Fusion Research Prevailed.* Here is what Footlick wrote:

> If the [scientific] presentation was weak, the press conference was a fiasco. The Salt Lake City newspapers and television stations were all represented, along with a sizable Los Angeles press contingent and some national science writers; as many as 150 reporters and a dozen television cameras jammed into a room much too small and hot.
>
> After a few timid questions, a physicist from the California Institute of Technology — a nonjournalist who had crashed the press conference — commandeered a microphone and began shouting loaded questions at Pons and Fleischmann. Soon everyone was grabbing microphones and interrupting one another; a number of people, some of them physicists cholerically denouncing the work, stood on chairs to shout.
>
> Pons and Fleischmann sat stone-faced in the television lights, perhaps stunned, certainly angry. After a few minutes, they announced that they would participate no longer, stood up, and walked out.

Except for the three scientists at the head table — Fleischmann, Pons, Jones — there was standing room only in the press briefing room for the rest of the electrochemists who had spoken that evening.

This was only one of numerous insults public and private. Physicists at Utah, pressured by their friends at other institutions, were demanding that the university disavow the research. In one bizarre case, a man who claimed to come from the Massachusetts Institute of Technology camped outside the door of Pons and Fleischmann's laboratory, trying to force his way inside when the door was opened. Telephone calls and faxes arrived from around the world, calling for more information or calling the experiment a fraud.

Footlick's account seemed out of place — no science journalist at the time had described the Electrochemical Society press conference this way. Footlick very kindly lent me his original audiotapes of the interviews on which he based his account. His source was Pam Fogle, the news director for the University of Utah, who was in the audience in Los Angeles. But I was intent on finding either original recordings or a second source to corroborate Fogle's memory.

I made multiple inquiries of historians, journalists and media outlets. Eventually, I hit pay dirt when I contacted Sam Hornblower, a producer with the CBS television program *60 Minutes*. He helped me locate 47 minutes of raw footage they had taken of nearly the entire press conference. It turned out that Fogle's account was generally accurate.

However, based on my review of the videotape, there are several corrections to make of the Fogle/Footlick account: The scientist from Caltech who crashed the press conference was not a physicist, but chemist Nathan Lewis. The footage shows people interrupting one another, but it does not show anybody standing on chairs or shouting. The video does not show any physicists denouncing anyone's work — although one physicist, Steven Jones, came close to doing so, though he remained in his seat the whole time and kept his composure.

Someone was, in fact, shouting barbed questions at Pons and Fleischmann, but it was reporter Tom Heppenheimer from *Science* magazine. Fleischmann and Pons made no announcement about ending their participation early. They did not walk out but stayed until the end.

In the next sections I will present some of the highlights — and lowlights — of the spectacle.

Fleischmann and Pons Respond

Gamma-Ray Measurement

Reporters asked Fleischmann and Pons about their claimed gamma-ray measurements. Fleischmann informally withdrew those results:

> Stan Pons and I had a long correspondence about the deficiencies of our gamma ray measurements. I'm well aware of the criticism which was made today. We are well aware that those measurements are deficient. I think it is not meaningful to continue with those measurements in that present form. We have to use other techniques.

Heat Burst

Several reporters asked Fleischmann and Pons about the new

information they had presented that evening about an excess-heat burst.

Pons explained to the reporters that the burst had shown 1,000 percent to 5,000 percent of heat in excess of the electrical input power. Pons said the burst was sustained for two days and the integrated energy was 4.2 megajoules.

Helium Measurement

Richard Harris, from National Public Radio, asked Pons about the forthcoming collaboration between the University of Utah and Los Alamos National Laboratory. Pons told Harris that "there is a formal document that is being examined by the University of Utah that's to be signed before an official collaboration."

Another reporter asked when Pons thought it might be signed. Pons sighed and paused. Fleischmann whispered something to him. Pons told the reporter, "It's just not under my control."

Such evasive and incomplete responses were guaranteed to raise the hackles of the reporters. Charlie Petit, from the *San Francisco Chronicle,* was the first to express his annoyance and to push for a more complete response.

> PETIT: Dr. Pons, please explain why you can't say anything at all about who is going to carry out this critical test for helium in the palladium. I mean, so far we've heard there's actually no evidence of fusion at all except by reduction. And you mentioned the Los Alamos collaboration. Why can't you tell us what group is going to be looking for that helium?
> PONS: We have an agreement not to do so.
> PETIT: But what's the motive? Is this part of the commercialization aspect of this?
> FLEISCHMANN: Yes. You must realize that we are not free agents to answer certain questions.

Both Fleischmann and Pons were withholding crucial information. But they certainly did not have any intention of irritating and frustrating the news media. They were restricted by a nondisclosure agreement. The University of Utah owned the rights to Fleischmann and Pons' work, and attorneys representing the university were calling the shots, as revealed by reporter Anne Burnett of KUER-FM.

Records I obtained from the Department of Energy under a Freedom of Information Act request explain what Fleischmann and Pons could not. The DOE provided a set of memos sent from Robert O. Hunter, the director of the Office of Energy Research, to Admiral Watkins, the Secretary of Energy. I have excerpted three entries from the memos:

> May 3, 1989: Los Alamos [National Laboratory] is working to collaborate with the University of Utah to assist in measurements and in any other way possible. However, the University of Utah has hired a law firm to protect their interests, which has slowed the process.
>
> June 14, 1989: In view of the lack of response by the University of Utah, LANL has decided to break off negotiations concerning verification of experimental results but remains open to a serious collaboration.
>
> June 28, 1989: [Pacific Northwest National Laboratory] was approached by the University of Utah to serve as a mediator in the analysis of Professor Pons' electrodes.

Appleby and Huggins Confirm Excess Heat

John Appleby, the director for the Center for Electrochemical Systems and Hydrogen Research of Texas A&M University, asked Chairman Deal whether he could contribute a comment. Deal agreed, and Appleby came to the podium and said that excess heat was the necessary proof of a nuclear reaction:

> The proof that we indicated this evening shows that, definitely, anomalous heat has been produced by palladium in heavy water. We've also done all the blank experiments which show that heat is not produced with platinum in heavy water, nor with palladium in light water.

Anthony John Appleby, speaking at May 1989 Electrochemical Society meeting press conference in Los Angeles

Appleby proposed a nuclear reaction between deuterium and helium-4 to give lithium-6. A reporter asked Jones, who may have been the only physicist in the room, his opinion. Jones mocked Appleby's proposal and provoked laughter among the reporters.

In response to another reporter's question, Jones gave what was by now his trademark spiel about nuclear products and national-debt-type numbers.

Bob Huggins, a professor in the departments of Materials Science and Engineering at Stanford University, politely asked Deal whether he could contribute a comment. Deal agreed, and Huggins came to the podium.

Huggins reinforced Appleby's view that excess heat had been proved. He aimed his next comment at the people who were in Baltimore — Lewis (who was in the room) and Meyerhof — who had suggested that Fleischmann and Pons had made a fatal mistake by not using a mechanical stirrer. "The question of stirring," Huggins said, "has been laid to rest both by the work in Utah and by work in our laboratory, where we've clearly been able to show that [stirring] is not important."

Jones Challenges Huggins

The decorum that had prevailed at the start of the press conference soon disintegrated. While Huggins was midsentence, Jones tried to interrupt, "Could I — could I, I wonder if I might —" Huggins politely held his ground. "Yes, just a second."

A moment later, as soon as Huggins finished speaking, Jones took his turn. He presented himself disingenuously as someone who thought it was "exciting" that people were seeing excess heat.

Jones tried to get Huggins to say whether he would permit third parties to inspect his cathodes for nuclear by-products. Huggins didn't take the bait, and Jones got aggressive; the tone of his voice got increasingly angry and frustrated. Huggins remained unfazed as Jones badgered him.

HUGGINS: Yes, when we're in a position to supply samples to people, we should be delighted to do that —

JONES: How soon do you —

HUGGINS: — I should point out that we're not using rods. We're making our own palladium samples, and the way we do this is probably a significant deviation from the way other people have made them —

JONES: But if you have heat, let's see if it's fusion.

HUGGINS: Alright.

JONES: We have an opportunity here!

HUGGINS: Alright.

JONES: We must take it!

HUGGINS: We'd be delighted to collaborate with people when we're in a position to do so.

Jones, obviously discouraged that he failed to get Huggins to agree to the test, abandoned his forced smile and shook his head in disgust. The test wasn't Jones' original idea; it was Harold Furth's. And it wasn't valid, either, as discussed in Chapter 12.

Lewis Commandeers Podium

Huggins stepped away from the podium. Just as Deal began to say, "Let's have Professor Bockris —," Lewis commandeered the podium. Lewis put his hand up toward Deal, and Deal stopped speaking. Lewis thanked Deal and began to speak. The free-for-all had begun.

Bruce Deal, president of the Electrochemical Society, 1989

Lewis began by disputing things that Huggins had said. When he stopped to take a breath, a reporter asked a question. But Lewis wasn't done with his monologue. He told the reporter, "Hang on!" and explained what excellent work he and his team had done at Caltech. He told reporters how strange it was that researchers in three laboratories — University of Utah, Texas A&M, and Stanford — had observed excess heat but, more important, he had seen nothing.

"So, presumably," Lewis said, "there are three samples that work, and none of the 50 that we tried, nor any of the ones the national laboratories have tried, nor MIT, nor anyone else has, is working. It is very difficult to believe that there are three sets of magic samples in the world."

It wasn't true. By April 29, excess heat had also been observed at the Indira Gandhi Center for Atomic Research, Korea Advanced Institute of Science and Technology, Portland State University, Case Western Reserve University, Tata Institute of Fundamental Research and Central Electrochemical Research Institute. (See Appendix C)

Bockris Calls Lewis Out

Lewis stepped away from the podium, and once again, Deal invited John O'Mara Bockris, the director of the Surface Electrochemistry Lab at Texas A&M, to the podium. Although Fleischmann and Pons refrained from lashing out at Lewis, Bockris did and said the obvious.

"It's important to [notice]," Bockris said, "the hints constantly made by Lewis that there's something extraordinary about this [just because] he can't repeat [it]. Really, it's he who's the only one [who can't do it]."

A reporter challenged Bockris. "How come you could not reproduce Dr. Fleischmann's experiment with the piece of the material provided by Professor Fleischmann?" Bockris squarely acknowledged the reproducibility problem. "There's a great deal of irreproducibility in the phenomenon," Bockris said. "I made a special point of that in my own lecture. At Texas A&M, I stressed that our overall count was one electrode in three which worked; 20 electrodes and only 5 or 6 of them have actually produced heat."

Heppenheimer Attacks

The next reporter to ask a question was Tom Heppenheimer, from *Science* magazine, published by the American Association for the Advancement of Science. Heppenheimer asked an argumentative question about calculations of the molar percentage of helium that, according to a scientist he spoke with, should have been evident in Fleischmann and Pons' results. Heppenheimer's tone was angry from the start.

John O'Mara Bockris during Electrochemical Society meeting press conference

Before he took a breath from his rapid-fire rant to allow Fleischmann or Pons the chance to respond, Heppenheimer switched from anger to outright hostility, almost yelling his question:

> I would add, since you have done experiments with apparatus that is just one step up from Emily Criddle's kitchen, I would recommend that such a very large amount of helium might be found with equipment that is just one step up from what we find in a high school chemistry laboratory. Why this coyness about finding helium in your electrodes?

Fleischmann and Pons were stunned and speechless. Five seconds of silence went by and were broken only by Heppenheimer, who demanded, "Where is it?!" Nobody in the room uttered a sound. Another four seconds of silence passed. Fleischmann was unwilling to respond to such rudeness. Pons finally found the words to speak:

PONS: We could not run that analysis yet.

HEPPENHEIMER: Why not?

PONS: It is being run now.

HEPPENHEIMER: We've heard that people promise a three-day turnaround. I can get my film developed —

PONS: I found out about that about an hour ago, OK? I'm not waiting for laboratories —

HEPPENHEIMER: See, you just dodged the most essential issue!

PONS: What?

HEPPENHEIMER: You just dodged the most essential issue. A couple of years ago, I was writing about a perpetual motion inventor, and he always dodged the most essential issue, also!

Stanley Pons listening to Tom Heppenheimer during Electrochemical Society meeting press conference

By now, Heppenheimer was yelling. Deal calmly de-escalated the conflict: "Could we go to the next question?"

Matthew Ellis from Reuters had the next question ready to go, and for the next few minutes, the discussion revolved around criticisms made at Baltimore.

Impromptu Press Conference

Following that, Lewis commandeered the podium and gave a full rundown on all the tests that he and his crew performed at Caltech.

In each case, they found nothing, nothing and more nothing. He didn't seem to stop to take a breath. Eventually, Lewis paused for a split-second, and Deal deftly jumped in and encouraged Lewis to wrap up: "So, in conclusion?"

"In conclusion," Lewis said, "we have no evidence in our laboratory with any of samples for fusion." Lewis was so furious that it apparently didn't cross his mind that, as Bockris said, Lewis' failure didn't mean that the others in the room hadn't succeeded but that the failure was his own.

Martin Fleischmann momentarily losing his composure at the Electrochemical Society meeting press conference

Deal asked for a statement from either Fleischmann or Pons in response to Lewis. Fleischmann began talking about helium and tritium. Lewis remained at the podium and, after a few minutes, interrupted Fleischmann.

LEWIS: We know the foreground; we don't know the background. I would like to —

FLEISCHMANN: The background, I beg your pardon! The background is available in the corrections to the paper.

LEWIS: That might be. I would like to specifically —

FLEISCHMANN: Well, then, please don't, don't —

DEAL: Could we go on to other questions, please?

The scene was broadcast by every television station that ran any report on the meeting. The scene has appeared in almost all video documentaries on the topic. It was the only time during the press conference that Fleischmann lost his composure.

University Stifles Helium News

John Van, from the *Chicago Tribune,* politely brought the conversation back to helium measurements. As Pons began to respond, another reporter interrupted and yelled at him about the delay.

Nevertheless, Pons kept his cool and made it clear that the helium analysis was out of their hands. Pons hinted that, if the matter were in their hands, things would happen faster. "That decision is not up to us," Pons said. "We have a prior commitment, and the decision is not up to us. We would get fast results. We would get these analyses done forthwith."

Even though they were being pressed by reporters, Fleischmann and Pons did not blame the University of Utah for dragging its feet in making arrangements with Los Alamos to test the cathodes for helium.

In hindsight, all the drama and challenges surrounding the idea of measuring the cathodes for helium appear to have been meaningless. Fleischmann and Pons were wrong that the phenomenon was a bulk effect. Rather, it is a surface effect. Many later years, surface and depth profiling with scanning electron microscopes and secondary ion mass spectrometers confirmed this. The helium measured in the experiments appears at the surface of the electrodes (and remains at the surface) and is captured in the off-gases.

Lewis Takes Podium Again

The discussion came back to heat measurements. Fleischmann responded to a reporter's question that referred to Lewis' accusation that Fleischmann and Pons incompetently neglected to stir their cell.

Nathan Lewis during Electrochemical Society meeting press conference

Fleischmann calmly explained their measurements — until Lewis jumped to the podium without permission and began arguing with Fleischmann. Lewis argued that he had precisely replicated their cell based on an "8½ x 11 glossy photograph" in which Pons was holding the cell. Fleischmann corrected Lewis, and they argued again:

FLEISCHMANN: I'm sorry, that is not —
LEWIS: That might be an incorrect cell.
FLEISCHMANN: It is an incorrect cell.
LEWIS: And I'll be happy to change it.
PONS: It's several times, many times the volume.

Lewis' cells were too large to make use of sparging, which Fleischmann and Pons had used to ensure a consistent thermal gradient in their cells. The next reporter began to ask a question, but not to Lewis. No longer the focus of the reporters' attention, Lewis quietly stepped away from the podium.

Appleby took a place at the podium without an invitation and began speaking. He said that Lewis' objections did not apply to the type of calorimetry that Appleby had performed in his group at Texas A&M:

> We are convinced that excess heat was evolved from the electrodes that we tested.
>
> I cannot tell you whether it's as a result of fusion. All I can say is that the amount of heat being evolved inside the palladium that we used would have been sufficient in one hour to volatilize completely that piece of palladium. So I think that rules out any chemical process.

The Human Side

Lee Siegel asked a question that allowed Fleischmann to cool down and regain his composure. "Given this controversy," Siegel said, "which is very hard for a layperson to get the detailed understanding of, are you willing to acknowledge any possibility at all that your observations are wrong and that you did not have fusion, or are you completely convinced that you had fusion?

"I have always been ready to acknowledge the fact that our experiments may be faulty," Fleischmann replied. "I said in the beginning, I said that throughout every meeting to the press, I have emphasized that you cannot prove something right; you can only prove it wrong. For that reason, one has to have very full publication of all the information. If we are wrong, I'll be the first to admit it."

Eventually a reporter asked an empathic question about the human side of the experience for Fleischmann and Pons. "How do you guys feel," the reporter said, "after what you've been through this last month? Let's step back from the science a minute. Do you have any regrets about the way — let me put that another way — do you wish now that you had just published this quietly instead of holding the news conference?"

"I don't think the result would have made any difference when the [preliminary note] came out," Pons said. "In the end," Fleischmann said, "I think there would have been a very marginal difference. But I wish we had been left to get on with our work at our own pace without the media attention."

Polarization in the Press

Like the Baltimore meeting, the Los Angeles Electrochemical Society "cold fusion" meeting and press conference left a record of its aftermath. Two *Los Angeles Times* journalists, with very different perspectives, reported the story.

Patt Morrison: "Hopes Outweigh Doubts"

As researchers, many members of the Electrochemical Society have doubts. But as chemists, they have hopes. If Pons and Fleischmann are right, not only would peace and plenty ensue in a world of limitless clean energy, but it would provide an earthier satisfaction for chemists: twitting what one called the "supreme arrogance" of physicists, who rank above chemists in the pecking order of science and consider fusion research their preserve.

Physical chemist Fritz Will of Schenectady, N.Y., is a past president of the Electrochemical Society. Last week, a Nobel laureate put a hand on Will's shoulder and told him, "Fritz, if cold fusion were ever to be discovered, it wouldn't be discovered by a chemist."

Thomas H. Maugh, II: "Fusion Genie's Still in Bottle, Scientists' Critics Say"

Suddenly, there is no more talk of a Nobel Prize. The two chemists are besieged by critics — castigated for their inability to explain their experiment, their failure to reveal full details of the research and their decision to circumvent the normal rules of science by announcing their results at a press conference before reporting them at a meeting or publishing them in a scientific journal.

Fleischmann and Pons were certainly to blame for failing to reveal full details of their experiment. But Maugh perpetuated a falsehood — one which was written into history — about Fleischmann and Pons' ethics. There was no circumvention. In the preceding six weeks, they had been, in fact, among the minority of scientists who submitted a manuscript for publication, not to mention received acceptance, before announcing their results to the press.

Thus, after Baltimore and Los Angeles, the battle lines emerged: optimists versus naysayers, believers versus non-believers, legitimate science versus pathological science.

If this doesn't work we're going to look pretty silly.

Cartoonist Joan Cartier's depiction of Princeton University thermonuclear fusion researchers

CHAPTER 23

India Advances Again

Massive Government Effort

As Nathan Lewis was whining in Baltimore and Los Angeles about deficiencies, some imagined and some real, in Fleischmann and Pons' preliminary note, groundbreaking research was taking place on the other side of the globe.

Researchers at the Bhabha Atomic Research Center (BARC) in Trombay, on the Indian Ocean coast near Mumbai, India, succeeded in their replications of the experiment — with no direct help from Fleischmann or Pons. Within a year, BARC researchers had observed, among other results, nine sets of experiments that produced both tritium and neutrons from six independent groups.

In fact, BARC researchers observed their first significant results on April 21, 1989, the same day that Lewis gave a seminar at Caltech and announced that, despite his monumental efforts and superior instruments, he couldn't get "cold fusion" to work. According to Lewis, Fleischmann and Pons had reported something illusory and led him on a wild-goose chase, or, if it was real, they were at fault for not giving him personal assistance.

Two groups at BARC, conducting independent experiments on April 21, observed neutron and tritium bursts. One group was a collaboration among the Heavy Water Division, the Neutron Physics Division, the Health Physics Division and the director of BARC. The other group was in the Analytical Chemistry Division.

Mahadeva Srinivasan, the director of the Neutron Physics Division and assistant director of the Physics Group, played a key role. Had he or anyone at BARC received any direct tips, support, information, or assistance from Fleischmann or Pons? "No tips," Srinivasan wrote me. "Entirely on our own."

By September, more than 50 scientists working in 10 groups at BARC had observed a dozen successful experiments. The researchers had the support of Padmanabha Krishnagopala Iyengar, the director of BARC between March 10, 1984, and Jan. 31, 1990. In December 1989, Iyengar and Srinivasan published the collective works in the BARC-1500 report. Most of the groups stopped their "cold fusion" research in the spring of 1990, after the subject had fallen into disrepute.

There are two key aspects to the groups' successes. First, they did not succumb to the mistaken *Dictum of Reproducibility*. Even though their experiments were not highly repeatable, the few experiments that did show positive results were unambiguous. Second, they accepted the validity of their experimental data at face value even though it appeared to conflict with prevailing theory.

Indigenous Talent

The Indian nuclear researchers had a long history of independent success in challenging areas of nuclear research and technology. By 1945, at the end of World War II, the United States, collaborating with Canada and the United Kingdom, had developed and dropped two nuclear fission bombs on populated areas in Japan.

Soon after, the Soviet Union began developing its own nuclear weapons arsenal; it and the United States began rapidly building their stocks of ever-more-powerful nuclear weapons and expanding their capacity to destroy each other, if not the entire world. Other countries began developing nuclear weapons as well.

By 1968, concerned nations developed the Treaty on the Non-Proliferation of Nuclear Weapons, more commonly known as the Non-Proliferation Treaty, or NPT. The three principles of the NPT were 1) nonproliferation of nuclear technology to states that did not have it, 2) disarmament among states that did have nuclear weapons, and 3) an agreement to support the peaceful uses of nuclear technology among nuclear-weapon states as well as non-nuclear-weapon states that had signed the NPT. A condition of the third principle was an agreement by signatories of the NPT that non-nuclear-weapon states could not develop their own nuclear weapons.

The first three signatories of the NPT were the United States, the Soviet Union and the United Kingdom. France and China also had nuclear weapons at the time, and the NPT recognized these five countries as nuclear-weapon states. At the time, India did not have nuclear-weapon technology, and did not wish to limit its ability to develop nuclear weapons, so India did not sign the agreement. This was an understandable position considering its nuclear-weapon neighbor to the northeast, China, and its neighbor to the northwest, Pakistan, which was also unwilling to give up its ability to develop nuclear weapons.

Journalist Raj Chengappa tells the detailed story of India's nuclear developments in his book *Weapons of Peace*. In an e-mail to me, Chengappa explained additional details about India and the NPT:

> There were other factors that discouraged India from acceding to the NPT. In 1962, India was defeated by China in a border war, and two years later, China did its first nuclear test. These developments meant that India was extremely vulnerable — China posed a clear and present danger. This was in addition to the fact that, by 1974, when India tested its first nuclear device, we had already fought three wars with Pakistan.
>
> India was and is a reluctant nuclear state. Its independence was built on the principle of non-violence and the fight for truth that Mahatma Gandhi so successfully preached. Nuclear weapons were anathema to the foundations on which India won its hard-fought freedom.
>
> The reason India finally went ahead with the development of its nuclear-weapons program was because the U.S. and U.K. refused our request to provide a nuclear shield if China attacked us. We were vulnerable to coercion by China, and developing our own bomb was a way to

restore the military imbalance. The threat that Pakistan would go nuclear also loomed large after India defeated it in 1971.

Moreover, India balked at the unequal terms that the NPT imposed — the haves including its main enemy, China, could keep their weapons, and the have-nots had to give up their rights to develop such weapons in exchange for technology to build nuclear power stations. But India already had such nuclear power technology, and it had little to gain by signing the NPT and plenty to lose by giving up the right to defend itself against a nuclear adversary.

Thus, India was cut off from assistance and materials from the nuclear-weapons states. This affected not only its ability to build nuclear weapons but also its ability to develop more advanced reactors. Additionally, its existing group of uranium-based reactors was severely affected when India was cut off from foreign sources of uranium after the NPT took effect. India has limited uranium resources although it has abundant sources of thorium, another important nuclear material.

The story begins in 1967, with Raja Ramanna, the director of the BARC Physics Group. Ramanna tasked a researcher in his group, Rajagopala Chidambaram, 31, with a key role for the development of the device, according to Chengappa:

> Ramanna wanted Chidambaram to carry out studies on the shock wave propagation on plutonium for a possible nuclear explosion. He was also to figure out the equation of the state of plutonium (essentially, how a plutonium sphere behaves under the extreme pressure and temperature conditions involved in such an explosion). (Chengappa, 2000, 118)

Both Ramanna and Chidambaram later became directors of BARC. Carey Sublette, the creator of a Web site on the history of nuclear weapons, described Srinivasan's role in the development of India's Purnima fission reactor:

Mahadeva Srinivasan developed a sophisticated physics model for criticality calculations in 1970 and later became the reactor's chief physicist. Construction began that year when sufficient separated plutonium finally became available. Purnima went critical on May 18, 1972. With Purnima as a test bed, the Indian physicists were able to refine their understanding of the physics of fast fission and fast neutrons. Although Srinivasan was not formally part of the later bomb development team, his expertise in fast critical systems underlay the nuclear design of the device. (Sublette, 2008)

Chengappa had more to say about Srinivasan's calculations:

Ramanna asked Srinivasan to simultaneously begin work on the physics of neutron multiplication for the explosive device. ... Srinivasan's technique for calculating the Purnima reactor's criticality was so effective that, years later, scientists at the famed Lawrence Livermore laboratory in the U.S. referred to it as the Trombay Criticality Formula and even used it to teach their students. (Chengappa, 2000, 124)

In 1974, the BARC team became national heroes when they successfully designed, built and tested their first nuclear-fission-based explosive proof-of-concept, called the Pokhran Nuclear Explosion. People in more-developed nations were shocked to find out that an underdeveloped country like India had achieved such a remarkable feat, particularly without help from the nuclear-weapons states, as Chengappa explained to me in an e-mail.

It validated India's ability to build nuclear weapons and caused shock waves across the world because it directly challenged the NPT that the big powers had pushed for. It announced India as a nation capable of making nuclear weapons and was a significant milestone and turning point both for India and the world.

Ramanna, Iyengar, and Chidambaram were the core team, in addition to other BARC scientists, including Srinivasan, who brought India into the exclusive nuclear-superpower club. Iyengar, Srinivasan and Chidambaram would also play important roles in the "cold fusion" conflict: the first two in support of "cold fusion," the latter against it.

The First Experiments

The sprawling BARC campus is home to India's most significant and largest center for nuclear research in the nation. When I visited the lab in 2008, it employed 15,000 people: 13,000 of them worked on-site, and 4,500 of those were researchers. The rest were support and administrative staff. BARC has a staff 50% larger than the U.S. Los Alamos National Laboratory, whose primary purpose is the design of nuclear weapons. BARC's mission, however, is primarily to support peaceful applications of nuclear technology, such as energy production and research for medicine and agriculture.

One of the benefits of such a facility is, as Srinivasan explained, having experts in almost any field of science within walking distance. Such multidisciplinary expertise proved immensely helpful and enabled the researchers to begin "cold fusion" experiments immediately and make rapid progress.

The News Arrives

Researchers in the Heavy Water Division had been working on a plasma focus-type fusion experiment when the news of "cold fusion" reached Trombay. Srinivasan told the story in a 1994 interview for *Cold Fusion Magazine.*

"On March 24, 1989," Srinivasan said, "we saw this little four-line news item in the *Times of India*, saying that neutrons were seen to be produced by a small battery and bottle experiment. I immediately got interested in it from the point of view of a neutron source."

Iyengar wasted no time, and at 10 that morning, he called a meeting of all interested scientists and engineers at BARC. Iyengar, Srinivasan said, was a person who was always interested in exciting new things. People with a variety of backgrounds, including neutron physicists, chemists, and chemical engineers, immediately got involved. Within six weeks, 12 groups at BARC were independently setting up cells and carrying out experiments. Ten of those groups got positive results. (George, 1994)

Experiments Begin

The researchers in the Heavy Water Division were able to start experiments immediately with an unusual electrolytic device they had on hand. It was called a Milton-Roy electrolytic cell and was a diffusion-type water electrolyzer designed to generate ultrapure hydrogen gas. The cathode comprised an array of 16 palladium-silver alloy membrane tubes with a total surface area of 300 cm^2. The cell was capable of running at a very high level of power: 100 amps peak and 60 amps continuous. It was possibly the largest "cold fusion" electrolytic cell ever used. (Iyengar and Srinivasan, 1990)

Their neutron detection system comprised banks of three types of detectors. The first set was a bank of BF$_3$ counters embedded in paraffin moderator blocks, which were sensitive primarily to slow neutrons. The second bank was composed of ^{3}He detectors for thermal neutrons, also encased in paraffin blocks. The third type of detectors was proton-recoil-type plastic scintillators that were sensitive to fast neutrons.

During the electrolysis experiments, the researchers typically placed a set of the BF$_3$ counters and proton-recoil counters close to the cell to detect the signal, and placed the ^{3}He detectors 1.5 meters away to detect the background.

Early Success

On April 21, they began the experiment with a current of 30 amps and slowly raised it to 60 amps. Soon, the neutron detectors near the cell started showing counts well above the background values. The cell produced a number of distinct neutron peaks in both the BF$_3$ bank and the plastic scintillator detector. These completely independent foreground detectors tracked each other nearly identically.

The researchers estimated that the total number of neutrons generated during the four-hour run was 4×10^7. They drew a sample of the electrolyte and sent it to the Tritium Division for analysis.

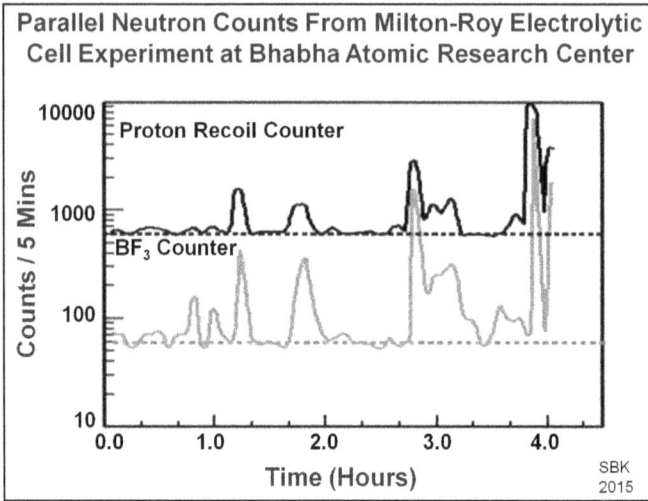

Parallel Neutron Counts From Milton-Roy Electrolytic Cell Experiment at Bhabha Atomic Research Center

Electrolysis of deuterium normally causes a separation of deuterium from trace levels of tritium in the deuterium and results in an enriched level of tritium after electrolysis. After taking into account the expected enrichment, they estimated that an excess of 8×10^{15} atoms of tritium were produced in that run. The starting value of tritium was 2.6 Becquerels/ml, and the final value was 55,600 Becquerels/ml, giving an increase of 21,000 times.

First Report

Iyengar reported the results for the first time on July 3, 1989, at the Fifth International Conference on Emerging Nuclear Energy Systems, in Karlsruhe, Germany. He wrote that the neutron counting rate, averaged over a five-minute interval, was also a couple of orders of magnitude larger than that of background count rates. (Iyengar, 1989)

The scientists were assuming that the process was fusion; therefore, they were expecting to see the normal branching ratio of equal counts of neutrons to tritium. However, this is not what they measured. Instead, the corresponding neutron-to-tritium ratio was 0.5×10^{-8} to 1.

As the researchers wrote, this was their first indication that the neutron-to-tritium yield ratio in "cold fusion" experiments was exceptionally low.

Variety of Experiments

The BARC groups carried out a variety of experiments employing both electrolytic and gas-loading of deuterium. They used palladium as well as titanium for the host metal. Many of the neutrons and tritium were also measured in bursts. Srinivasan was fascinated by the bursts because he thought they could imply chain reaction events, as he wrote in a 2009 review article: "Approximately 20% of the neutrons produced could be attributed to high multiplicity events wherein more than 20 neutrons are generated per burst." (Srinivasan, 2009)

Srinivasan made a key point about typical neutron detection. Most detectors pick up a very small percentage of neutrons emitted from a source. For example, he wrote, when neutron-detection efficiency is as low as 1%, even if 100 neutrons are emitted in a single sharp bunch, the detector will still detect it only as a single neutron event. If 10 counts are registered during a one-minute interval, it could imply that the source produced either a) 1,000 single neutron emission events or b) 10 burst events, each of which emitted 100 neutrons. Some of the experiments suggested that the neutrons and tritium were being generated simultaneously.

As Srinivasan explained, the correspondence between the neutrons and the tritium is highly likely despite the fact that their graph shows a latency. They detected the neutrons first and the tritium later because electrolyte samples were taken only periodically for the tritium assays. The neutrons and tritium could have been produced at the same time, or the products could have closely followed each other. (Srinivasan, 2009)

When Srinivasan and Iyengar normalized the yields of neutrons and tritium across all groups, based on cathode surface area, they found a systematic pattern. The neutron-to-tritium ratios were all within the range of 10^{-6} to 10^{-9}. On average, one neutron was emitted for every 10 million tritons.

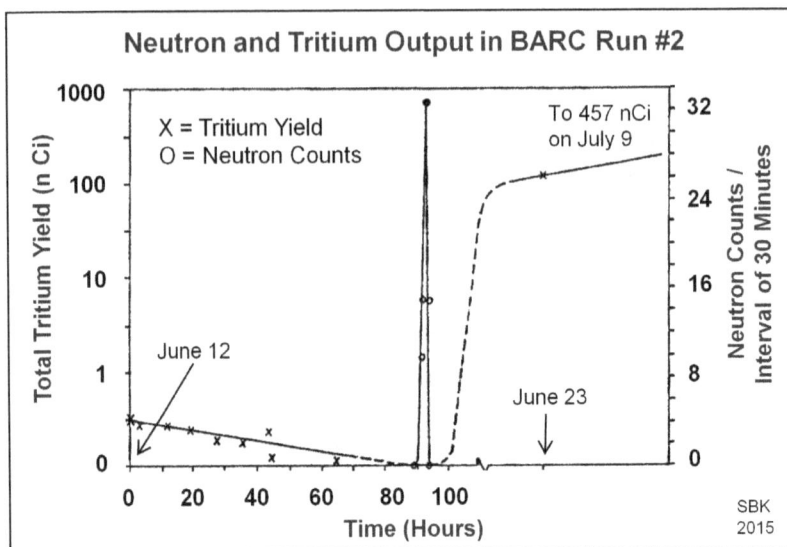

Neutron spike and tritium yield during run #2 of Milton-Roy cell at BARC, June 12, 1989

Surface Phenomena

They made an intuitive guess that the phenomena were based on cathode surface area rather than cathode volume. Fleischmann and Pons, on the other hand, had assumed that the effects were based on volume. More than a decade later, depth profile analyses provided extensive evidence for the phenomena as surface, rather than volume, effects.

Unusual Pattern

The BARC researchers also saw another unusual feature in eight of the electrolytic cells. The burst of neutrons and tritium in the cells occurred on the first day of electrolysis, after only a few hours of charging. After continued electrolysis, for reasons they didn't understand, all the cells that had produced neutrons stopped emitting neutrons.

Fleischmann told me about something similar in a June 4, 2004, letter. They had found that the generation of neutrons was confined to the initial stages of the charging of the electrodes.

Summary of Results From Groups Reporting Tritium and Neutrons in BARC Electrolysis Experiments							
Serial # (Date(s))	Cathode	Shape	Area Cm^2	Anode	Neutron Yield	Tritium Yield	n/T Ratio
1 (May 21/89)	Ti	Rod	104	SS pipe	$3x10^7$	$1.4x10^{14}$	$2x10^{-7}$
2 (April 21/89)	Pd-Ag	Tubes	300	Ni pipe	$4x10^7$	$8x10^{15}$	$5x10^{-7}$
3 (June 12-16/89)	Pd-Ag	Tubes	300	Ni pipe	$0.9x10^7$	$1.9x10^{15}$	$5x10^{-7}$
4 (May 5/89)	Pd-Ag	Disks	78	Porous Ni	$5x10^6$	$4x10^{15}$	$1.2x10^{-9}$
6 (April 21/89)	Pd	Cyl.	5.9	Pt Mesh	$3x10^6$	$7.2x10^{13}$	$4x10^{-8}$
7 (June-Aug./89)	Pd	Cube	6.0	Pt Mesh	$1.4x10^6$	$6.7x10^{11}$	$1.7x10^{-6}$
8 (Jan-Apr/90)	Pd	Pellet	5.7	Pt Mesh	$3x10^6$	$4x10^{12}$	$1x10^{-5}$
9 (July/89)	Pd	Ring	18	Pt Mesh	$1.8x10^8$	$1.8x10^{11}$	$1x10^{-2}$
							SBK 2016

Among the nuclear products for which the BARC researchers analyzed, tritium was the dominant nuclear signature. But they realized that both the neutron production and the tritium production were minor compared with the heat. Meanwhile, they received confirmation of their results from Kevin Wolf and John Bockris at Texas A&M University, Tom Claytor at Los Alamos, and other researchers. It was very clear to them that neutrons and tritium were only secondary phenomena.

Political Disruption

The groups at BARC, like everyone else in the world, had reproducibility issues. Like many other researchers, they too initially thought that "cold fusion" was so simple. They soon realized they could not reproduce the results on every attempt. Then, in July, the news reached BARC that the U.S. Department of Energy had deemed "cold fusion" evidence "unconvincing." Following that, scientists at BARC — some of them senior physicists who doubted the results — became more vocal. These doubters challenged the successful researchers to demonstrate the results on demand, but the researchers couldn't do so on demand. (George, 1994) Chidambaram, who was the director of the Physics Group at the time, was one such doubter.

Another Very Cold Fusion Experiment

However, the BARC researchers were also successful until early 1990 with other approaches, including a clever variation of the Frascati experiment, as Srinivasan explained:

> We took the deuterated titanium chips and dropped them directly into a container of liquid nitrogen. Then we took out those pieces and monitored them individually for tritium. It was a tough problem because we had 1,000 small chips with a total of about five grams. We divided them into lots of 20 and put them into a windowless beta detector. Some of the lots gave significant counts. Finally, we were able to show that four out of 1,000 chips had very high activity at the microcuries level.
>
> Those chips are still preserved by us — and they still give this signal. For instance, when Douglas Morrison visited us around August 1990, I showed them to him. The moment we placed one of those chips into the detector, the count rate indicated a very high level of activity, giving a beautiful beta (electron energy) spectrum. I asked him to speculate as to where this beta spectrum could have come from. I even gave him copies of the spectrum. He has never talked about it anywhere or mentioned it in any of his writings.
>
> The more exciting thing about that particular titanium-chip experiment is that not only do only four out of the 1,000 chips have that high activity, but even in those four chips, there are very small hot spots, showing that what is happening is happening very selectively. There is clearly something very special about those sites. This is telling us something very important [about the mechanism], because theoreticians immediately imagine a [metal] lattice which is fully loaded with deuterium, and that what is happening is happening everywhere in the whole lattice. I suspect that it is not the case. ... We came to the startling conclusion that each of those hot spots is the result of a micro-nuclear explosion. (George, 1994)

Michael Ravnitzky, an editor of this book, recognized the value of a set of results like the four out of 1,000 active chips:

> This gives a road map for understanding the microcrystalline or crack structure that produces this effect: Make a large number of chips. Identify which ones produce the effect. Then analyze those spots on those chips to identify the microstructure that best exhibits the effect.

Expert Neutron Measurements

In the summary of the BARC-1500 book, Srinivasan and Iyengar provided a concise summary of their Frascati replication:

> The neutron detection system comprised 24 He^3 counters arranged in a well-like array and had a counting efficiency of 10%. The neutron count rate reached a peak value of 10^5 per 40 seconds as compared to initial background levels of 60 per 40 seconds. The neutron emission phase lasted for several hours at times.

The BARC researchers also analyzed the experiments to better understand the nature of the neutron bursts, Srinivasan explained:

> There is another interesting experiment we did by measuring the probability distribution of neutron counts. We did this to answer the question: In all these cold fusion experiments wherein we see neutrons, are these neutrons being emitted by the sample one at a time or in bursts of two, three, four or more at a time?
>
> In other words, was the neutron emission following Poisson statistics, or was it non-Poisson? This is basic to the mechanism behind it. So we devised an experiment to look for neutrons in 20-millisecond intervals because that is as far as we could go down with our setup. All we did was feed the data out to a personal computer and chop it up into 20-millisecond blocks. We did a statistical analysis and showed

that about 15-20 percent of the counts were coming in bursts of several tens of neutrons at a time. (George, 1994)

Srinivasan had heard that some scientists doubted their claims and had imagined, but not specified, some kind of error. Srinivasan had this response:

> In our experiments, we had two banks of neutron detectors looking at the source, and a background detector bank away from the experiment. We trust our measurements, unless both of these banks near the sample simultaneously decided to behave in a crazy manner or respond to cosmic-ray-caused spallation neutrons, not to the background bank.
>
> In our experiments, the background counter was absolutely stable. I am also aware of the argument that cosmic rays can cause spallation neutrons only in the Pd cathode and so do not give a signal in the background bank. But the size and duration of the detected neutron episode does not support that argument. (George, 1994)

In the summary of the BARC-1500 book, Srinivasan and Iyengar made several important points. They emphasized that many cells gave null results, and those were not included in their tables with the positive results. On the whole, their results clearly demonstrated to them that "cold fusion" was essentially aneutronic in nature. They also concluded that "cold fusion" in electrolytic cells appeared to be essentially a surface phenomenon.

In contrast to some of their nuclear physics peers in the West, they did not allow their preconceptions rooted in prevailing theory to weaken their confidence in the data that lay before them.

CHAPTER 24

Spaghetti Fusion

Success at Italian Nuclear Lab

In April 1989, dramatic news had come out of Italy. The ENEA-Frascati lab had reported "very cold fusion" by loading deuterium into titanium shavings, enclosed together in a stainless-steel cell. Ugo Bardi, a professor of physical chemistry at the University of Florence, told me that he was in Berkeley when he first heard about it. The Americans there, he wrote, gave it a nickname, "spaghetti fusion," alluding to the Italian-made Western films known as "Spaghetti Westerns."

The ENEA-Frascati management quickly organized a press conference and issued a press release, and this news had an immediate impact around the world. It caused the U.S. Department of Energy to take "cold fusion" seriously.

The three lead researchers on the ENEA-Frascati experiment were physicists Francesco Scaramuzzi, Antonella De Ninno and Antonio Frattolillo. Like the researchers at BARC, the Frascati group received no help from Fleischmann or Pons, De Ninno told me. Their story runs from April 1989 through March 1990.

Here are the technical details. The Italians had come up with a different concept to load deuterium into a host metal, one that eliminated all the complexities of electrochemistry. Their idea was to load deuterium gas into titanium shavings and subject it to changes in temperature.

Episode 1

On Friday, April 7, 1989, around 4 p.m., the Frascati researchers began the experiment. A technician had already machined fresh titanium shavings, which had been de-oxidized with a thermal treatment. The cell vessel and gas manifold had been vacuum-tested. They placed 100 grams of titanium shavings into the cell, filled it with deuterium gas, and placed the Dewar flask around the cell.

About 20 cm away, they had placed a BF_3 detector to measure neutron emissions. Unlike the BARC researchers, the Frascati researchers had only a single detector, which made them vulnerable to criticism from people who doubted their observations.

Around 8 p.m., they filled the Dewar with liquid nitrogen and went home. It took a couple of seconds to fill the Dewar, which dropped down to 77 degrees Kelvin. De Ninno told me that the following morning her home phone rang:

> It was Franco. He was in the lab and said that the cell was producing a lot of neutrons. I thought what every scientist thinks when nature fits with his/her ideas: "Wow! I knew it!" I told him, "I'm coming. I will be right there."

They had detected neutron counts of 20 +/- 4 per hour, with a background of 2.3 counts per hour. (See graph) After seeing the startling results, they prepared the experiment for a second run. By 11 a.m., they loaded the Dewar with fresh liquid nitrogen. They waited an hour, but nothing happened, so they filled the Dewar with more liquid nitrogen. At 1 p.m., the second run began producing neutron counts. By Saturday evening, the rates doubled to 40 counts per hour. At 8 p.m., they filled the Dewar with fresh liquid nitrogen again and went home while the detector continued to register neutrons. The gap between hours 31 and 33 indicates a break in the data acquisition.

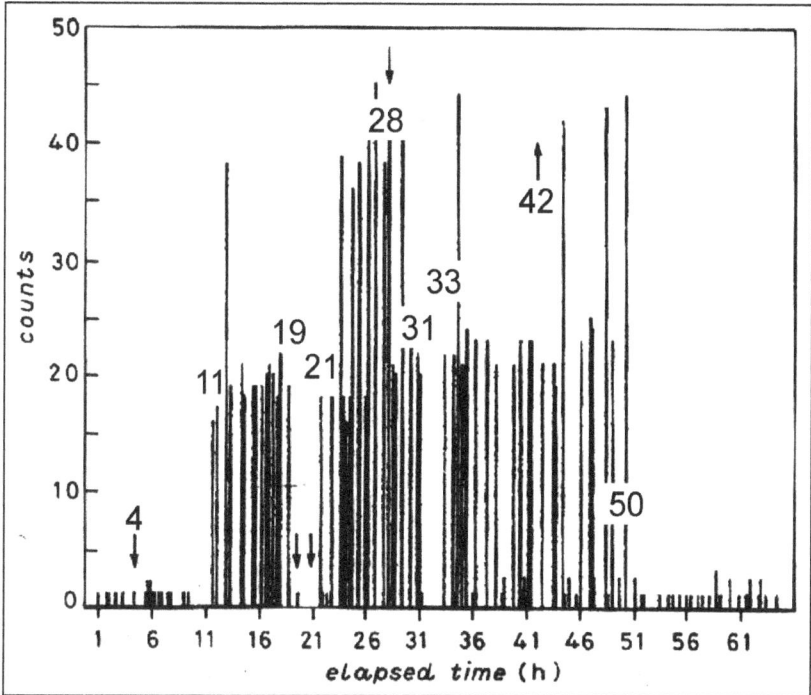

ENEA-Frascati deuterium-titanium experimental episode #1, April 7-10. In the graph, the downward pointing arrows at hours 4, 19, 20 and 28 indicate when the researchers filled the Dewar with liquid nitrogen. The upward pointing arrow at hour 42 indicates when they removed the Dewar from the cell.

On Sunday morning, they were still detecting neutron emissions. They removed the Dewar from the cell and allowed the cell to rise to room temperature. As it did, they continued to measure neutron counts until 6 p.m. They stopped taking data at 8 a.m. on Monday morning.

Although she was optimistic, De Ninno still wasn't sure whether the neutron signals were real. She went to Marcello Martone, the head of the Neutron Detection Division, and asked his opinion. "If you can repeat it," Martone said, "they are real."

Episode 2

So they prepared for their next round of experiments. By April 15, they were ready to begin. They cooled the cell down to liquid-nitrogen temperature, then removed the Dewar and allowed the cell to rise toward room temperature. (See graph) After about three hours, they started seeing neutron counts, but this time they observed a very different emission pattern. The count rate steadily increased for seven hours, at which point the counts peaked and steadily declined.

About 15 hours after the start of the experiment, they moved the detector away from the cell, and, as expected, they saw a sharp drop-off in count rates, followed by a small tail. Based on the efficiency of the counter, they estimated that the cell emitted more than 5,000 neutrons per second during this run.

The results looked convincing, and the effect seemed repeatable. At this point, they prepared a manuscript with their findings. Lab management prepared a press release and organized a press conference.

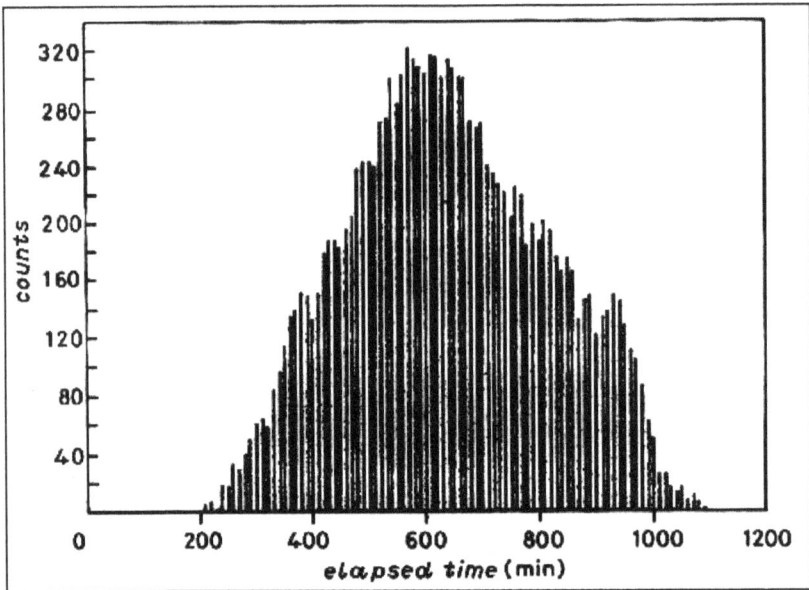

ENEA-Frascati deuterium-titanium episode #2, April 15-16

Episode 3

A few weeks later, they ran another set of experiments and, on May 8, got a third round of positive results but not as strong as the April 7-10 results. They ran tests and checked that neutron detector for several months to see whether they could observe any erratic behavior or malfunctions, but it ran perfectly.

Episode 4

A fourth episode was never reported publicly. Nathan Lewis and physicist Charlie Barnes, at Caltech, knew about it. David Goodstein, a physicist and the vice provost of Caltech, had come to Italy to inspect Frascati's "very cold fusion". Goodstein and Scaramuzzi were old friends.

In a June 30 letter to Lewis and Barnes, Goodstein explained the Frascati experiments in detail. One of the most fascinating facts revealed by his analysis was that each titanium sample produced bursts of neutrons only once. This implied that a crucial unknown material or preparation factor was responsible for the neutron phenomenon.

According to Goodstein's letter, the fourth episode took place during the week of June 19. The Frascati group used a set of four BF_3 detectors, two near the cell and two several meters away. They registered neutron emissions 100 times greater than background.

Garwin Imagines Errors

The initial results in April were so profound that, according to De Ninno, Italian as well as international scientific visitors streamed through the laboratory to learn more and speak with the researchers. One of them was Richard Garwin. On June 9, 1989, after Episode 3 but before Episode 4, Garwin visited the lab and met with Martone and De Ninno. In a three-page follow-up letter to Martone, on June 12, Garwin expressed his doubts about the results even though he didn't point out anything wrong with the experiment.

Garwin wrote that he was by no means sure that the measured bursts were caused by neutrons. He offered no suggestions for improving the repeatability of the experiment.

Garwin began by quoting a newsletter from Douglas Morrison, who, in turn, had quoted a statement given by Yale physics professor Moshe

Gai, who said that all BF_3 counters are unreliable and subject to false counts. Garwin also suggested to Martone that he look at the weather data from April 15-16 to see whether there was unusual humidity that day that might have caused condensation on the wires of the counter and then caused a fault. Here's what De Ninno wrote in 2015 when she first saw the letter Garwin had written to Martone:

> Garwin's criticism about the moisture is nonsense. The cables of the detector are coaxial RG-59 cables with an outer plastic sheath, an inner copper shield, a dielectric insulator and a copper core. They protect against electromagnetic interference and are often used outdoors, as well.
>
> BF_3 detectors have been used for a long time on the Frascati Tokamak machine, and their performance has been recorded for months. They are periodically re-calibrated with an Am-Be source.

Garwin also wrote in his letter to Martone that the Frascati researchers were "obligated" to run specific tests that he (Garwin) decided were required.

De Ninno sent me a September 1990 internal ENEA report in which they performed extensive tests, including those Garwin identified as "obligatory." The biggest concern, of Garwin's as well as of the Frascati researchers, was that the pattern of neutron signals was caused by an electronic-based artifact. They ran all the conceivable tests they could, including checking calibration of the detector with a burst source of neutrons from a microtron accelerator, but nothing showed any malfunction. All facts led to the conclusion that the original signals, which appeared as trains of high-frequency pulses, were from neutron bursts that had saturated the detector.

They also performed tests to check for the influence of temperature and moisture variation, radio frequency generation, ultrasound generation, and instability of the power supply. None of those tests gave false-positive results.

Goodstein Finds No Errors

A few weeks after Garwin went to Frascati, Goodstein went there. Despite the fact that Caltech scientists Lewis and Koonin had, according to the newspapers, disproven "cold fusion" in May, Goodstein was not so quick to dismiss the entire idea.

Furthermore, Goodstein's extremely detailed and technical letter was also devoid of the imaginary errors that filled Garwin's letter to Martone. Lastly, Goodstein's letter also reveals that Lewis had been attempting unsuccessfully to replicate the Frascati experiment. "I'm trying to convince Franco," Goodstein wrote, "to come to Caltech in late July to speak with you directly. He would be at Caltech from July 28 to Aug. 2. Will either or both of you be there to speak with him? Can you arrange for him to give a seminar?"

In 2015, I asked De Ninno whether she remembered anything about such a visit. She did:

> It was 26 years ago!!! I remember our attempt to use more than one neutron detector. This document is very important! It shows that in June there was a genuine scientific interest around this experiment. Later, something began to go in the wrong way: the irreproducibility, of course. But, as a matter of fact, almost all the groups which had tried to reproduce the experiment decided to quit in a couple of months. What had happened to all the groups which have had good results as shown in the hand-drawn neutron graph you found? (Chapter 29)
>
> In March 1990, Franco and I went to Caltech to visit Goodstein. He was quite doubtful about "cold fusion" but very warm with Franco; they have been friends for more than 20 years. Goodstein organized a meeting for us with Koonin. Koonin was very, very doubtful and almost rude with us. He said that "cold fusion" was impossible and the experiments are certainly wrong. On March 20, we were invited at the Los Alamos Laboratory by Howard Menlove

where I gave a seminar. I remember the date precisely: It was my 29th birthday, and I was very excited! There we met Steve Jones. I had the impression that he was not interested in our research, but I can't explain why.

Goodstein has been careful throughout the years to avoid making dismissive comments about "cold fusion" research. He has called it a "pariah field," but that term does not necessarily place blame.

He has been careful to avoid judging his Caltech colleagues for their behavior in Baltimore. Goodstein has often recognized Koonin and Lewis for their role in casting "cold fusion" out of the scientific establishment. It was, and still is, a politically safe position, no matter how the story ends.

In Search of Understanding

The Frascati researchers attempted more experiments later that summer, but the researchers couldn't repeat the four sets of results they had seen earlier. In the first three experiments, they had used a neutron detector that had worked reliably for a decade, and the neutrons had been measured by experts. In the fourth experiment, they used a set of four detectors. There was no obvious reason the results could not be reproduced. There was also no obvious reason the results might have been false.

In October 1989, the Frascati researchers began a second round of experiments, which went until March 1990. They used a detector very similar to a detector used by scientists at the Los Alamos National Laboratory.

The Frascati researchers made a series of 19 runs with a variety of parameters and materials. Ten of them showed a total of 19 neutron bursts. The nine runs that did not produce any neutron bursts did not contain deuterium:

Titanium shavings loaded with deuterium (4 runs, 6 bursts)
Titanium alloy with deuterium (3 runs, 2 bursts)
Titanium sheet with deuterium (2 runs, 9 bursts)
Deuterium without titanium (1 run, 2 bursts)

They could not explain the neutron bursts in the experiment without titanium. I spoke with theorist Lewis Larsen about the last experiment. He and I had spoken several times in the last few years about other experiments performed in Italy by Francesco Piantelli in stainless-steel chambers with deuterium gas. Larsen suggested that the fraction of nickel within the stainless steel could have been responsible for the reactions.

In the second round of experiments, the Frascati researchers also searched for tritium and, in a few cases, found as much as three times the amount of tritium in the final values than in the starting values.

The April 1989 Frascati experiments with deuterium gas and titanium had shown real nuclear evidence. The researchers had not only confirmed the existence of a new class of nuclear phenomena that occurred in metal hydrides but also revealed that the phenomena were not limited to electrolysis or palladium.

Some people believed that, because the experiments could not be repeated on every attempt, the successful experiments had to be illusory. De Ninno saw it differently:

> I believe that "cold fusion" research is like the finger pointing at the moon: Only silly people look at the finger; wise people look at the moon. This phenomenon sheds a new light on the physics of condensed matter and can really be the gate for a new revolution in physics. Don't believe that physics is just the challenge in measuring with higher accuracy what is already known.

Francesco Scaramuzzi (2007) Photo: S.B. Krivit

CHAPTER 25

Nothing to See — Move Along

"Statistical Illusions"

By May 1989, Secretary of Energy Admiral Watkins' three initiatives were well under way. Watkins was getting weekly reports on "cold fusion" activity from the 10 participating Department of Energy national laboratories.

The Department of Energy's Energy Research Advisory Board had a panel of 20 experts who had begun looking at "cold fusion." And the DOE-sponsored "Workshop on Cold Fusion Phenomena" took place in Santa Fe, New Mexico from May 23-25, 1989.

Chrien Report on Santa Fe Workshop

Robert E. Chrien, a physicist at Brookhaven National Laboratory, sent an 18-page report about the Santa Fe workshop to only five of the 20 members of the DOE "cold fusion" panel.

Only a handful of "cold fusion" confirmations were reported at the conference, and most presenters reported failures to confirm. For most of the attendees, who had paid $400 for admission, the journey was a disappointment. Chrien described the dramatic historical perspective of the events leading up to the meeting:

> The ["cold fusion"] claims have produced a flurry of excitement and activity such as has not been seen in nuclear physics since the discovery of fission. ... The experimental

335

situation is, at best, chaotic. Positive and negative evidence abounds. It is quite clear that many experiments have been hurriedly assembled and are meaningless. ... The Texas A&M results are illustrative of this chaos. They are, in fact, interpretable as the best evidence *against* fusion as an explanation of the phenomena.

Appleby at Texas A&M reports excess heat at the level of 40 milliwatts/gm of Pd, and this agrees with the rates of Pons and Fleischmann.

Kevin Wolf at Texas A&M reports a neutron rate of 4/sec ... and tritium production at 3×10^{14} atoms of tritium after 10 hours of electrolysis. ... These reported rates of 10^{-10}, 10^{-13}, and 10^{-23} fusions/deuteron pair/second differ by many orders of magnitude and are strongly inconsistent with a single process as their origin. It seems to me to be fruitless to search for a single cause for such data.

The most intriguing reports concern the observations of bursts of neutrons, detected by moderated ^{3}He detectors. Howard Menlove, of LANL, describes the most convincing evidence, using a set of ^{3}He detectors embedded in polyethylene, developed for the materials safeguards program. Bursts are detected from cylinders containing Pd and Ti under 40 atm of D_2 gas. The cylinders are cooled to liquid-nitrogen temperatures and allowed to warm up.

LANL Confirmation of Frascati

Howard Menlove, a physicist and nuclear engineer working at Los Alamos, reported a replication of the Frascati "Spaghetti Fusion" experiment at the workshop. Menlove was recognized as an expert in neutron and fission physics and gamma-ray spectroscopy. His confirmation now meant that experts in some of the top national laboratories in the world —the United States, India and Italy — had observed evidence for new, unexplained nuclear reactions taking place in low-energy environments.

However, the BARC experiments were not known outside of India until July 3, 1989, when P.K. Iyengar gave a paper at the Fifth International Conference on Emerging Nuclear Energy Systems, in Karlsruhe, Germany. Of the eight papers presented there, most of their authors worked outside the U.S.

Iyengar's friend Hiroshi Takahashi, at Brookhaven National Laboratory, was in Karlsruhe, but the news of the BARC experiments didn't hit the U.S. physics community until Iyengar faxed Takahashi a copy of his Karlsruhe paper on Aug. 23, 1989; he, in turn, faxed it to David Goodwin at the DOE.

Menlove submitted a paper to *Nature* on June 12, but as far as I can tell, it didn't publish there. Instead, it published in the *Journal of Fusion Energy*.

Considering the acrimonious public comments of *Nature* Editor-in-Chief John Maddox and Richard Garwin, *Nature* would have been the last journal to publish any paper that confirmed "cold fusion." The history of the topic includes many instances of papers submitted to *Nature* that were rejected by the editors and not sent out for peer review.

Menlove, however, was able to see past the inconsistency between experiment and theory, as shown by a Reuters news article that quoted him. Reuters reported that Menlove told reporters that he had "no doubt" that a nuclear reaction of some sort had occurred. Menlove said that neutrons had been released, which proved that a nuclear reaction had occurred, although he said the mechanism that produced the neutrons had not been identified.

Menlove, a neutron expert, knew there had to be a nuclear explanation, but he didn't know what it might be, as he explained to Anne Burnett of KUER-FM. "You've got to explain the emissions of neutrons," Menlove said. "We don't know if it's cold fusion, hot fusion, or something other, but I don't know any chemical processes that would give the neutrons."

Texas A&M Confirmations

Electrochemist John Appleby, at Texas A&M University, presented his confirmation of Fleischmann and Pons' excess heat in Santa Fe.

Appleby and Robert Huggins, a materials scientist at Stanford, may have been the only people at the workshop who reported excess heat.

Although Appleby's experiment was far more sophisticated than Huggins', his results had little impact on the consensus at the workshop. Since the 1930s, nuclear science has been the domain of physicists, and only physicists have been considered credible and qualified to discuss nuclear science. A handful of chemists was not going to change this no matter how good their data were.

To their credit, neither Appleby nor Huggins claimed that their excess heat was from fusion. The same was true of Bockris, who was also was at the Santa Fe workshop. As Jerry E. Bishop pointed out in the May 23, 1989, *Wall Street Journal*, only Fleischmann and Pons were attributing the heat to fusion, although none of the chemists could identify any chemical reaction that could account for the level of excess heat. When pressed for explanations, Appleby and Huggins refused to speculate on what sort of mechanism produced the excess heat.

The following day, in a *Los Angeles Times* article titled "Lack of Evidence Fails to Sway Claim on Fusion," Thomas H. Maugh II reported that nuclear chemist Kevin Wolf, at Texas A&M, had measured large quantities of produced tritium in experiments at Texas A&M and a small quantity of neutrons in a cell that produced excess heat. The tritium had been confirmed by researchers at General Motors Corp. and at the Los Alamos National Laboratory.

Without knowing it, Wolf had confirmed the neutron-to-tritium ratio that the groups at BARC had seen. After several hours of electrolysis, Wolf said, the amount of tritium jumped 100 to 1,000 times above the starting level. Wolf, like everyone else, could not reproduce the results on demand.

Illusions and Verdicts

Bill Broad wrote up the event for the *New York Times* on May 24. He wrote nothing about Wolf and his presentation. The next day, Broad published another article on the workshop and reported the general discouraging attitude among the attendees:

Several teams of scientists said at a conference today that their experiments on low-temperature nuclear fusion had produced a few neutrons, suggesting that something curious was going on but nothing that would revolutionize the world's production of energy. [Critics] immediately challenged the reported observations, saying they were statistical illusions.

Harwell Abandons Hope

Three weeks later, on June 15, the U.K.'s Harwell laboratory held a press conference to announce the termination of its cold fusion research. Press conferences were standard practice for everyone involved in "cold fusion." The lead scientist on the project was electrochemist David Williams, a former student of Fleischmann's. According to Williams, none of their results was positive. However, according to Bockris, who had had a conversation with Williams, they did, in fact, obtain positive results. Anne Burnett, of KUER-FM, recorded Bockris as he recalled and narrated the conversation in his testimony before the Utah Legislature:

BOCKRIS: I hear you are getting better results because Stan Pons has told me you're seeing lots of stuff.

WILLIAMS: I don't know where you got that idea from.

BOCKRIS: Well, aren't you then? You're not seeing anything?

WILLIAMS: No, neither heat, nor neutrons, nor tritium.

BOCKRIS: Have you never seen any heat?

WILLIAMS: Well, I mean, only bursts of heat, nothing constant.

BOCKRIS: Well, did you ever see any bursts of neutrons corresponding to the heat?

WILLIAMS: Well, only for short time, but that must be an artifact.

BOCKRIS: Well, did you see any tritium?

WILLIAMS: I used tritiated water.

BOCKRIS: Well, it's curious to start with tritiated water if you want to see new tritium.

WILLIAMS: Yes, well, I started with 17 kilo-Becquerels of tritiated water, and I got 36 after 10 hours.

BOCKRIS: Well, the tritium went up.

WILLIAMS: Oh, yeah, it sort of moves up a bit.

Bockris, like Menlove, did not deny the credibility of the successful experiments' results just because they were poorly repeatable, as he explained to the Utah Legislature. "On his side," Bockris said, "if his bosses are asking him, "Can you consistently produce neutrons?" the answer is "No." And it's "No" for all of us.

In a June 6, 2004, letter to me, Fleischmann wrote something similar about Williams' experiment. He explained that a visual inspection of Williams' heat data reveals evidence for bursts in the palladium-deuterium system but no such bursts in the palladium-hydrogen system.

In 1992, two other scientists also examined the Harwell heat data and confirmed that there were 10 instances of excess heat in their "cold fusion" experiments. (Melich and Hansen, 1992)

I sought out Williams' side of the story and sent Bockris' quote to him. Williams wrote back that he couldn't remember speaking to Bockris about it. He also took great offense at Bockris' statement.

"The supposed quote is a complete misrepresentation of our results," Williams wrote. "Where did this come from?"

I explained where it came from and sent him the audio recording of Bockris speaking to the Utah Legislature. I also sent him part of Fleischmann's letter to me.

Williams sent me a lengthy and testy e-mail in response. Most of it was his explanation of his group's experimental and analytical process from his published paper. (Williams, 1989)

Williams did not deny the comments ascribed to him by Bockris. So I wrote him back and asked for straightforward answers about whether he saw any possible heat bursts, neutron bursts, or heat and neutron correspondences. Williams wrote that he believed the data showed nothing.

CHAPTER 26

Hired Hands

Fleischmann, Pons Versus the University

In the fall of 1990, local newspapers reported Stanley Pons missing, his phone disconnected and his house up for sale. This chapter reveals the precipitating circumstances that took place in the summer of 1989. They include Pons' relationship with the University of Utah and the establishment of a "cold fusion" institute by the university without his and Fleischmann's consent.

As summer 1989 approached, the "cold fusion" conflict faded from view. Through May 1989, "cold fusion" showed up almost daily in the newspapers. From June through August, coverage fell to weekly stories. Fleischmann and Pons had tired of the public spotlight. The Electrochemical Society press conference was the first and last press conference for either of them since the now-infamous moment at the University of Utah, aside from an invitation-only meeting with reporters at a science conference in Salt Lake City in March 1990.

Fleischmann sought refuge at his home in England, but Pons was stuck in Salt Lake City. The personal attacks hit Pons particularly hard; his skin was not as thick as Fleischmann's. He told reporters that, if other scientists continued attacking him, they would hurt his feelings.

For the Benefit of the University

But conflict was hard for Fleischmann and Pons to avoid. In May and June, their troubles came from people in Utah. The local "cold fusion" events in the summer of 1989 were tracked by reporter Anne Burnett, of KUER-FM, who produced a weekly update from April 21 through August 25. The tapes have been preserved at the Cornell Cold Fusion Archive.

Throughout May, the public and the scientific community waited eagerly for the collaboration between the University of Utah and the Los Alamos National Laboratory to begin. The first planned activity of the anticipated collaboration was the test to detect helium in their post-experimental cathodes. This idea had been suggested and then demanded by some physicists who said that this would provide proof of "cold fusion." They, like Fleischmann and Pons, erroneously assumed that the reactions took place within the bulk of the cathodes; therefore, they thought that assays of the cathodes would reveal newly created and retained helium.

In her May 19 broadcast, Burnett interviewed an attorney from the University of Utah. The attorney explained that the negotiations with Los Alamos were going slowly because of the university:

> I can tell you that the patent attorneys are working with the inventors. By the patent policy of the university, the inventions are the property of the university, and they have been assigned to the university by the inventors.
>
> The lawyers right now are busy developing the history of the invention and firming up our patent position.

The university wanted to protect what it thought was a potentially billion-dollar technology.

There is one uncertainty about the ownership of the technology: It concerns the $100,000 paid by Fleischmann and Pons. It is unclear how the University of Utah proposed to handle this with Fleischmann and Pons. A document written by Dietrich K. Gehmlich, a professor emeritus at the University of Utah College of Engineering, says that Peterson committed the university to repay the $100,000 to the two electrochemists. However, Gehmlich's document contains several errors so I am not sure how reliable his information is.

The division of profits appeared to follow a standard protocol, according to science writer JoAnn Jacobsen-Wells, of the *Deseret News*. Thirty-three percent of any royalties would go to the university, 33 percent to the Department of Chemistry, and 33 percent to Fleischmann and Pons. Jacobsen-Wells wrote that Fleischmann and Pons committed to placing any royalties they earned from the patented research into a charitable trust.

A March 23, 1989, University of Utah press release was less equivocal about ownership. It said that "the fusion technology is owned by the University of Utah." However, during the press conference that day, university President Chase Peterson seemed uncertain. Here is the key sentence: "With the ownership, if it turns out to lie with the university of Utah, as we think it will, then we would do all in our power to have this exploited, by ourselves and others, for the benefit of cheap energy with little cost to the world's ecology."

The university was active in courting industrial partners. In Burnett's May 19 radio broadcast, James Brophy, the University of Utah vice president for research, said that 45 to 50 companies had signed nondisclosure agreements with them. The primary and initial benefit to the companies was access to the "cold fusion" patent applications.

Pam Fogle, the university's news director in 1989, told me that, in addition to in-house attorneys at the university, the administration also was represented by lawyers from the state of Utah's Office of the Attorney General, and additional legal counsel from the East Coast, the West Coast, and Texas.

Perhaps there were too many cooks in the kitchen. The University of Utah was unable to come to terms with Los Alamos. On June 14, Los Alamos gave up its attempt to deal with the University of Utah because of the university's lack of response. Los Alamos wasn't necessarily difficult to deal with; the lab had no problem developing a research collaboration agreement with Texas A&M and with Stanford University.

Finally, on June 29, 1989, the University of Utah established its first "cold fusion" research collaboration, with General Electric Corp. GE soon signed a formal financial agreement with the university, but nothing ever materialized from the collaboration.

National Cold Fusion Institute

In the days following the University of Utah fusion press conference, president Peterson began publicly discussing the idea of an off-campus research facility — the National Cold Fusion Institute — to develop and commercialize Fleischmann and Pons' research. But by June, any hope of a national cold fusion institute funded by seed money from the federal government was lost.

State bureaucrats in Utah — members of the newly formed Fusion/Energy Advisory Council — were still hopeful. Administrators from the University of Utah were scrambling to figure out how to leverage state money and capitalize on "cold fusion." Administrators asked the state for $5 million to build and staff the NCFI center.

The administrators assumed that they could rely on Fleischmann and Pons to cooperate with them to bring wealth to the state. As it turned out, that assumption was mistaken.

By May 16, 1989, the University of Utah had filed its seventh U.S. patent application on Fleischmann and Pons' work. The university likely assumed that the information in the patent applications was sufficiently detailed to allow commercial exploitation.

On Dec, 2, 1993, the *Deseret News* reported that the university sold the rights to the patent applications to ENECO, a private Salt Lake City company. In 1997, ENECO gave up its attempts to prosecute the applications and returned the rights to the university, at which time the university abandoned the applications. ENECO filed for bankruptcy in 2008.

The closely held secret of the university's highly publicized National Cold Fusion Institute concept was that neither Fleischmann nor Pons wanted any part of it. Pons did his best that summer not only to remain out of the media spotlight but also to avoid being seen on campus. Electronic messages from John Gladysz, the University of Utah chemistry professor who was exchanging information with chemists in the Caltech Chemistry Department, hinted at the tensions in May and June:

Pons is rumored to be on vacation in England and to have an extensive vacation itinerary this summer. The university here keeps on putting out incredible propaganda.

The state "fusion council" still hasn't released the $5 million. I wouldn't be surprised if several of the members have been strongly influenced by the reports from Caltech, etc. The council had one hearing today, and they will have another on July 11, supposedly with Fleischmann and Pons testifying. Fleischmann and Pons are rumored to be in England writing fusion paper #2, but I think they are basically in hiding, possibly here.

Someone is logging in as Pons on the computer, but this might be someone else. The council wants Pons and Fleischmann to "open the books," but interestingly, the governor is saying, "Dammit all, let's just get on with it and give them the money before Texas A&M beats us out." So it will be interesting to see if Fleischmann and Pons cooperate.

As far as Gladysz's comment about propaganda, the University of Utah administrators were telling Fogle only what they wanted her to know and asking her to disseminate to the news media only what they wanted the media to know. Another electronic message from Gladysz to Caltech confirms this.

"I continue to pick up more intelligence about how all of this happened," Gladysz wrote. "Apparently, Pons, Brophy, Peterson, and the lawyers were the inside ring from time zero for many of the important decisions, press conferences, etc. Nobody outside the circle was ever consulted. It will be quite a while before I think I have a complete picture, however."

At the end of May, local and national newspapers reported that Fleischmann and Pons turned down an invitation to meet with Utah state officials who were considering giving the $5 million to the university. The university administrators must have been livid.

Meanwhile, the support from General Electric had been in concept only. Hugo Rossi, the dean of the College of Science, said in Burnett's Aug. 18 broadcast that the corporation had stipulated that Fleischmann and Pons must be able to demonstrate the production of 10 watts of heat continuously before it made a financial contribution.

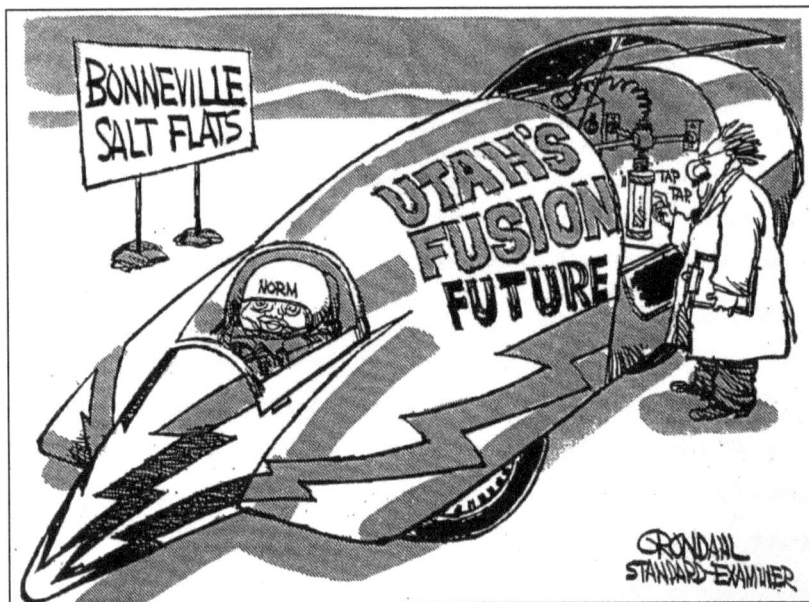

Utah Governor Norman Bangerter depicted in the driver's seat, by Cal Grondahl

On Aug. 7, according to the *Deseret News*, the state Fusion/Energy Advisory Council authorized $5 million of Utah public money for "cold fusion." The motion had been brought to the council by Mitchell Melich, a Salt Lake City attorney who had run for governor. His son, Michael Melich, would soon follow in his father's footsteps as an avid promoter of "cold fusion," though the younger Melich did most of his work behind the scenes. He also engaged in a long-term strategy of gathering information, sometimes overtly and sometimes surreptitiously, about "cold fusion" research. I had many interactions with Michael Melich once I began reporting on the topic in 2000.

The doors opened to the new institute in a nearby off-campus building. But in what must have been an embarrassment to university and state officials, Fleischmann and Pons chose to remain where they had been, in the basement of the university chemistry building. If there was any knowledge or expertise that had not been disclosed in the patents, it remained with them in their basement lab.

For the first director of the NCFI, the university appointed Rossi. University officials had hoped that more than 40 researchers would work there, ideally running 30 to 100 "cold fusion" cells simultaneously.

Directly Opposed

Long before I had researched any of this background, Fleischmann had given me a cryptic clue about the prickly nature of the relationship between the university and Fleishmann and Pons. I had met Fleischmann for the first time at the 10th International Conference on Cold Fusion in Cambridge, Massachusetts, at the Royal Sonesta Hotel.

People told me I could probably find him at the bar, which I did. Between sips of beer, he told me that he and Pons were treated by the university as "hired hands." It sounded demeaning, and I wasn't sure what he had meant. I asked him to clarify but he refused to say anything more then or ever to me about their relationship with the university.

Pons, as far as I know, has also never openly criticized the University of Utah. Nor did he complain to me when I met him. Again, it is John Gladysz who, in an electronic message to Michael Sailor at Caltech, offers insight, through a second-hand perspective from Pons:

> I sat next to Pons at dinner. Over the course of the next hour, you might say I took Stan to the mat. I learned that there is a lot of "institutional responsibility" for all of the screw-ups. That is, the things they came across as poor public relations were frequently attributable to university administrators. There are some real hard feelings toward certain colleagues.

In 2012, I received a copy of a letter that Fleischmann had sent to his close friend electrochemist John Bockris on Feb. 7, 1991. It confirms many of the details I have since independently confirmed and provides additional understanding.

In the following excerpt from his letter, Fleischmann discusses the 1989 period of the NCFI and mentions the first director, Hugo Rossi, as well as the second director, Fritz Will. Fleischmann also mentions the Johnson Matthey company, with which he and Pons had made private arrangements to obtain and use the company's precious metals in their experiments. Fleischmann also mentions Gary Triggs, Pons' attorney:

> While Stan and I were deciding on our course of action, the university made a strong bid to establish an institute. At that time, Stan and I discussed the competence of various people in the university to do the research. It was our belief that the university had no special competence in physics, engineering or calorimetry and that our competence in electrochemistry was not necessarily relevant to the research needs.
>
> The only people who we thought were likely to make a really original contribution were Milton Wadsworth's materials science group, but, against that, we felt that the immediate needs were more likely to be met by collaborating with Johnson Matthey. The line taken by the people in the university was strongly oriented toward engineering, and Stan in particular tried to put a block on that. For one thing, he pointed out that I had probably more competence in the engineering sector than did the engineers, but, of course, this was all to no avail.
>
> As you know, we advised Hugo Rossi against a bricks-and-mortar institute, and our advice was ignored. Stan felt so strongly about this that he wrote a letter to Hugo, and I believe this was the first letter which was not answered. Since then, and as I have again told you on several occasions, we have had no reply to any letter. Here, Fritz is misleading you. The matter has nothing to do with Gary Triggs.
>
> In our letter to Hugo, we told him that we did not want to take part in the institute but that we would nevertheless support his efforts. We told him that this would delay our work by at least six months (that is our rough-and-ready

development toward a demonstration) since we would have to do very careful measurements. Needless to say, Hugo ignores past history.

At various periods, Hugo asked for advice, ignored it — no, in fact, countermanded it. Running up to October 1989, we were appalled by the research effort in NCFl, and we naturally realized that we would be blamed for its failure. We decided that, in that case, we might as well join the effort, and we devised a program together with Milton Wadsworth's group. We were strongly influenced in this by our belief that they are the only people in the university likely to be able to make an original contribution, and we have always been especially attracted to the notion of working with [Raj K.] Rajamani. He strikes us as one of the few people with the necessary background in engineering and mathematics to make a thorough job of the data evaluation, and we devised a strategy that he should make an independent evaluation of the experiments. Hugo Rossi resigned as director, and all this happened round about the time of the NSF-EPRI meeting.

The NSF-EPRI meeting took place in October 1989, but we'll get to that in Chapter 31. During the summer, a few more important things happened between the university and Fleischmann and Pons.

Patents and Perpetual Motion

It was no wonder the University of Utah had such difficulty getting patents for Fleischmann and Pons' "cold fusion" ideas. A June 5, 1989, memorandum from Kenneth L. Cage, the director of Group 220 at the United States Patent and Trademark Office, sheds light.

In the memo to his staff, he alerted them that a large number of applications relating to "cold fusion" were starting to come in.

"There is a possibility that a few applications may slip through," Cage wrote, "without being identified. If one of your examiners should receive an application relating to cold fusion, he or she should check to make sure the words "COLD FUSION" are stamped on the file wrapper.

He also directed his staff to route any action on any of those applications through his office and the Office of the Assistant Commissioner for Patents.

Many years later, on March 27, 2006, five group directors in the patent office distributed another memo to the patent office Technology Center 2800 managers. The gist of the memo was similar to that of the 1989 memo, but this one, a reminder about the "Sensitive Application Warning System," was more detailed. Here are some of the categories identified by the memo:

> Perpetual-motion machines
> Antigravity devices
> Room-temperature superconductivity
> Free energy
> Other matters that violate the general laws of physics
> Applications containing claims to subject matter which, if
>   issued, would generate unfavorable publicity for the
>   USPTO

Both memos are available on the *New Energy Times* Web site.

Falling Apart

In early August, Fleischmann and Pons were making plans to attend the September meeting of the International Society of Electrochemistry, in Japan. Recall that, in April, Utah Governor Norm Bangerter, had made a trip to the Far East, soliciting business investment in Utah from companies in Japan, Korea and Taiwan. His efforts came back to haunt him. The Asian companies went directly to Fleischmann and Pons.

Toyota Motor Corporation, the local Utah newspapers reported, had offered to fund Fleischmann and Pons' travel expenses, on the condition that the two of them speak at a special closed session at Toyota. Pons turned down the Toyota offer.

"Under the circumstances," Pons said, "I don't feel it would be right to accept Toyota's invitation because we are already in industrial collaboration in this country with General Electric Corp. I think it would be a conflict of interest if I were now to accept an invitation to speak at a closed meeting of another industrial company." The idea of Pons getting cozy with the major Japanese corporation must have been disconcerting to people in Utah.

NCFI was going nowhere. By the end of August, only two researchers were on staff. Pons, of course, had no involvement in NCFI at the time, and it was widely reported that he remained cloistered in his basement laboratory on campus. A quote published in a *Los Angeles Times* news story summarized the situation.

"Stan will have a presence and an office here," Rossi said, "but he and Fleischmann are not really interested in the kind of programmatic research we intend to do. We want to find out if there is commercial viability in cold fusion. Stan and Martin are more interested in experimentation."

The *Los Angeles Times* quote confirms one of the essential points in Fleischmann's letter: He and Pons wanted to better understand the science before considering engineering any practical applications.

By the end of September, the press was reporting that the NCFI had failed to duplicate any aspect of the Fleischmann-Pons experiment. Rossi was losing his confidence.

"We have a conference coming up here next February," Rossi said. "If we don't have any papers to present, then this place will be closing up shop. I'm not saying I will do that. I'm just saying I think that is what would happen."

The failure of the institute to produce results by then and the lack of involvement by Fleischmann and Pons must have enraged the Utah legislators who had agreed to fund the institute. The obvious question was whom to blame.

The university had done the heavy promotion and had sold the idea to the Utah Legislature and to the Utah public, all while knowing that Fleischmann and Pons had never agreed to participate. If Utahns believed they had been swindled they were right. If they wondered whether Fleischmann and Pons never had the fusion breakthrough that they had claimed, they were justified in their suspicions. Fleischmann knew how it would turn out: He and Pons would be blamed.

In earlier months, Utahns stood proudly by Fleischmann and Pons in the wake of Nathan Lewis and Steven Koonin's hostile behavior toward them. But now, that support was wearing thin.

In a 1994 documentary film produced by Jerry Thompson for the Canadian Broadcasting Corporation, Pons' wife, Sheila Pons, explained how school had become unbearable for her family.

"Our daughter was in school. At the time, she was 11 or 12 years old, a very sensitive age," Pons said. "Some of the children said to her, 'Your dad's a fraud. Why did he do this to us? Why do we have Utah smeared like this because of your dad?'"

State pride had given way to disappointment and shame, trickling down from adults to their children. The puffery of the NCFI by the university and the NCFI's subsequent failure led to other, more nefarious consequences. Some people concluded that Fleischmann and Pons knew precisely in 1989 how to fully control the experiment (they didn't) and that the two chemists held the secrets to developing "cold fusion" into a golden commercial opportunity.

When the university failed in its attempt to transfer Fleischmann and Pons' knowledge out of the university and into the state-owned institute, some people assumed that Fleischmann and Pons were deliberately withholding the secret to untold riches and the salvation for Utah's financial struggles.

Fleischmann and Pons did not lack funds for their own research by this time, thanks to their longstanding relationship with the Office of Naval Research, an offer from the Department of Energy, and offers from wealthy industrialists.

Some people, insistent on obtaining Fleischmann and Pons' knowledge, resorted to heavy-handedness. Pons began receiving threatening phone calls from people who demanded that he share his assumed secrets.

While Fleischmann and Pons did their best to remain out of the spotlight during the summer, scientists representing but not working for the DOE were doing their best to put an end to "cold fusion."

CHAPTER 27

A Tale of Two Telescopes

ERAB Panel Faces Competition

Two independent U.S. groups, both sponsored by the federal government, evaluated "cold fusion" research in 1989. One group viewed "cold fusion" as the biggest scientific mistake of the century, and the other group saw it as a possible new field of science.

The first group of scientists was sponsored by the Department of Energy. The second group of scientists was sponsored by the National Science Foundation (NSF) and the Electric Power Research Institute (EPRI).

The three-day NSF/EPRI workshop took place in October 1989. Although the workshop was well-documented in its 711-page proceedings, published copies are extremely hard to find, and only 100 were printed. Several significant aspects of this workshop have never been mentioned in historical accounts of the "cold fusion" controversy.

The Department of Energy "cold fusion" review took place between April and November of 1989. These chapters represent, to my knowledge, the first public discussion of the inner workings of that review. Chapters 27-32 will explore the history of these two group efforts.

These chapters provide a rich perspective from which to view human reactions in the context of a potentially major scientific breakthrough.

DOE ERAB Cold Fusion Panel

On April 24, 1989, Admiral James Watkins, the U.S. secretary of Energy, had directed the Department of Energy's Energy Research Advisory Board (ERAB) to convene an independent panel of experts to evaluate "cold fusion." Watkins had his own source of information about "cold fusion": the 10 DOE national laboratories participating in "cold fusion" research. But his goal with the ERAB board was to authorize an independent evaluation of "cold fusion" on behalf of the Department of Energy, as announced in the department's April 21 press release (Chapter 14).

The "cold fusion" panel, formed specifically for that purpose, was managed by the ERAB, a standing committee that advised the Secretary of Energy on energy-related matters. Until 2015, only the members of the panel and some DOE staff members knew a crucial fact about the ERAB "cold fusion" panel report: The conclusions and opinions expressed in the report did not represent the DOE; they represented only the majority of the panel members.

Two of the panel members were DOE staff members: David Goodwin, the panel technical advisor, and William Woodard, the secretary of the panel. Among several hundred documents about the ERAB panel in the *New Energy Times* Richard Garwin Cold Fusion Archive, all documents from Woodard are perfunctory. Nothing from Woodard contains any scientific opinion or evaluation or indicates that he had any effect on the panel's opinion, process or outcome. Woodard, however, was enthusiastic about "cold fusion."

Admiral Watkins showed no sign of doubt or cynicism about "cold fusion." To the contrary, his April 24 letter shows that he took the phenomena very seriously. His academic training in mechanical engineering rather than in nuclear physics may have contributed to his open-mindedness. He was also a colleague and close friend of Admiral Hyman Rickover, the man who led the Navy's pioneering nuclear-powered submarine program during the 1950s-1970s.

Watkins was a pragmatist who had been exposed to and was familiar with the real-world applications of nuclear energy. He was certainly no ivory-tower academic. His open-mindedness may also have been motivated by his desire to ensure that neither the Department of Energy nor, by implication, the White House would be surprised by a potentially new energy or defense development.

For 26 years, the public perception of the 1989 DOE "cold fusion" review has been that the panel's negative conclusion represented the DOE's opinion: "cold fusion" was worthless as a possible new source of energy. This is not the case.

The DOE, at least as represented by Watkins and Goodwin, was genuinely interested and hopeful. It was the majority of the panel members, particularly the physicists such as Garwin, who countermanded Watkins' goal of an objective analysis of the subject.

DOE Santa Fe Cold Fusion Workshop

Watkins' April directive to Los Alamos to organize a workshop on "cold fusion" in May in Santa Fe, New Mexico, had little impact on the "cold fusion" controversy partly because the Santa Fe workshop came on the heels of the devastating Baltimore and Los Angeles meetings by only a few weeks.

By the end of May, most scientists had failed in their own replication attempts and arrived in Santa Fe frustrated and angry. The few who reported successes were drowned out by the overwhelming numbers of failures.

The opening comments on May 23, 1989, from Norman Hackerman, a chemist and Scientific Advisory Board chairman of the Welch Foundation, characterized the nature of the "cold fusion" debate:

> In sponsoring the present workshop, the Department of Energy hopes to provide a forum in which information can be exchanged among active researchers in the field of cold fusion: physicists, chemists, materials scientists, metallurgists, etc. In fact, it is this multidisciplinary facet of cold fusion research that makes the field so fascinating and, at the same time, so prone to controversy.

Watkins' Directives to the Panel

On April 24, 1989, Watkins sent a letter to John H. Schoettler, the chairman of the ERAB. He charged Schoettler with three tasks:

1. Review the experiments and theory of the recent work on cold fusion.
2. Identify research that should be undertaken to determine, if possible, what physical, chemical or other processes may be involved.
3. Finally, identify what research and development direction the DoE should pursue to fully understand these phenomena and develop the information that could lead to their practical application.

Watkins did not charge the ERAB with deciding whether the research might lead to practical sources of energy. Watkins did not ask the panel members to decide whether the evidence for the new nuclear process initially termed "cold fusion" was persuasive.

The ERAB panel, directed by nuclear chemist John Huizenga performed task #1. However, it did not perform tasks #2 or #3. Instead, the panel rejected the scientific validity of the phenomena. Once that happened, the panel didn't need to recommend to the DOE how to pursue research and development that could lead to practical applications.

Nuclear chemist Glenn Seaborg, as mentioned in Chapter 13, said in a 1995 lecture that he told President George H.W. Bush how to respond to the question of "cold fusion."

"You're going to have to create a high-level panel that will study it for six months," Seaborg said, "and then they'll come out and tell you it's not valid," and that's what he did.

The Selected Experts

After ERAB chairman Schoettler received his orders from Watkins, Schoettler called Huizenga and asked him to chair the panel. Like Seaborg, Huizenga was a nuclear chemist. He was a professor at the University of Rochester, a major U.S. thermonuclear fusion research center. Like Garwin, he had worked on the hydrogen bomb program. Huizenga died in 2014, and, according to his obituary at the University of Rochester Web site, he was best remembered — proudly — for his role in leading the ERAB panel to "conclude that there was no convincing evidence to support the claims of achieving cold fusion."

Huizenga was a natural choice because he was already a member of the ERAB. He also shared the same perspective about "cold fusion" that Seaborg did, as Huizenga wrote in his 1993 book *Cold Fusion: The Scientific Fiasco of the Century*." My initial feeling was that the whole cold fusion episode," Huizenga wrote, "would be short-lived and that it would be wise to delay appointing such a panel."

The panel's threadbare four-page interim report, which Huizenga thought would suffice, is evidence that he thought that a review of "cold fusion" would be a waste of time.

By the end of May, most of the panel members had been selected, according to Robert Park, the spokesman for the American Physical Society. For many years, Park wrote a weekly newsletter about topics of interest to the membership of the society. His newsletters were always witty and entertaining, and often sarcastic and caustic. "Cold fusion" was a popular topic for Park, who called it "voodoo science." In his May 19, 1989, newsletter, he told readers about the newly formed ERAB panel and wrote that Watkins wanted an interim report by August. "The final coroner's report," Park wrote, "is due Nov. 15."

Despite Huizenga's reluctance, Seaborg and Schoettler convinced Huizenga to form the review panel and carry out its task. Huizenga's first choice for a co-chair, as he explained in his book, was Mildred Dresselhaus, also an ERAB member, and a professor at MIT, another major U.S. thermonuclear fusion research center. But her time limitations did not allow her to accept the invitation as co-chair.

Nevertheless, Huizenga and Dresslhaus together selected the other candidates for the panel.

John Huizenga image courtesy University of Rochester Libraries, Rare Books and Special Collections.

For the co-chair, Huizenga selected Norman Ramsey, a professor of physics at Harvard and a Nobel prize recipient in physics. Ramsey agreed to serve as co-chair but only during the initial months, because his travel plans in the late summer conflicted. Of the documents in Garwin's archive, none is from Ramsey. He was not part of any of the topical sub-groups, but he did attend the final panel meeting on Oct. 30-31, 1989. Clearly, despite the equal titles, Huizenga was directing the panel.

The panel included other notable scientists. One was William Happer Jr., a professor of physics at Princeton University, another major U.S. thermonuclear fusion research center. Happer was chairman of the Advisory Council of the Princeton Plasma Physics Laboratory and had said of Fleischmann and Pons, "just by looking at these guys on television, it was obvious that they were incompetent boobs." (Happer, 1993)

Another member was Allen J. Bard, a professor of chemistry at the University of Texas, who made one brief attempt to replicate the Fleischmann-Pons experiment and failed. The panel included Steven Koonin, who delivered the brutal public "incompetence and delusion" comment in Baltimore about Fleischmann and Pons. Mark Wrighton, professor and head of the MIT Chemistry Department, was on the panel, too. As discussed in Chapter 9, he had tried the Fleischmann-Pons experiment for just a few days, failed, and immediately announced his failure to the press. A total of 22 scientists were on the panel, apart from the DOE staff members. Four of them, chemists Bard, Larry Faulkner (University of Illinois), Barry Miller (AT&T Bell Laboratories), and Mark Wrighton (MIT), had expertise in electrochemistry, according to Elton Cairns, who responded to my e-mail inquiry. Cairns was the president of the Electrochemical Society in 1989.

Appendix F contains the full list of panel members, the sub-groups, meeting dates and site visits. Of eight laboratory visits, five were conducted by single members of the panel or a pair of members. Panel members visited only two laboratories as a group: the University of Utah and Texas A&M University. For unknown reasons, some members visited the Caltech laboratory as part of the ERAB panel, despite the fact that Caltech had found nothing. Information on the laboratory visits comes from both the final ERAB report and documents in the Garwin archive.

NSF/EPRI Workshop Planned

Independent of and parallel to the ERAB cold fusion panel, the National Science Foundation (NSF) and the Electric Power Research Institute (EPRI) were also preparing to assemble a group of scientists to assess the "cold fusion" phenomena.

EPRI had been providing grants to some of the most successful groups working on "cold fusion" from the very beginning. EPRI staff members were well-informed about the details of the experiments and the results.

Unlike academicians working in federally funded thermonuclear fusion research, EPRI's staff members and member constituents were more interested in new ideas. EPRI's mission was all about practical electric power, and if there was a better way to make it — or if there was a possibility of a disruptive technology — they and their 600 utility-industry member companies wanted to know.

When the news of "cold fusion" broke, EPRI told several of its sponsored researchers that they could spend unlimited funds on "cold fusion" research. A corporate video from May 24, 1989, called "Technology Update — Cold Fusion," featuring Joe Santucci, in EPRI's Nuclear Power Division, shows that, when most scientists were sniggering at "cold fusion," EPRI was taking it seriously.

A May 23, 1989, press release from EPRI summarized the results of its four groups of sponsored "cold fusion" researchers at Texas A&M. Excess heat had been produced by nine of 20 palladium rods used in calorimetry experiments. Tritium had been produced at levels up to 10,000 background. Bursts of neutrons had been detected intermittently in one cathode.

Two weeks later, on June 9, 1989, Thomas R. Schneider, a senior science advisor for exploratory research at EPRI, sent a letter to Paul Werbos, the program director at the National Science Foundation, and agreed to Werbos' request to co-sponsor a research and development workshop with the NSF on "cold fusion."

They selected physicist C.W. "Paul" Chu, at the University of Texas, Houston, and John Appleby, an electrochemist at Texas A&M University, as co-chairmen. Chu was famous for his and his colleague Maw-Kuen Wu's ground-breaking discovery of high-temperature superconductivity in 1987. Schneider and Werbos wanted to avoid media coverage and did not make any efforts to get media attention in advance of the workshop. They had hoped to schedule the workshop for the end of August; however, scheduling conflicts among the invited participants delayed the workshop until Oct. 16-18, 1989.

Meanwhile, the ERAB "cold fusion" panel members had begun their task. The panel's meeting took place during the DOE Santa Fe workshop in May.

Charting Their Own Course

ERAB Panel Disregards Directives

The ERAB panel members knew of confirmations that had been reported at the Santa Fe workshop, as discussed in Chapter 25. They heard about other confirmations there, as well.

Charles D. Scott, a senior corporate fellow in the Chemical Technology Division at the Oak Ridge National Laboratory, and his colleagues reported that one of their "cold fusion" experiments produced excess heat. "There has been one apparently anomalous neutron flux measurement," the researchers wrote, "and periods of up to 12 hours of apparent excess energy. None of these results have been reproduced, nor can they be explained by conventional nuclear or chemical theory."

On May 23, the first day of the workshop, panel members also heard the news (from a press release) from the Manne Siegbahn Institute for Physics in Sweden. They had confirmed neutron counts 10 times greater than background in their Fleischmann-Pons-type experiment.

First ERAB Panel Meeting

On the evening of May 24, 1989, a majority of the ERAB panel members sat down for two hours and determined how they would accomplish their task. Should they perform site visits? Where and by whom? Should they separate into sub-panels?

The Garwin archive includes a notable document, a description of this meeting written by Dave Goodwin, the DOE technical advisor to the panel. Goodwin's academic training was in physics and engineering. At DOE, his task was to assist the associate director of high energy and nuclear physics. He, like Watkins, also had experience in the business world, having worked for three commercial nuclear power plants.

Here's why Goodwin's summary is so profound: He assumed that the reported experimental confirmations and data were real. With rare exceptions, all other documents from panel members in the Garwin archive assume that all the results were mistakes.

In addition to listing the successful replications reported from several countries, Goodwin questioned the failures and listed eight possible explanations for them. No other document in the Garwin archive contains any instance in which Garwin or his fellow panel members questioned failed experiments. Rather, the panel members assumed that null results were the correct answer to the question of "cold fusion." Consequently, they focused on searching for errors or artifacts that would explain away claims of positive results. In addition to receiving Watkins' instructions, the panel received Goodwin's instructions.

They were to:
1. Provide a general summary of the basic experimental
   research
2. Identify details of the experimental and theoretical
   research
3. Recommend a general level of basic research to be
   considered for the next fiscal year.

Crucially, there were no instructions from Watkins or Goodwin that directed the panel to:
1. Decide whether the reported results were valid
2. Judge the experimental results based on any assumed
   theoretical process
3. Attempt to disprove the claims of the successful
   experiments

That is the course of action, however, that the panel took. Watkins' April 21 press release made no assumption about the underlying process:

"The origin of any heat released has not been established, be it nuclear, chemical, mechanical or another process. Similarly, a mechanism for production of a fusion reaction, if any, at room temperature in solids has not been established."

By the time the panel began its work, most of the scientists who had reported successful confirmations were not claiming that the underlying process was fusion. The vast majority of news stories in April and May of 1989 show that the scientists who had reported positive results refused to speculate on the underlying process, let alone attribute them to fusion. In contrast, the ERAB panel members approached their task with an assumption that the underlying process was, as Fleischmann and Pons had claimed, deuterium-deuterium fusion.

The panel members either failed to recognize or acknowledge that the experimental data looked nothing like fusion. They ignored an alternative possibility, that the collection of experimental data constituted evidence for a new, previously unrecognized type of nuclear process that looked nothing like the results of fission or fusion.

When I later asked panel member Garwin why he didn't acknowledge the results as evidence for an unexplained nuclear process, he told me that he judged the data according to the theory of fusion simply because Fleischmann and Pons had called it "fusion."

All documents from Goodwin and Watkins make clear that the two men were open-minded and eager to explore "cold fusion" as a possible new source of energy. It was not the Department of Energy that dismissed "cold fusion" in 1989. Rather, it was the panel members who directed, managed and determined the outcome of the review.

In a perverse way, Watkins got what he asked for: a group of scientists to independently review the subject. Not only did they independently review the topic, but they also independently established their own objectives and took the review in another direction.

Menlove's Neutrons Withstand Mockery

Even though there was evidence of a new kind of nuclear process, most of the panel members had difficulty seeing the data before them.

Douglas Morrison, although not on the panel, provided a simple example of this blindness in his June 1, 1989, newsletter. He wrote about Menlove's neutron results, which were well known by the panel members:

> Menlove's normal random neutron emission is 0.05 to 0.2 n/s while the bursts are from 10 to 300 n/s. The bursts last less than 100 microseconds. The neutron detector consists of 18 helium-3 tubes and has a high efficiency of 34%. *The dummy cylinder gave no counts over two hours.* The active cylinder was followed for longer periods, up to 25 hours, but the burst only occurred in the first 40 to 80 minutes. [emphasis added]
>
> The result looks very significant, but the experiment has a major problem. I asked if they had done the control experiment with H_2 in place of D_2, and the answer was, "No!" Now most people have said that one should not accept experiments where the most elementary checks have not been made.

Morrison, as was typical of him, had used the rhetorical device of mockery ("most elementary") in his last sentence to marginalize the result and distract readers from the actual data. Menlove, as Morrison so clearly stated, had performed a parallel experiment with an empty dummy cylinder, which gave no neutron counts. This was a control experiment, although it was not the control that Morrison would have used. Morrison was fixated on the deuterium-deuterium fusion theory; Menlove was not.

Morrison's dismissal was groundless, and here's why: If Menlove did an experiment with H_2 and obtained neutrons, Morrison would then have been left with the extraordinarily difficult task of explaining what sort of freakish act of nature could have generated the neutrons.

Any neutrons measured from the metal shavings in the active cylinder, whether filled with hydrogen or deuterium gas, would have indicated a novel scientific phenomenon so long as Menlove did not place a miniature nuclear reactor or a radioactive isotope in the cylinder.

Menlove recognized Morrison's shortsightedness, but he ran tests a few weeks later with hydrogen gas, anyway. He got no neutrons with hydrogen and reported this at the NSF/EPRI workshop in October.

Happer Visits University of Utah

A few days later, on June 2, panel member Will Happer went to the University of Utah to see Pons' work. Pons did not have a cell that was generating excess heat at the time. Happer made no effort to hide his disdain in his report to the panel:

> I would be happy to endorse cold fusion if I were given the opportunity to see a working cell and verify the calibration of the heat flow measurements. I think this might take about a week of my participation in the experiments. Should the excess heat really be present and due to nuclear fusion (that is, not of chemical origin), I volunteer to join my committee members in carrying Pons in a sedan chair to Stockholm for his Nobel Prize, and I would rejoice to see him become a trillionaire for a discovery of such benefit to mankind. On the other hand, I object to whitewashing a religious cult under the aegis of ERAB.

Garwin Tempers His Bet

Garwin, however, as shown in his June 9 message to his IBM colleague James Ziegler, had a brief moment of uncertainty about the new science even though he had placed his bet against it in April. He wrote to Ziegler that he heard about the ENEA-Casaccia replication of the ENEA-Frascati experiment. Scaramuzzi had mentioned it at the Santa Fe workshop.

The Casaccia researchers performed their titanium-deuterium gas experiment differently from the Frascati researchers. They also got much higher neutron counts. "They had neutrons for five minutes," Garwin wrote, "with a peak rate of 7,000 counts/minute. With an efficiency of the counter of about 0.5%, this was a source rate of 1 million per minute, peak."

The Casaccia researchers ran the experiments a second time and got 1,800 counts/minute. A third run registered 7,000 counts. This is the only instance I have ever seen in which Garwin was motivated to do a "cold fusion" experiment. "This is sufficiently quick and clean that I think we should do it," Garwin wrote. No records in his archive mention whether IBM made such an attempt. According to a rumor Garwin heard on July 27, the Casaccia researchers' BF_3 detector had malfunctioned and they had retracted their claim. I have been unable to locate additional details on the rumored retraction.

Garwin had criticized "cold fusion" as unscientific because the experiments were poorly reproducible and the results didn't match prevailing theory. Nevertheless, when he saw an experiment that appeared easy to do, suddenly the research was real enough for him to consider attempting a replication.

Wolf and Bockris: Only a Nuclear Process

Members of the panel were aware of the tritium reported by Kevin Wolf at the Santa Fe workshop. Wolf was a nuclear chemist who worked in the Cyclotron Institute at Texas A&M University. His encounters with the strange phenomena of "cold fusion" took him on a convoluted journey. He initially reported seeing tritium, but he later attempted to discredit the validity of his own data.

Electrochemist John O'Mara Bockris worked in the Texas A&M Chemistry Department. He and Wolf worked together on some "cold fusion" experiments, and Wolf ran some experiments independently.

Among the documents in Garwin's archive is a June 12 fax from Hoffman that apparently went to all the panel members. The fax contained a June 9 manuscript from the Texas A&M researchers, the summary of which could not been clearer.

"We describe the observation of tritium in seven out of 11 electrochemical cells," the Texas researchers wrote, "at levels which could not be produced by any process other than a nuclear one."

The experiments produced tritium at levels hundreds to thousands of times greater than background. The final tritium values in these experiments compared with the starting values were, respectively, 380x, 49,000x, 1,200x, 37,000x, 330x, 7,600x, and 630x. Anything above 5x was scientifically significant. In a June 9 letter that Bockris wrote to his colleagues, he contrasted the difficulty of dismissing the tritium with the ease of dismissing the excess heat. "As far as the chemical explanations of the heat are concerned," Bockris wrote, "of course we are just waiting for someone to make one. They will not explain the tritium."

Bockris sent a copy of manuscript to Garwin, and the panel made plans to visit Texas A&M on June 19. Bockris accurately anticipated the response from the physicists. In advance of the panel's visit, Bockris sent the panel members a letter on June 12 conveying his thoughts on the nature of scientific discovery, new paradigms, and human behavior.

Despite his reputation for being a maverick, his letter reveals not only his wisdom about the nature of scientific revolutions, but also his cautious approach to the "cold fusion" research. Few written accounts from 1989 describe the character and context of the "cold fusion" controversy better than his letter. I have reproduced most of it here:

To "Cold Fusion" Visitors:

In the following, we seek to avoid misunderstandings about the work on electrochemical Cold Fusion which is going on in three sub-groups at Texas A&M University.

1) We are interested in the experiments reported by Fleischmann and Pons, and by Jones, which mention cold fusion obtained by electrochemical confinement.

2) We take the attitude that the presence of cold fusion in the experiments carried out by these workers is unproven.

3) Our attitude is to stress experiment. We seek to find out whether there *are* neutrons evolved from palladium electrodes under certain circumstances; whether tritium *is* produced during deuterium evolution at palladium electrodes; and whether the sometimes observed excess heat can be replicated in our laboratories.

Of course, we are interested in attempting to bring the reproducibility under better control.

4) When we have obtained reproducibility in the region of >50%, and can instruct others how to do the experiments with the same success rate, then we shall investigate the dependence of heat evolution, neutron production and tritium evolution as a function of the variables such as overpotential, metal substrate, D/Pd ratio, dislocation density, dendritic promontories, etc.

When we have established some of these dependencies, perhaps in a year, we shall then have a basis on which to decide if the new phenomena originate in nuclear processes.

5) We are particularly unenthusiastic in the discussion of the application of present theories of fusion in plasmas to the idea of fusion in electrochemical confinement because we think that the difference of conditions, particularly in respect to screening by electrons of deuterium-deuterium interaction, is an extreme one, and that it has not yet been properly investigated theoretically. Our attitude is that we may be in an emerging area of science and that, in such situations, experiment usually molds theory to fit it.

Historically, when new science is emerging, it is often reviled and denigrated until the new paradigm is accepted. It is too early to say whether this is the situation in this field.

6) At the time of writing, the phenomenon is less than three months old. Two or three years ([after research with] 5-6 Centers, 100 people) will be the right sort of time to think of in order to make a decision as to whether it is worth big money. The idea that a number of meetings are already planned, and even decisions made up on the basis of happenings at them at this time, appears to us to be unwise, partly because of the emotional outbursts by physicists which have occurred at some of them and the great negativity widely shown, but mainly because of the small degree of knowledge among us all.

Although we welcome criticism, we believe that

spending a great deal of time in angry condemnation of the phenomena we are investigating is not a good way to further understanding of new phenomena which understandably exist. We would rather tell you in a relaxed way about our results and compare them with the positive results of others in various parts of the world. We believe it is agreement among scientists, particularly between those in various countries, which eventually decides what is regarded as "truth" for a few decades in a field.

We think the new (and shaky) "facts" should be isolated from comparison with the older theories until the facts are firm and agreed upon — at least to a good degree.

8) About negative results: We think that, in attempts to verify newly claimed phenomena, negative results have much less value than positive ones. Negative results can be obtained without skill and experience.

It has always been the anomalies which can be seen in science which give rise to the new ways of thinking which cyclically invade the sciences. The constant reiteration of the old way, particularly with the great anger and emotion as we see among our colleagues and visitors, has not been the way that changes in scientific attitudes have come in the past.

Therefore, when persons tell us that they have carried out the electrolysis of deuterium evolution in palladium and see nothing new, particularly if, as is usual, they are furious about it, have spent little time on it, and have little experience as to how to do experiments of the type named, we tend to discount their contribution.

This is particularly so because the phenomena under consideration are undoubtedly elusive. Added to this is the fact that the effects — when they indeed turn on — are difficult to find in electrodes as small as 1- and 2-mm diameters (quickly chargeable), and can only easily be detected (when they display) in most calorimeters when the size of the electrode is something in the region of 4-6 mm. However, a 6 mm electrode takes 72 days to charge before the experiment can begin.

Thus, as we are now less than 72 days from the announcement, to start experiments it will be necessary not only to charge electrodes but to gather equipment of various kinds both electrochemical and nuclear, to say nothing of super-pure Pd rods. It is remarkable that those who were not already working in electrochemistry before the announcement was made could have made experiments at all, let alone gotten results upon which the national policy in funding is to be founded.

Most of the experiments in which negative results have been obtained have come from laboratories which have little record of research in physical electrochemistry or, when in a tiny number of cases the laboratories were electrochemical, little experience in nuclear measurements.

The remaining pages of Bockris' letter listed his summary of the most common errors that he thought researchers had made in their experiments. Although he was clear that that he didn't know all the required parameters, properties, and conditions, he precisely described a dozen common errors that were guaranteed to produce null results. His letter reveals many subtleties and complexities in electrochemistry that no physicists would typically have known. "Finally," Bockris wrote, "there is no doubt that irreproducibility is the bane of these experiments. We are looking increasingly toward the concept that the phenomenon occurs at the surface rather than in the interior."

The idea of "cold fusion" as a surface reaction was remarkably astute so early in the field's development; this topic comes up later. A few days later, on June 20, Bockris wrote to David Thompson, a scientist at the Johnson Matthey Technology Center. Fleischmann and Pons had an agreement with Johnson Matthey and were getting their precious metals from the company. Bockris described to Thompson his perceptions of the underlying causes for the general fear of "cold fusion." He identified three general categories: 1) reluctance to disturb established lines of federal fusion funding, 2) fear of non-conformity, and 3) paradigm paralysis (discussed more in Chapter 33).

Even though researcher Charles D. Scott presented results at the Santa Fe workshop, someone else at Oak Ridge had also seen positive results. Bockris explained this to me in a May 5, 2004, letter:

> One of the scientists at the Santa Fe workshop, who was from a government lab, asked if he could have a private word with me. We went into an empty office and shut the door. He produced graphs from his briefcase and said, "You see, I have results which are positive and prove the effect. But I showed them to my boss on the way over in the plane, and he told me I mustn't present anything like this. I would have to destroy the results and never mention them."

The Panel Investigates Texas A&M Claims

The panel members knew they had to go to Texas A&M University to see what was going on with its "cold fusion" experiments. The visit was scheduled for June 19. As the date approached, Bockris and Garwin engaged in a duel of words by fax machine. They were aware that any significant levels of tritium above background, taking into account natural enrichment that occurs during electrolysis, would be unequivocal evidence of a nuclear reaction. Tritium exists in trace amounts in heavy water. During electrolysis, a greater proportion of deuterium separates from heavy water than does tritium and leaves a disproportionately greater quantity of tritium in the electrolyte.

Electrolytic Tritium Enrichment

Panel member Bigeleisen explained this in detail in his Sept. 7 draft of the tritium section of the final ERAB report. "The amount of enrichment," Bigeleisen wrote, "is primarily a function of the amount of water electrolyzed for a given type of cathode. It can reach a factor of five when 95% of the initial charge of water is electrolyzed." The Bockris group had observed levels of tritium hundreds to thousands of times greater than background by this time, in seven experiments. There was no question about these results; they were well beyond the levels that electrolytic enrichment could attain.

Garwin's Problems

Garwin found no problems with the tritium measurements. Nevertheless, he was determined to find a way to challenge the tritium data, as his June 14 fax to Bockris revealed. "I have two major problems with your June 9 paper," Garwin wrote."

His first problem was that Bockris' group had not observed the correct number of neutrons relative to tritium, according to conventional theory. His second problem was that he knew of no mechanism that could provide enough energy in the room-temperature experiments to overcome the Coulomb barrier. "That has *always* been the problem, and it remains the problem," Garwin wrote. "Please let me know your views on these remarks."

Bockris admonished the panel members to momentarily set aside older theories. "I don't think these matters will be resolved for a year or two," Bockris wrote. "After the facts are known and these nuclear particles are connected with the heat (if such a connection is made), that is the time to think about fusion and theorize."

Bockris told Garwin that they had considered the possibility that the tritium had been placed in the cells illicitly. However, Bockris made it clear that the probability that someone had placed tritium — for each data point — in all seven of the cells, over a period of weeks, was so far-fetched that Garwin's speculation was approaching a conspiracy theory.

Garwin's explanation to me for his dismissal of the experimental data just "because it was called cold fusion" doesn't stand up to scrutiny. Garwin knew on April 12, when Fleischmann spoke at the Erice Fusion Forum workshop, that the results were inconsistent with the accepted theory of deuterium fusion. Fleischmann could not have been any clearer about the disparity. "This is an experimental observation," Fleischmann said. "If you predict what the neutron rate should be, it's between 10^8 and 10^{10} times higher than the rate which we observe. If we are right about this, ... then there must be another decay channel."

The results didn't match the D+D fusion branching ratio; however, anybody who understood fundamental concepts of nuclear science knew that neutrons and tritium could be produced only by nuclear reactions, no matter what Fleischmann and Pons called it. Any neutrons or tritium produced in electrochemical cells represented new science.

Closed Minds

On June 21, Bockris wrote to panel member Jacob Bigeleisen. "I have experienced quite a few commissions of inquiry in my time," Bockris wrote, "but I never found one which had its mind made up beforehand as much as this one."

Another report from Goodwin, distributed to the panel members on June 21, summarized the evidence the members had seen in Texas: excess heat, various control experiments, measurements of neutrons and tritium. The tritium samples had been re-measured at General Motors and at three national laboratories: Argonne, Los Alamos, and Battelle.

The most fascinating thing Goodwin mentioned in his report was a secondary ion mass spectroscopy (SIMS) analysis of Charles Martin's and John Appleby's palladium cathodes at Texas A&M. The SIMS analysis had been done at Lawrence Livermore National Laboratory. In Martin's palladium cathodes, they found increased levels of carbon, silicon and oxygen. In Appleby's palladium cathodes, they found iron, chromium, magnesium, calcium and silicon.

Despite these anomalies, the confused and upset Martin left the field thinking that none of his experiments had been successful. In the end, he concluded that his good friend Pons had misled him, and he turned on Bockris and accused him of faking his tritium.

On July 5, Bockris wrote to Phillip M. Stone, the director of the DOE's Office of Science and Technology, and complained about the composition of the committee. To Bockris, the panel was unfairly weighted by 11 physicists and only three electrochemists. He wrote that it was unreasonable to include in the committee people who had expressed their negative opinions publicly. Bockris also told Stone about the experiences of other "cold fusion" scientists when panel members visited them:

> "I have never seen a set of men before with their minds so made up before they started."
> "I have never been treated like this since I was a junior lieutenant in the Army."
> "It was as though they were investigating a crime. I was afraid to say anything positive."

In a May 5, 2004, letter, Bockris wrote to me about the visit from the panel members.

"I was a principal objective of their dislike," Bockris wrote. "They treated me as though I was a prisoner and had committed some awfulness which they were trying to discover."

Thermonuclear Fusion Funding at Risk

On June 15, just one week before the ERAB panel was to draft its interim report, researchers in the magnetic confinement fusion field received bad news from Congress. The Subcommittee on Energy and Water of the House Committee on Appropriations proposed cutting $68 million from the $349 million that had been budgeted for fiscal year 1990. Further to their disappointment, just a week earlier, on June 7, the Department of Energy had proposed a 10-year competition between the magnetic confinement fusion and inertial confinement fusion fields.

On June 21, representatives of the magnetic fusion field sent a letter to Admiral Watkins. They were strongly opposed to the idea of competition. In addition, the $50 million in funds that was proposed to be removed from the magnetic fusion program, they said, would cause a major disruption to the progress of their research. "Out of concern for the long-term health of fusion development," the 14 representatives wrote, "we strongly urge that recommendations for major changes in the magnetic fusion program ... be delayed until their technical merits and impact on the affected programs have been thoroughly reviewed."

The $50 million targeted for removal had been designated for the next-generation fusion test reactor, called the Compact Ignition Tokamak, to be built at the Princeton University Plasma Physics Laboratory. On June 29, Harold P. Furth, the director of that lab, sent a memo telling his staff that the House of Representatives approved an amendment restoring $25 million of the proposed $60 million cut. Inevitably, there wasn't enough money to build the Compact Ignition Tokamak, and the idea faded into history.

Sweep It Under the Rug

ERAB Panel Deals With Data

ERAB panel members met for the second time on June 22 in Washington, D.C. Woodard gave them a 60-page document that summarized "cold fusion" research at the national laboratories. Most of the labs had reported no positive results. The report mentions the curious occurrence of "false positives" at many of the laboratories. The report doesn't explain how the researchers could distinguish between real positives and false positives.

The Interim Report: What They Knew

Sandia National Laboratory reported "unexpected false positive signal neutrons and neutron bursts." Argonne National Laboratory reported their neutron measurements as "false positive, up to eight times background, not reproducible." Researchers at Lawrence Berkeley Laboratory (home of panel member Hoffman) wrote "false positive bursts." The LBL researchers wrote that a "mechanical and/or electrical interference" was responsible for the artifacts. Yet they had no idea what was responsible for the interference.

Researchers at Lawrence Livermore National Laboratory wrote that they "appear to have duplicated ... bursts of a few hundred neutrons." John Nuckolls, the director of LLNL, wrote to Woodard that they had repeated the Frascati experiment and "might have seen bursts of a few neutrons, but are not yet convinced. ... There are some curious observations that should be explained, but at a low funding level." The accompanying technical report from LLNL, written by an unidentified scientist, was more direct: "We appear to have duplicated the Frascati results."

It's no wonder that many of the national labs reported their results as "false positives." By this time, June 1989, Fleischmann and Pons had been publicly accused of incompetence, delusion and fraud by prominent scientists in the United States. After Baltimore, the impression in the scientific community was that "cold fusion" was bunk. Scientists who suggested otherwise were now risking their reputations.

The summary at the top of Page 41 of the document, possibly written by Michael Saltmarsh, says, "None of the groups has confirmed the production of excess heat." On the same page, in the section for the experiments performed by Donald P. Hutchinson and his colleagues, the researchers confirmed the production of excess heat: "One cell indicated an apparent energy imbalance of about 10% for several hours, and this apparent imbalance has been shown to disappear if the experiment is run at 5° C instead of 13° C. This effect is not understood."

The temperature-dependency variation reported by Hutchinson was well-known by Fleischmann and Pons. They reported this in 1993 and it became known as the "positive-feedback effect." As cells become warmer, a feedback mechanism amplifies the heating effect. (Fleischmann and Pons, 1993)

In the document, E.L. Fuller, at Oak Ridge, reported a cathode that showed anomalous changes such as hardness, fractures and blistering. Fuller knew that the low power applied in Fleischmann-Pons-type cells could not explain these features.

Charles D. Scott, at Oak Ridge, reported "two short periods of apparent excess [power], although these did not exceed 20% of the [power] input or last more than a few hours." He followed that immediately with this disclaimer: "These observations have not been reproduced and do not confirm the claimed cold fusion effects. However, the apparent unusual phenomena are interesting, and they will be the basis for additional investigation."

Many people in the U.S. national labs had reproduced aspects of the Fleischmann-Pons and Frascati experiments, but few of them had the courage to suggest that the observations were "reproducible."

The Interim Report: What They Wrote

On June 23, the day after the panel meeting, Garwin wrote the first draft of his group's analysis for the nuclear particles section of the panel's interim report. Of the four paragraphs in his draft, two are the most interesting:

> Although many experiments report no neutrons, some have reported on the order of 1 neutron per second. ...
>
> Numerous experiments have sought tritium production in electrochemical cells. No experiments with Pt anodes and Pd cathodes report excess tritium beyond the level expected from electrochemical enrichment of the tritium in the original heavy water. One group of experiments reports tritium production of some 10^{12} tritium atoms per second only for Pd cathodes with Ni anodes. These experiments need to be verified by independent investigators.

Where did Garwin get the 1 neutron per second number? He knew that Menlove had seen bursts up to 300 neutrons per second. Garwin knew that Frascati had seen bursts up to 5,000 neutrons per second. He ignored the burst rates and considered only the steady emission rates.

Garwin also knew that two groups at Texas A&M had observed tritium seven times at levels hundreds to thousands of times above background. The tritium had been independently verified by five laboratories. The Texas A&M experiments were independent verifications of the Fleischmann and Pons tritium. There was no scientific reason that those particular experiments "needed" to be verified further. Garwin's summary was not an example of scientific skepticism; it was an example of doubt and denial.

Storms Seeks Publicity

Also on June 23, two researchers at Los Alamos, Edmund Storms and his wife, Carol Talcott, went directly to the news media and told JoAnn Jacobsen-Wells, a reporter at the Salt Lake City *Deseret News*, that they, too, had found tritium in their "cold fusion" experiments.

According to Jacobsen-Wells, Los Alamos was the first government lab in the world to announce the finding of tritium. Actually, the BARC lab in India was the first. But more important, the news given to Jacobsen-Wells was not an official announcement by Los Alamos.

Jacobsen-Wells quoted Los Alamos spokesman Jeff Schwartz in the article. Schwartz came very close to denying Storms' claim, saying that it had not been confirmed or peer-reviewed. Storms' publicity-seeking was not appreciated by everyone at Los Alamos. William Johnson, at the laboratory, posted an electronic message to the sci.physics.fusion newsgroup on July 12. Johnson wrote that Storms' use of the news media was "inappropriate and a breach of the normal standards of conduct for professional scientists."

Wishing It Were Dead

"END OF COLD FUSION IN SIGHT": That was the headline in *Nature* on July 6, a week before the third meeting of the ERAB panel. John Maddox, the editor-in-chief of *Nature*, detested "cold fusion" and wanted to see it buried.

"Although the evidence now accumulating," Maddox wrote, "does not prove that the original observations of cold fusion were mistaken, there seems no doubt that cold fusion will never be a commercial source of energy."

Maddox had subtly backtracked. On April 27, he had written that "there were no confirmations of the Fleischmann-Pons experiment." He tried to use a failed replication attempt to bolster his position.

"It seems the time has come," he wrote, "to dismiss cold fusion as an illusion of the past. At the outset, the suggestion that deuterium nuclei can be made to fuse together at ordinary temperatures seemed a brave leap of the imagination. The article on Page 29 of this issue by Moshe Gai et al., of Yale University, is merely another nail in the coffin of the idea. The Yale group has done its best to replicate the conditions of the original experiments but has failed to replicate their results. Similar outcomes have been reported from other laboratories. So what has been learned from these hectic months?"

As meticulous as Maddox was about so many aspects of science, on this matter, he was misguided. Gai's failure meant only that Gai failed, nothing more.

Draft of the Panel's Interim Report

Maddox was not alone in his aversion to the new science. Members of the ERAB panel met on July 11-12, 1989, in Washington, D.C. Here, they prepared a draft of the interim report that had been requested by Admiral Watkins. When the meeting ended, Woodard faxed a copy — a scant four pages — to all of the panel members.

The lead paragraph explained that some labs had reported positive results and some had reported null results. There was no consensus about the science and thus no reason to consider the reported phenomena seriously:

> Although the panel's task is not yet completed, the panel finds that the experiments reported to date do not present convincing evidence that useful sources of energy will result from the phenomena attributed to cold fusion. Indeed, evidence for the discovery of a new nuclear process termed cold fusion is not persuasive. Hence, no special programs to establish cold fusion research centers or to support new efforts to find cold fusion are justified at the present time.

Watkins had not charged the ERAB panel with deciding whether the research might lead to practical sources of energy. Nor did he ask the panel members to decide whether the evidence for the new nuclear process termed "cold fusion" was persuasive. He had directed them to identify what research should be undertaken to better understand the phenomena in the hopes that it could lead to a new source of energy. The panel members had hijacked the mission. They crafted the following paragraph in order to accommodate the few positive results that they could not summarily dismiss:

> However, there remain unresolved issues and scientifically interesting questions stemming from reported cold fusion efforts. Some of these are relevant to the mission of DOE and should be handled by carefully focused and cooperative efforts within current programs by normal mechanisms for project selection.

Among those "unresolved issues" were Menlove's neutrons, Frascati's neutrons, Casccia's neutrons, Wolf's tritium, Bockris' tritium, Appleby's excess heat, etc., etc. How was this denial of science possible? Simple: It was inconceivable to all but a few scientists.

Direct to the News Media

Woodard had asked panel members to review the draft and submit comments by July 21. He marked it clearly as a preliminary draft that needed to be reviewed by all panel members.

Nevertheless, within hours, someone on the panel gave Bill Broad, of the *New York Times,* a copy of the July 12 interim draft report, which had not been reviewed or approved for release, let alone delivered to Watkins. Broad broke the news on July 13: "Panel Rejects Fusion Claim, Urging No Federal Spending." He had spoken with Huizenga.

"I think we're pretty unanimous on the content," Huizenga said, "but people may want to change a word here or there." Huizenga told Broad that the interim report would be "the gist of the final report, unless something surprising happens between now and then."

Something surprising did happen. More than a word or two changed between the interim report and the final report. Not only did it grow from a skimpy four pages to 60 pages, but also the panelists embarked on a massive systematic effort to collect information about experimental results from everyone they knew who was working on "cold fusion" domestically and internationally. Nothing of the sort had been done before the release of the interim report. Someone, perhaps Watkins, had found their interim report woefully inadequate and had directed them to do a better job.

A week later, on July 20, the panel finished and released the final version of the interim report. It was almost identical to the July 12 draft except for one change. The July 12 draft said, "Although many experiments report no neutrons, some have reported on the order of 1 neutron per second."

The July 20 final version said, "Although many experimenters report no neutrons, some report as many as 0.1 neutron per second." In addition to their complete denial of the neutron bursts, Garwin and the panel members, without any apparent scientific basis, reduced the random neutron rates by an order of magnitude.

Memorable Mistake in Science History

As of July 12, "cold fusion" was dead, well, mostly dead. Its execution did not wait for the November release of the final report, let alone for the July 20 official interim report. All that was missing, to borrow Bob Park's words, was the "final coroner's report" to be delivered in November.

Journalist Robert Bazell, on NBC's *The Today Show*, got the point across on July 12. "Pons and Fleischmann hinted they may have found a cheap new source of energy for the world," Bazell said. "Not so, the panel concluded in its draft report. The report said there is no convincing evidence that what is called cold fusion will ever produce significant amounts of energy."

Park, as usual, provided an entertaining quote. "It's a story the American people love," Park said. "They love the story about the manager of the New York Yankees who spots some hillbilly throwing rocks at squirrels, then takes him to New York to win the World Series. But, of course, it never happened."

"It now appears," Bazell concluded, "that cold fusion will be remembered as one of the memorable mistakes in the history of science."

From this day, the stigma was securely bound to the research for decades. Thomas F. Droege, an independent researcher and an independent thinker, sent Huizenga a letter on Sept. 9. Droege had been in no rush to resolve the "cold fusion" dilemma. His group had spent three months preparing their equipment before they started their first experiment. Droege sent Huizenga a letter scolding him for the opportunity his panel had destroyed:

> What you are doing is driving the work elsewhere. This means that, in the end, you will not know what is going on [with the science]. If it proves to be an important discovery, then the DOE will look like fools and will not have any of the stars in their stable. I recently participated in a summer study at Breckenridge, Colorado, where more than 350 high-energy physicists from all over the world participated. I made it a point to talk to as many as I could about "anomalous heat." You would think I was ringing a bell and crying, "Leper, leper."

From then on, any scientist, particularly those funded by the federal government, who suggested that the research might be credible would be putting his or her reputation at risk, let alone becoming a target for ridicule and mockery. Any further inquiries sent out by the ERAB panel after July 12 were virtually pointless.

Panel's Perspective Discourages Bockris

Bockris did not anticipate the level of denial in the interim report. He was dismayed at its conclusions. He wrote to Huizenga on July 17 that they left out the most essential part in the report – the tritium.

Actually, they didn't leave it out, but they made it appear as if the null results were as significant as the positive results. "Tritium levels above normal," the panel wrote, "have been reported in some cells following electrolysis but not in others." The panel also marginalized the tritium at Texas A&M with the following sentence: "One group reports finding up to 10^{14} tritium atoms (neglecting losses to the gas phase) in each of several cells."

Reporting the yield in numbers of atoms was a tricky way to obscure the positive results. An honest way to report the Texas A&M tritium might have looked like this: "Two groups at Texas A&M reported a total of seven experiments in which tritium levels were hundreds to thousands of times greater than background.

Bockris predicted in his June 9 letter to his colleagues that the ERAB panel would not find a way to explain the tritium. However, he didn't imagine that they would find a way to sweep it under the rug. Neither did he imagine that, months later, reporter Gary Taubes and *Science* magazine would insinuate that Bockris and his colleagues had spiked their cells to give the fraudulent appearance of tritium production. Despite Garwin's reputation for being outspoken, Bockris was conservative about "cold fusion," as he wrote in his July 17 letter to Huizenga.

"Even were the correlation between the tritium and the heat to be established," Bockris wrote, "and reproducibility attained, there is no evidence to indicate that a great new technological pathway to energy has been found. There is also no reason to believe that it has not." Huizenga sent this response to Bockris:

> As you can see by the panel's interim recommendations, more experiments need be designed to check the large amounts of tritium that you and your colleagues have reported. When possible, it would be desirable to measure also the neutron production in the same cells producing tritium. What is the present status of Storm's reported tritium? I have heard that he was working in a laboratory in Los Alamos where high-level tritium experiments were performed in the past.

From this point, for these critics, there would never be enough successful experiments, performed by scientists of their liking, published in peer-reviewed journals of their preference, to satisfy them. Huizenga, Garwin, Maddox and others were not demonstrating scientific skepticism; they had become deniers.

From this point, when provided with new confirmatory results, the deniers would always require larger effects, or lower backgrounds, or some other new demand.

On July 3, Bockris sent a letter to panel member Norman Ramsey, the only Nobel Prize winner on the panel. In the letter, Bockris explained some of the latest insights that he had uncovered about the research. Bockris also explained new insights he had obtained that indicated that the phenomena were surface-based, rather than volume-based. Bockris immediately connected the dots and realized the fortuitous nature of a surface-based effect: "The adventitious nature of the observations increases the possibility that we shall be able to use cheaper electrode materials." No longer would large pieces of palladium be required.

Bockris saw it: "A new field, 'nuclear electrochemistry,' has thus been born." He knew nothing about weak nuclear interactions and thus could only attribute the results to a nuclear fusion process. He gave Ramsey specific ideas about what research should be undertaken to determine a better understanding of the phenomena — exactly the type of information that Watkins had originally requested.

Unresolved Issue: Menlove's Neutrons

Immediately following the interim report, the biggest "unresolved issue" panel members thought they had was Menlove's neutrons. The many reports of failures did not concern the panel members. They didn't care why those experiments failed because, for them, failure *was* the correct result.

The responsibility to draft the neutron section of the final ERAB report fell to Garwin. His neutron section was to be incorporated into the nuclear products section. Physicist John Schiffer was the group leader for the nuclear products subgroup. Schiffer had been the associate director of the Physics Division at Argonne National Laboratory for 25 years, the director of the lab for four years, and an editor of the journal *Physics Letters B.* Jacob Bigeleisen was a member of the subgroup and had had a role in the historic Manhattan Project nuclear weapons program.

The Frascati neutrons, which had prompted urgent activity by the U.S. Department of Energy on April 18, were relatively easy for Garwin to cast doubt on; the Frascati researchers had used only one detector, and it was poorly repeatable. Garwin had written to Martone, the head of the neutron measurement group at Frascati, on June 12, making several suggestions about how the detector used in their experiments could have experienced faults.

Garwin did not prove any fault; nor did the Frascati researchers, who looked hard for possible artifacts. Nevertheless, in a letter Garwin wrote to Menlove and copied to Woodard, he dismissed the results as a matter of disbelief: his own. "The neutron 'bursts' from Frascati could hardly be believed," Garwin wrote.

But Menlove's replication of Frascati was much harder to write off because, instead of a single BF_3 detector, Menlove used multiple helium-3 detectors. Garwin, in a July 12 letter, in his official capacity as a member of the ERAB board, issued demands to Menlove and sent a copy of the letter to the Department of Energy. His demands were the scientific equivalent of a subpoena.

"So now I have to ask for your data showing simultaneous counts in redundant counters," Garwin wrote. "Furthermore, I have to ask for your data showing counts in one counter set and no counts in the counters that might be expected to be influenced by the neutron bursts."

Five days went by without a response. Garwin was getting anxious. After several attempts, Garwin finally made telephone contact with Menlove on July 17. Garwin described the details of his call in a July 24 letter to Woodard. Menlove had responded to all of Garwin's questions, particularly about a well-designed set of control experiments. "In fact," Menlove said, "about half as many dummy samples as real ones are run. We have never gotten a signal from a dummy."

Menlove immediately followed up the phone conversation with a detailed four-page letter on July 25, answered all of Garwin's questions, and responded to all of Garwin's demands. Menlove was an expert in neutron counting for treaty verification, and he was not intimidated by Garwin.

On Sept. 12, Garwin cast aspersions on the Frascati team's results. He sent a letter to Woodard telling him that the Frascati researchers had not been able to repeat their "beautiful results of April on neutron bursts" and that they had stopped running experiments. Garwin did not think that the Frascati results were "beautiful." At the end of his letter, Garwin relayed rumors about Menlove's results.

"I have heard informal comments," Garwin wrote, "that several of the people who work with Howard Menlove do not believe that they are counting neutrons."

Menlove continued his experiments for at least the next two years and got a few more strong results, as well as many null results. His primary interest was searching for the mechanism that was triggering the neutron emissions. He was not surprised at his low rate of repeatability because he had tried a large variation in sample types and procedures. The final sentence in his paper speaks to the heart of the problem in the research then, as it does today: "The number of experimental variables far exceeds our capacity to investigate the parameters." (Menlove, 1991)

Neutrons All Over the World

The Garwin archive contains a hand-drawn graph of some reported neutron measurements in "cold fusion" experiments. The graph seems to be from July or August 1989. Garwin and his fellow panel members knew in 1989 that either low-level or burst neutrons from "cold fusion" had been measured all over the world.

The y-axis is neutron source rate in neutrons/second and is a log scale. The x-axis does not appear to be used; the lab names do not seem to be placed according to date of experiment. The values appear to be a mixture of both average (random) and burst neutron rates.

News archives correlate some of the neutron reports, but some of the other labs shown on the graph had not publicly reported neutrons.

Countries represented include the United States, Germany, Argentina, Japan, Sweden, Canada, Italy and France. India is not listed here, which suggests that the graph was drawn before September 1989.

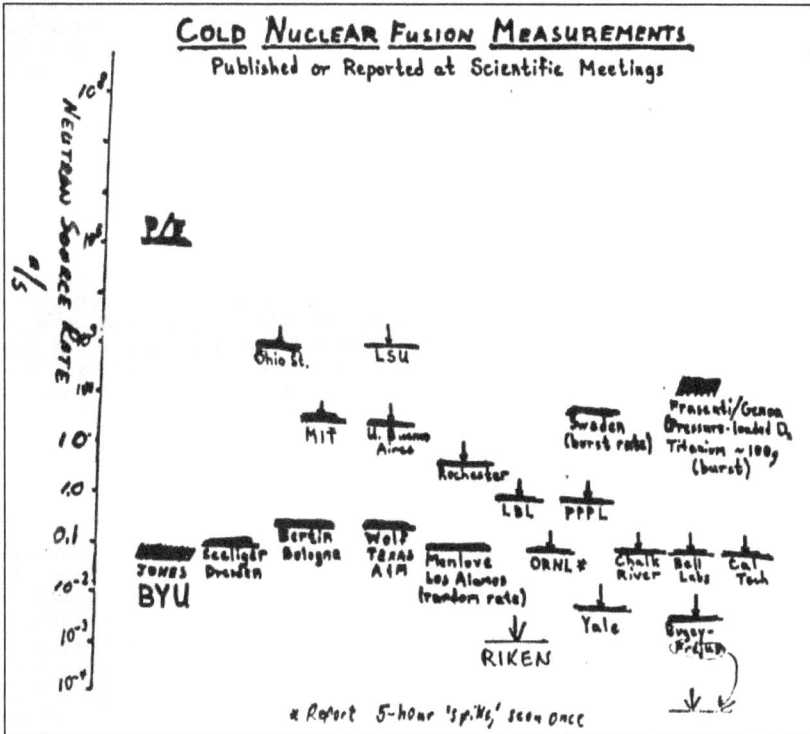

Hand-drawn graph of worldwide neutron measurements in "cold fusion" experiments from the Garwin archive, author unknown

The most fascinating listings are MIT, the University of Rochester (Huizenga's affiliation), Lawrence Berkeley Laboratory (Hoffman's affiliation), Princeton Plasma Physics Laboratory (Happer's affiliation), Chalk River, and Bell Labs. All researchers from these laboratories had claimed no evidence for "cold fusion."

Garwin handled the "unresolved issue" of the Casaccia neutrons in a July 27, 1989, letter to the Department of Energy. Garwin told Woodard that he had met Ugo Valbusa, a solid-state physicist from Genoa, Italy, who spoke with him about problems he had heard second-hand about the Casaccia work. Garwin's letter to Woodard included no specific technical details; nevertheless, he was enthusiastic.

"The Casaccia group," Garwin wrote, "has now discovered that the problem is due to a difficulty in their BF_3 counter! They have no positive result. Science marches on!"

Artist: Don Wright

CHAPTER 30

Could We Do Without This?

ERAB Panel Ignores More Data

John Huizenga, the chairman of the ERAB "cold fusion" panel, had told Bill Broad of the *New York Times* that he didn't expect more than a word or two to change between the interim report and the final report. Nevertheless, Huizenga and his panel regrouped and put significantly more effort into the task. After the panel obtained reports of more successful results, they looked for more ways to dismiss them.

Panel Conducts Two Broad Surveys

Between July 13 and July 20, the panel sent out a survey called "Request for Summaries of Past and Present Cold Fusion Research." The panel distributed this survey to national laboratories and domestic and international universities and industrial laboratories. The surveys went out along with a copy of the draft interim report. This was a strange thing to include. The panel was asking for the labs' research status while suggesting the predetermined conclusion. On August 9, the panel sent out a second survey called "Request for Tritium Production Results." In response to these two requests, the panel received a large body of research summaries.

One of the tritium results in the survey responses in the Aug. 28 report from Los Alamos was that of Edmund Storms and Carol Talcott. Their response was now an official result. This data now became another "unresolved issue" for the ERAB panel members.

Among the surveys, another researcher at Los Alamos, Thomas Claytor, told the panel that he, too, had detected "tritium well above background in one cell that had also shown neutron activity."

He used a novel method: a solid-state metal-insulator-semiconductor junction and deuterium gas. The method was later used by researchers Tadahiko Mizuno, in Japan, and Jean-Paul Biberian, in France. His other cells showed tritium levels identical to that found in the deuterium gas in the supply cylinder.

There are no letters between Claytor and the panel members in the Garwin archive. Claytor didn't publicize his results. The panel seems to have left him alone.

More "Unresolved Issues" — Tritium and Neutrons From India

On Aug. 23, 1989, a new "unresolved issue" appeared. P.K. Iyengar, the director of the Bhabha Atomic Research Centre (BARC) in India, sent his July 3 Karlsruhe conference paper to his friend Hiroshi Takahashi at Brookhaven National Laboratory. Takahashi immediately sent the paper to Goodwin at the Department of Energy. On Aug. 28, Woodard sent the Iyengar paper to everyone on the ERAB panel.

Until this time, there is no indication that Garwin and the other panel members knew anything about the extensive effort at BARC, India's largest and most significant nuclear research laboratory. As reported in Chapter 23, more than 50 scientists in 10 groups had been working on "cold fusion" at BARC. Most of the groups had succeeded and observed significant levels of tritium and neutrons, often in the same experiments. The BARC researchers saw burst rates of neutrons in the range of thousands per second. Like the researchers at Texas A&M, they saw tritium at hundreds to thousands of times greater than background starting values in several experiments. The tritium results, of course, were far beyond the level possible by electrolytic enrichment. Nobody on the ERAB panel dared question the competency of the BARC experimentalists. Therefore, on Sept. 19, Garwin, as usual, disputed their experimental results based on nuclear fusion theory.

Livermore: Pops Like a Firecracker

Back on April 12, 1989, the *San Jose Mercury News* had reported an explosion of a "cold fusion" experiment at the Lawrence Livermore National Laboratory. The Sept. 8 report from Livermore, in response to the panel's survey, revealed fascinating details about this explosion. The Livermore researchers were working with a replica of the Fleischmann-Pons cell with slightly smaller dimensions. Here is the researchers' description of the explosion:

> The first run was made at 4 amperes and 12 volts, with a current density of 0.4 amperes per square centimeter on the Pd electrode. After about 1.3 hours, the Pd rod suddenly glowed red, then white for a brief moment, and the cell exploded with a force similar to a small firecracker. The Pd electrode was discolored, except where it had been covered by the Teflon holder. The Teflon was not distorted or melted where it had been in contact with the Pd rod. Most of the force of the explosion was probably due to the recombination of electrolytically produced hydrogen and oxygen which was ignited by the hot Pd surface. The discolored layer was 0.5 micron thick and contained Na, K, Ca, and Li. All the metals except Li probably were leached out of the glass vessel by the action of the LiOH solution.
>
> The explosion was possibly a surface reaction not involving the bulk of the Pd rod, or a reaction between the electrolyte or hydrided Pd and a Li-Pd alloy formed on the electrode surface. This may be the same phenomenon reported by Fleischmann and Pons. The apparatus was rebuilt and more experiments run without explosions.

It is difficult to know how similar or dissimilar their event was to the Fleischmann-Pons event. However, based on the description given by the researchers, the cathode was in the cell, undergoing electrolysis, at the time of the explosion. The researchers did not write whether cathodes used in blank cells also contained Na, K, Ca or Li.

No recombination-based explosion, whether it occurred in the headspace or among bubbles in the electrolyte, could have caused the sudden heating of the cathode, because the cathode heating took place before the explosion.

Author Frank Close speculated that the effect was simply what is called the "cigarette lighter effect." This doesn't explain the explosion because the cathode was in the cell, in contact with the electrolyte.

A more feasible explanation, according to Lewis Larsen, is the following: 1) hotspots began reacting on the cathode surface, 2) the positive-feedback heating process began and rapidly increased the local temperature surrounding the cathode and 3) the electrolyte on the cathode surface flash-boiled and vaporized the electrolyte. 4) When water converts to steam, it expands volumetrically 1,600 times. Large quantities of steam were generated suddenly at the cathode surface and expanded outward in all directions.

A Mistaken Assumption?

After a site visit to Pons' lab on June 2, William Happer wrote to panel members that, because he saw a wire that was part of the calorimetry system in Pons' lab that was "straight, unscratched and unbent," Fleischmann and Pons could not possibly have calibrated most of their cells. Happer's sophomoric detective work was contradicted by extensive documentation by Fleischmann and Pons about their calibration, not to mention their ability to run blank cells for extended periods with an accuracy of $+/-1$ mW.

On Aug. 9, 1989, Huizenga wrote to Pons requesting more information about his and Fleischmann's experiments. After having hosted Happer and other panelists in his lab, and after seeing the panel members' interim report, Pons did not appreciate what the panel was doing, as he wrote to Huizenga on Sept. 10:

> We are still much concerned by the very nature of this biased committee, and find it difficult to understand how

any sensible selection procedure could come up with a list dominated by members who had already publicly declared (and viciously so) their verdict and opposition to this research. Nevertheless, we have on several occasions tried to comply with its requests since it determines the availability of badly needed research funds for others working in this important area.

You can therefore imagine our irritation with the subsequent statements made by members of your committee regarding the accuracy of our calorimetric measurements. We established the maximum error limits for our experiments before we submitted our preliminary publication, and we did indeed give these in the paper. In our more recent work, we have further established that we in fact somewhat overestimated the magnitude of the errors. These latest results will be published shortly, and we strongly suggest that you wait until this paper is available to the general scientific community before making ill-considered statements regarding the most important signature of the processes. Failure to do so will inevitably do further severe damage to the general public's opinion of science in general and the assessment procedures in particular in this country.

The panel members knew that their conclusions and recommendations would affect the expenditure of federal research dollars that might go toward "cold fusion" research or to thermonuclear fusion research. The panel members knew that the impact of their conclusions and recommendations would affect "cold fusion" research in the United States. In fact, the impact of the panel's conclusions affected "cold fusion" researchers worldwide, because foreign governments accepted the U.S. panel's conclusions.

If Watkins assumed that scientists affiliated with institutions that received federal funding for thermonuclear fusion research could participate on this panel without overt bias, he was mistaken.

Los Alamos Confirms Tritium

Although Edmund Storms had told the *Deseret News* on June 23 that he and Carol Talcott had seen the production of tritium in their electrolytic cells, the results of their experiments had not been approved for release and possibly not internally reviewed. On Sept. 13, 1989, Reed Jensen, the deputy associate director for research at Los Alamos, responded to Huizenga's tritium survey request. Jensen explained to Huizenga why he thought the research was important:

> Despite the frustration from the non-reproducibility of experimental results from all aspects of cold-fusion phenomena, Los Alamos researchers continue to work very hard to resolve the controversy over whether cold fusion is a reality or not. The prospect of producing tritium by such a simple technique is too important for us to abandon this research before examining all avenues in an objective and systematic manner.

Tritium is a key component of many nuclear weapons, and one which must be replenished over time because of tritium's relatively short half-life. A year earlier, in 1988, the Department of Energy's only source of tritium, the "K Reactor" at the Savannah River Site in South Carolina, was shut down for safety reasons.

For the next 16 years, the decommissioning of large portions of the nation's nuclear weapons stockpile allowed the extraction and purification of tritium for reuse in the remaining weapons, until a new production facility could be constructed in 2004.

Jensen explained that Storms and Talcott had operated 22 electrochemical cells simultaneously and that they had made considerable effort to increase the loading ratio by the use of an alloy and special surface-coating techniques. Two of their experiments, #29 and #30, showed a "60-fold increase in tritium content over the concentration of the stock solutions of electrolyte." Jensen put a disclaimer on the data and told Huizenga that the electrolyte had not been assayed in the cells before charging. Therefore, Jensen said, contamination could not be ruled out, despite the fact that they had taken precautions to avoid that possibility. The fact that the other 20 cells showed normal levels of tritium made the contamination scenario extraordinarily unlikely.

In the coming months, Storms and Talcott continued to obtain more positive results. On March 2, 1990, according to the *Wall Street Journal* — which some annoyed scientists cynically called the *Journal of Wall Street* — Talcott reported that they had measured excess tritium in seven of nine new "cold fusion" cells, up to an 80-fold increase.

Science Marches On, in India!

On Aug. 28, 1989, Garwin had received a copy of the BARC paper, presented by Iyengar at the Karlsruhe conference. The final ERAB report was due soon, and Garwin had to deal with the positive results from India somehow. He sent a Telex message to Iyengar on Sept. 7 and tried to identify inconsistencies and weaknesses in their work.

On Sept. 13, Iyengar sent a Telex back, happily reporting that, since his presentation at Karlsruhe, the BARC researchers had observed more successful experiments, even with other detection techniques. Iyengar speculated about a theoretical mechanism for the inconsistency between the theoretically expected neutron-to-tritium ratio.

"Cold fusion is essentially aneutronic in nature," Iyengar concluded. In an earlier letter, when Garwin had told Iyengar that some Italian researchers had retracted their claims, Garwin had written to Iyengar with enthusiasm, "Science marches on!" Iyengar couldn't resist taking Garwin's words and throwing them back. "Let me also endorse your observation 'Science marches on,' in spite of good and bad reports. With best wishes, sincerely yours, P.K. Iyengar."

Garwin wrote to Woodard the same day, asking him to distribute the response from Iyengar to the full panel. Garwin's only comment to Woodard and the panel was in reference to Iyengar's idea that the branching ratio in "cold fusion" was essentially aneutronic. "I do not believe the argument of the second paragraph," Garwin wrote. Belief, again, had overtaken Garwin's scientific objectivity.

When it came time to write the final report, the panel followed Garwin's lead and dismissed the BARC results because the data did not agree with what they knew and understood.

Although Garwin never stood in front of television cameras denouncing Fleischmann and Pons, he had perhaps the most significant effect of anyone on the nascent field.

Years later, on Dec. 2, 2010, I discussed science philosophy with Garwin and asked him about his dismissal of "cold fusion" based on contradiction with theory. "It was never my position," Garwin wrote, "that the experimental results could not be right because no theory could account for them. Experiment is a principal approach to making discoveries and changing our theories!"

Yet this is exactly what he did. His dismissal was based on two arguments: a) the fact that the experiments were not fully reproducible and b) the fact that the results data didn't match fusion theory.

Excess Heat at Oak Ridge

An apprehensive scientist from Oak Ridge (Chapter 28) had revealed to Bockris at the Santa Fe workshop that he had seen some positive results. A different Oak Ridge researcher, Charles D. Scott, had reported his group's positive neutron and excess-heat results in Santa Fe.

Oak Ridge sent three documents in response to the ERAB panel's survey request. One of them, dated Sept. 1, is Scott's and his seven co-authors', from the Chemical Technology Division. The lead author on the other document, dated Sept. 14, was Donald P. Hutchinson, from the Oak Ridge Physics Division. He had four co-authors: two from the Physics Division and two from the Nuclear Weapons Division.

In the Hutchinson paper, one of four electrolytic cells produced 2-3 watts of excess heat for 300 hours. The heating anomaly began after the cathode had been charging for a week. The excess heat was 10% of the input power. The researchers wrote that, except for a short initial period of negative imbalance, the cell exhibited a sustained positive heat balance. The other three cells were in power balance for the entire duration. The researchers did not conclude that nuclear processes were responsible for the incident because they detected no excess neutrons or tritium from the experiment.

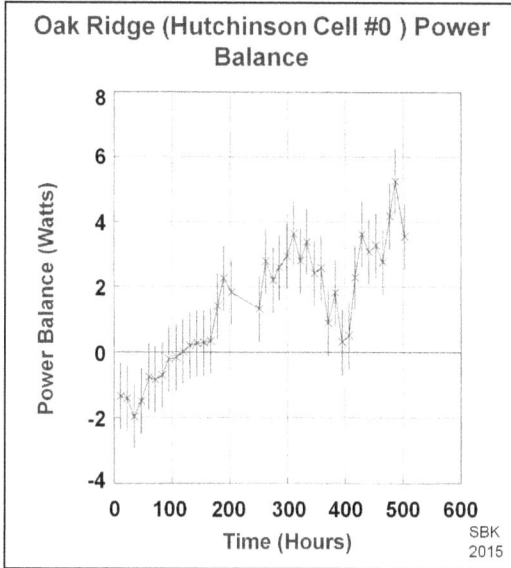

Power balance plot for Hutchinson cell in Oak Ridge "cold fusion" experiment

I discussed the "negative imbalance" with Melvin Miles, an electrochemist who is one of the most knowledgeable experts on Fleischmann-Pons-type electrolytic cells. He had seen this kind of initial negative imbalance reported by other researchers. As he explained, energy is going into the cell at the beginning to warm up the cathode and the water inside. If researchers don't account for this, as many did not, they see a false endothermic period.

Kevin Wolf Is in Trouble

On Sept. 15, 1989, panel member Jacob Bigeleisen, a distinguished professor of chemistry at the State University of New York at Stony Brook, sent a two-page letter to panel members John Schiffer and Garwin about the tritium. One page was the Sept. 13 Telex from Iyengar, in which Iyengar enthusiastically reported more successful

BARC experiments. Bigeleisen made no comment about the BARC tritium. Instead, he directed his fellow panel members to the "unresolved issue" at Texas A&M.

"It comes to the same conclusion: Kevin Wolf is in trouble," Bigeleisen wrote. "His neutron and tritium measurements are self-inconsistent." Wolf was indeed in trouble but not because of his data. Colleagues of Bigeleisen's at Stony Brook and the neighboring Brookhaven National Laboratory were preparing an attack.

On Sept. 17, Nicholas P. Samios, the director of Brookhaven, sent the survey response for his lab to Huizenga. It revealed the efforts of 16 scientists, only one of whom, Kelvin G. Lynn, reported an anomaly. One of eight of his electrolytic cells showed a four-sigma increase in tritium. Lynn thought that it could be explained by contamination in the deuterium gas that was used to pre-charge the cathodes. He didn't explain why the other seven cathodes didn't become "contaminated."

Vicious Opposition

Concurrently with Samios' report to the ERAB panel, he and Robert P. Crease, an assistant professor of philosophy at the State University of New York at Stony Brook and a historian at Brookhaven National Laboratory, launched a personal attack against Fleischmann and Pons. By inference, it was also an attack on every other scientist doing "cold fusion" research. Samios and Crease submitted an article to the *New York Times,* and it published on Sept. 24 in the *Times'* magazine section. They accused Fleischmann and Pons of self-deception. Here is an excerpt:

> Pons and Fleischmann apparently fell victim to the experimental scientist's worst nightmare. Usually, self-deception is quickly and relatively painlessly cleared up, either through the scientist's own labors or those of a neighboring lab bench. What made the case of Pons and Fleischmann different was their meteoric and public rise to celebrity, and their equally spectacular and public downfall. ... Cold fusion may someday rank among such notorious

scientific nondiscoveries as the Martian canals, N-rays and polywater. But it is more than a classic example of pathological science; it also provides a lesson about how experimental science works and why it is enormously difficult.

Letters to the editor, published on Nov. 5 in response to Crease and Samios, were more open-minded. Here's one by Stephen Ward, of Abington, Pennsylvania:

> I was amused, then not so amused, by "Symptom No. 2: The 'discoverers' are outsiders." The ridicule of outsiders in their quest for truth is the hallmark of orthodoxy. The function of orthodoxy, however, is to preserve the status quo. Creativity comes from outsiders, like two bicycle mechanics from Ohio or a lonely patent clerk in Switzerland.

Here's a letter from Robert Bernstein, of East Lansing, Michigan:

> Your authors contend that "good" scientists always try to "kill the discovery." That is straight out of Karl Popper's philosophy of falsification, and it simply doesn't describe the behavior of real scientists. No scientist kills his own discovery — you verify it to your own satisfaction, publish it and let the [critics] do their worst. If the discovery holds up, well and good. If not, tough. Only a very small fraction of published discoveries hold up to long-term scrutiny.

Letter writer Wade Roush, from Cambridge, Massachusetts, observed something subtle:

> The "mixture of excitement and [doubt]" with which Crease and Samios say they followed the cold fusion story no doubt stemmed in part from anxiety over new and possibly powerful competition. Crease and Samios wrote, "There were scientists who had seen, as early as the original press

conference, that something was amiss." This smug we-knew-it-all-along tone betrays their relief that, as it seems, cold fusion has fizzled.

Samios did not mention in the article the numerous failed attempts by researchers in his own lab. This was a serious omission and misrepresented his neutrality. Samios did not explain why, if "cold fusion" was so obviously the result of self-deception on March 23, 1989, so many researchers in his laboratory attempted to replicate it.

The Crease-Samios article was significant for several reasons. First, it became the model — persisting for at least the next 26 years — for the publicly reported story of "cold fusion." Second, it gave social permission for scientists to publicly refer to Fleischmann and Pons as, in a word, idiots — if not directly, at least by inference. Third, it locked down the stigma against the new science so tightly that no scientist of any stature would dare publicly question the myth.

The myth was that no experimental evidence existed for new nuclear phenomena and nobody had independently confirmed the Fleischmann-Pons experiment. "Cold fusion" had become forbidden science.

Runaway Heating in Belgium

On Sept. 25, thermonuclear fusion researchers from the Nuclear Research Center in Belgium, also known as SKC-CEN, responded to the ERAB panel's survey and submitted their paper "Experimental Evidence of Erroneous Heat Production in Cold Fusion Experiments." The title is deceptive in that it suggests that they found evidence for errors in Fleischmann and Pons' excess-heat measurements. In fact, they did not examine Fleischmann and Pons' calorimetry.

The Belgian researchers ran two electrolytic cells in parallel, one with D_2O and the other with H_2O. They kept the level of electrolyte in the cell constant by the regular addition of D_2O or H_2O, respectively. On June 16, a sudden temperature rise occurred in both cells. The cells had been stable for 46 hours. The authors explained what happened after both cells had been running at these elevated temperatures for an hour:

When decoupling the function generator, the H_2O cell suddenly normalized. This had no effect on the D_2O cell. Replacement of the potentiostat had also no result. The D_2O system was no longer controllable, and the cell had to be cooled by means of a cold air blower. For the D_2O cell, this abnormal situation lasted for about two hours, so that the dissipation may be estimated at 100 to 200 kJ.

They didn't detect any hard radiation, neutrons or gamma. They didn't detect a significant increase in tritium. For these reasons, the researchers concluded that the excess heat was not produced by a nuclear fusion reaction, so they searched for a hidden source of energy input. Here's their conclusion:

Excess energy has been observed during a Fleischmann and Pons-type cold fusion experiment. This phenomenon may be explained on the basis of a hidden input of electrical energy, which is not revealed by the standard DC measurements done in this type of electrochemical experiment. The hidden electrical energy consists of a low frequency AC power which is superimposed on the applied DC power. It may be generated by the occurrence of an oscillation instability in the electrical circuitry of the cell in case certain faulty conditions build up.

The fact that the Belgian researchers resorted to cooling the cell with a blower suggests that they had a runaway heating event. If they did remove the input power, they failed to explain how the cell produced heat for two hours. Nor did they explain why, if they continued to power the D_2O cell, it reacted to the hidden power so profoundly differently from the H_2O cell. They also didn't explain, if there was a hidden source of power, why it turned on only after 46 hours.

Six Impossible Things Before Breakfast

Toward the end of September, panel members were scrambling to complete their sub-sections of the final report. By Oct. 4, Schiffer had written several drafts of the nuclear products section and presented them to his group. In a cover letter to Garwin, he wrote that Garwin had focused too much attention on the BARC report, giving it too much weight and leaving too much room for readers to make up their own mind. Garwin, as he told his colleague Douglas Morrison on Oct. 10, thought that the BARC results were "the most credible evidence" for "cold fusion." Garwin's private comments to his friend Morrison were inconsistent with what Garwin said to members of the panel about the BARC results.

Schiffer cut the BARC section back, but he left in the tables listing all of the BARC results and the data as Appendix B. Schiffer suggested to Garwin, "I think we should probably remove them, as well as the Bockris tables, if you agree." The tables disappeared.

```
APPENDIX B

????COULD WE DO WITHOUT THIS????

Reproduce BARC tables

Reproduce Bockris tables

APPENDIX C

CONSIDERATIONS IN TRITIUM CONCENTRATION.
```

Schiffer's request to Garwin to omit the BARC and Bockris data

Schiffer, writing the draft conclusion for the nuclear products group, established the primary basis for denial:

> A number of careful experiments have been carried out
> to search for the expected products of cold fusion. _N_o_n_e

have seen these products at anywhere near the level that would be expected from the heat production reported in electrolysis, by many orders of magnitude. Some experiments report neutrons or tritium at a much lower level; however, the rates of these two fusion products, measured in the same experiments, are inconsistent with each other, again by large factors. *[Ed: Emphasis is in the original]*

That was the first big problem for the panel members: An unexplained process was generating enormous levels of energy but doing so without the dangerous radiation ordinarily expected by a nuclear reaction. The second big problem was that the neutron bursts were not reproducible by all experimenters. There was a third problem:

If there _w_e_r_e such a process as room-temperature fusion, it would require not only (a) the circumvention of fundamental quantum mechanical principles ... but also (b) drastic modifications of branching ratios in the D+D reaction and (c) the invention of an entirely new nuclear reaction process. *[Ed: Emphasis is in the original]*

An entirely new nuclear reaction process. Inconceivable. Schiffer ended his summary with a quote from a children's book:

```
A quotation from Lewis Carroll seens appropriate:

    'Alice laughed. "There's no use trying," she
said: "one can't believe impossible things."

    "I daresay you haven't had much practice,"
said the Queen.  "When I was your age, I always did it
for half-an-hour a day.  Why, sometimes I've believed
as many as six impossible things before breakfast."'

                        from 'Through the Looking Glass'
```

Schiffer's proposed ending to the nuclear products section of the ERAB report

Schiffer eventually deleted the quotation from the final report, thanks to the wisdom of Darleane Hoffman in response to the Sept. 27 draft. "As to the quotation," Hoffman wrote, "I am very fond of it, but our committee has already been accused of being biased against cold fusion, and I'm afraid this will make us appear snide as well — although perhaps it doesn't really matter!"

BARC-1500 Report Arrives on U.S. Soil

Another ERAB panel meeting took place on Oct. 13 in Chicago. No records of this meeting are available. In advance of the meeting, on Oct. 9, Garwin spoke with Iyengar, who was visiting his friends at Brookhaven. Iyengar called Garwin and told him that BARC was preparing a comprehensive report on all of BARC's "cold fusion" experiments, to be published as the BARC-1500 report.

The next day, Garwin sent the notes of his phone call to Woodard. Garwin marginalized the BARC achievements by focusing on the disparity between experiments and theory. "Of course," Garwin wrote, "I do not accept this conjecture."

According to an Oct. 16 letter from Garwin, Iyengar gave a copy of the draft BARC-1500 report to Martin Blume, the associate director of Brookhaven. Blume then sent a copy of the report to Garwin. It contained more than 100 pages of reports of successful measurements of neutrons and tritium, often produced concurrently, in "cold fusion" experiments.

Garwin also sent a letter to Iyengar on Oct. 16 thanking him for the report. Then Garwin wrote, "I don't know whether you have seen the *Salt Lake Tribune* article of 09/26/89. ... The Utah National Cold Fusion Institute is saying that the Institute's more than 20 electrochemical cells 'have produced no excess heat and no fusion products such as neutrons or tritium.'" Garwin was implying that the NCFI failures cast doubt on BARC's success. It was a scientifically meaningless comparison.

CHAPTER 31

Showdown in the Scientific Corral

Opinion Gums Up ERAB Panel's Plan

In the middle of October 1989, three scientific meetings furthered the development of the new science. In Hollywood, Florida, two dozen scientists gave presentations at a special session on "cold fusion" held at the fall Electrochemical Society meeting. Many of them reported positive findings: excess heat, neutrons, and tritium.

One of them was Richard Oriani (1920-2015), a physical chemist and professor at the University of Minnesota. Oriani, along with Fleischmann and Pons, was one of the first scientists who reported not only excess power but also excess integrated energy. He and his team observed 2.2 MJ/cm^3 of palladium during an 11-hour period. "Such magnitudes," Oriani wrote, "are very difficult to rationalize in terms of chemical reactions. At present, we have no evidence of nuclear reactions."

Oriani, like many other "cold fusion" researchers in 1989, did not assert any correlation between experimentally observed excess heat and the theory of deuterium-deuterium fusion, or between any nuclear products and the theory.

The second meeting occurred in Tokyo, Japan. Fifteen researchers presented "cold fusion" papers at the fall meeting of the Atomic Energy Society of Japan. The third meeting, the NSF/EPRI workshop on "cold fusion," was held in Washington, D.C.

NSF/EPRI Scientific Workshop

Despite the political attacks against "cold fusion" and the intense controversy, even within the National Science Foundation, Paul Werbos, Thomas R. Schneider, John Appleby, and Paul Chu pulled off a successful scientific workshop.

The NSF/EPRI Workshop on Anomalous Effects in Deuterated Materials brought together 50 scientists, including the world-famous hydrogen bomb development project physicist Edward Teller. The group included a half-dozen physicists and a half-dozen materials scientists; the rest were electrochemists, physical chemists, surface chemists and general chemists. Garwin was not invited to the workshop. He and Teller had known each other from Los Alamos; Teller had selected Garwin to develop the working design for a proof-of-concept of Teller and Stanislaw Ulam's concept for a hydrogen (fusion) bomb. (Chapter 7) At the NSF/EPRI workshop, Teller encouraged exploration of the new science. On the ERAB panel, Garwin was denying it.

Thirty scientists gave formal presentations at the workshop, and another dozen invited guests came there to listen. The goals of the workshop were to encourage greater cross-disciplinary communication, identify key required factors for the phenomena, and produce published proceedings of value to the research community.

The proceedings included an edited transcript of the discussions that took place after each presentation. These transcripts provide an invaluable narrative of these historic discussions. Four years later, the proceedings — a 711-page book — were finally published by EPRI, long after nearly all the participants had given up on the new science.

Werbos and Schneider were pleased with the success of the meeting and wrote that the participants demonstrated a "high degree of professionalism" and that "both [critics] and advocates engaged in constructive discussions."

Werbos and Schneider didn't seek advance publicity for the workshop, nor did they keep it a secret. In addition to inviting researchers who had reported positive results, they invited three members of the ERAB panel: Allen Bard, Steven Koonin, and Mark

Wrighton. They invited non-panel members who had been critical of the research, including Nathan Lewis, Moshe Gai, Steven Jones, Richard Petrasso, and Walter Meyerhof. Of those, Lewis, Bard, and Petrasso were present, according to the list of attendees.

Most of the workshop took place at NSF headquarters in Washington, D.C., and some of the sessions took place at the Wyndham Bristol Hotel.

Wolf on the Fence

On Oct. 13, the same day that the ERAB panel members met in Chicago, journalist Robert Pool wrote an optimistic news story for *Science* that discussed the forthcoming NSF/EPRI workshop. No press release appeared, but Pool heard about the workshop through the grapevine and made an inquiry to EPRI. Pool featured the Wolf tritium findings.

A few weeks earlier, Bigeleisen had written to his fellow panel members that Wolf was "in trouble" because his neutron and tritium measurements were "inconsistent." Wolf's real trouble was not with his data but with pressure from his peers. Because he was a nuclear chemist running "cold fusion" experiments in the Cyclotron Institute at Texas A&M University, his credibility was difficult to dismiss, and his tritium data was hard to refute. It was another major "unresolved issue" for the panel members. Wolf was in an awkward situation. Pool wrote that Wolf did not want to believe that the tritium he was observing was coming from a nuclear process.

"I'm doing everything I can think of to prove that it doesn't," Wolf said, "but we just keep getting more and more evidence all the time." By this time, the two groups at Texas A&M had run 13 experiments, with tritium levels that ranged from 100 to 1 million times background level.

Pool reported that another half-dozen laboratories had confirmed other phenomena associated with "cold fusion." Wolf had considered the contamination scenario, according to Pool, and had tried to remove and isolate every possible source of error. Nevertheless, Wolf could find no source of contamination.

Wolf's final tests, Pool wrote, would be to perform tests with light water instead of heavy water. If light-water experiments, which otherwise used the same components as the heavy-water experiments, showed tritium, Wolf would be convinced that it was real. It was unambiguous to Wolf that the nuclear reaction, if real, was not fusion.

"If tritium is being produced," Wolf said, "it certainly is not from conventional deuterium-deuterium fusion. Either it's tritium contamination, or it's some unknown nuclear reaction."

His insight was brilliant; the experimental data did not look like fusion. Teller and a handful of other people also got it. Wolf had spoken with Pool several weeks before the article went to press. He had hoped to complete the light-water tests in time for the NSF/EPRI workshop, and he did. At the workshop, in addition to discussing his tests with light-water experiments, he also reported a unique attempt to measure tritium in the headspace as well as in the electrolyte.

He fitted three cells with a separate recombination catalyst. He found that, "in all cases, the concentration of tritium was much higher in the recombinant D_2O compared to the electrolyte, which is the opposite to that expected from separation factors." His cell #C-8 produced 300 times more tritium in the gas phase, both in concentration and in an integrated activity. (Wolf, 1993)

At the workshop, he reported his latest set of experiments: the results of six D_2O cells and six H_2O cells. None of the H_2O cells showed a significant increase in tritium. Neither did five of the D_2O cells. But cell D-6 showed massive amounts of tritium both in the headspace and in the electrolyte. Wolf remained indifferent despite saying a few weeks earlier that null results in the H_2O cells would convince him. He began making up reasons to deny his own data. Now, he explained, he would be convinced only if the number of positive results in the D_2O cells were equal to the number of null results in H_2O cells.

"Due to the statistics of the matter," Wolf wrote, "one D_2O cell does not prove the case one way or the other." He did everything he could think of to rule out contamination. "Laboratories and equipment have been checked thoroughly," Wolf wrote, "and it is clear that no widespread contamination is present. Samples of palladium and nickel have been checked at Los Alamos, and no tritium was found. Similarly, samples of materials have been dissolved and counted in the present study, but not in the numbers necessary and not with proper sampling techniques for a model based on spot contamination."

Try as he might, he could not explain the tritium in cell D-6. The only remaining imaginable scenario was that the palladium used in cells D-6, C-8 and C-9 — and only those three — had been impregnated with high levels of tritium from a mysterious source.

After Wolf's presentation, the discussion finished with comments from Teller and Fleischmann, who reflected on the unexplained variances between the experimental data and the theory of fusion.

Teller came to the workshop with his own idea of how a nuclear reaction could lead to the anomalous results. He proposed that the key might be an as-yet-undiscovered neutral particle that would act as a catalyst to transfer neutrons between nearby nuclei. He lightheartedly called it the meshuganon, a combination of meshugana, the Yiddish word for crazy, and the suffix for subatomic particles.

Détente at the Scientific Summit

Unlike the struggling researchers in the National Cold Fusion Institute during the summer, Fleischmann and Pons, in their basement laboratory, ran more experiments and continued to produce positive results. At the NSF/EPRI workshop, they reported that 23 of 28 experiments had produced excess heat greater than 0.01 watts, with a precision of 0.001 watts. (Fleischmann and Pons, 1993)

Nathan Lewis participated in the discussion that followed their presentation. He was respectful, and Fleischmann and Pons responded to the few questions he asked. The Nathan Lewis who, in May, proclaimed to the world's physicists that he was an expert in the Fleischmann-Pons experiment, who "knew" that they had screwed up their calorimetry, who "knew" how they made their errors, who took Fleischmann and Pons to task based on "facts" he had read in newspapers, was gone. His complaint about how Fleischmann and Pons had failed to stir their cells and how they miscalculated their excess-heat measurements had vanished. Lewis the lion had turned into a lamb.

Here, in Washington, Lewis was able to question Fleischmann and Pons to his full satisfaction, yet neither he nor any of the other 49 scientists there identified any major errors with Fleischmann and Pons' heat measurements. But Lewis had so embarrassed himself in Baltimore that, even if he wanted to, he could not reverse the damage to Fleischmann and Pons' or his own reputation. In 1992, three years later, Lewis lamented to author Frank Close that, in his travels, people remember him best as "the person who killed cold fusion."

In the workshop discussion, Paul Chu asked Fleischmann and Pons about their best case of integrated energy. Pons said they measured 16 MJ over a 13-week period. Bard asked Pons whether he had used an oscilloscope to check for hidden alternating current effects. Yes, Pons had, and he found only 4 mV of oscillation. Menlove asked whether they had monitored for nuclear products. Pons said they had seen a maximum tritium enhancement of eight times. Pons was reluctant to state any neutron data because they still had only simple neutron detection equipment.

Rafelski asked whether they looked for helium. Pons said that, in their search for helium in the cathodes, they were having "difficulties" that were "tantalizingly ambiguous." Pons and Fleischmann were convinced that the nuclear effects were coming from within the bulk of the cathodes. Bockris asked Fleischmann whether he and Pons were sticking to the bulk idea, and Fleischmann, in a roundabout way (typical of Fleischmann), said yes.

Joe Santucci, with EPRI, asked Pons whether they had searched for helium in the headspace of the cells. Pons said it was not impossible but extremely difficult. Back on April 18, Pons had said in a lecture at Los Alamos that he and Fleischmann reviewed some of their earlier data and, to their surprise, saw a large amount of helium. However, Fleischmann and Pons stopped talking about helium soon after physicists began demanding assays of their cathodes.

David Thompson, from the Johnson Matthey Technical Center, said that, in his group's metallurgical analyses of all Fleischmann and Pons' cathodes that produced excess heat, there had been a consistent granular structure that was smaller than the structure on the cathodes that didn't produce excess heat.

Amazing Isotopic Shifts

The most significant report was presented by Debra R. Rolison and William E. O'Grady, researchers in the surface chemistry branch at the Naval Research Laboratory, in Washington, D.C.

Rolison and O'Grady had a difficult time accepting their data, but they reported what they measured. It was the first time any researchers in the field had reported — anomalous isotopic shifts. Isotopic shifts occur when the abundance of nuclei of one isotopic species either increases or decreases after an experiment, relative to the starting abundance. Like the production of tritium, these phenomena can only be the result of nuclear reactions. (Rolison and O'Grady, 1993)

Two of the deuterium/palladium electrolysis experiments performed by Rolison and O'Grady showed a significant increase, 45% and 85%, of the measurements at mass/charge 106, presumably palladium-106. For the measurements of mass/charge 105, they found a decrease of atoms. The sum of the peaks at 105 and 106 measured after the experiments corresponded well to the sum of the initial natural abundances of the isotopes. They analyzed the cathodes using time-of-flight secondary ion mass spectrometry. They also used inductively coupled plasma atomic spectrometry to look for contamination, but their samples were clean.

Rolison and O'Grady did their best to analyze for other isotopic species or compounds that could mimic the mass/charge of 106, but nothing else fit. The other tantalizing fact they learned was that the reaction zone of the anomalies appeared on the surface or near-surface area of the cathode, within the first 1,000 angstroms, rather than within the bulk of the cathode. They also saw significant reductions in palladium-108. The results defied then-current theoretical explanations.

"I continue to be concerned about the Pd-108 peak," Teller said. "If you cannot account for this via molecular impurities, its presence is very difficult to explain. To produce [this effect] would require energy input, so the laws of physics would be violated."

Another national lab had also observed shifts in isotopic abundances. Schneider, in an Oct. 30 summary of the workshop, explained that only after NRL reported its isotopic shifts did Teller disclose that researchers at Livermore had seen an isotopic anomaly — a depletion of lithium-6 — that defied current theoretical explanation. Schneider explained his interpretation of these data:

> The measurement of the isotope shifts suggests that some nuclear transformation is occurring which "dissolves" the deuterium and frees the neutrons which subsequently reacts with the surface of the electrode. Substantial energy could be released (similar to fusion energy) by such a reaction. However, this still leaves a huge gap in our understanding. Together with other experimental data, this strongly suggests that the deuterium-deuterium fusion is not the normal reaction which is involved.

Schneider got it, too, and was heading in the right direction. According to Goodwin's report to the ERAB panel on Oct. 19, Rolison and O'Grady had measured up to 100% increase in Pd-106. Goodwin also wrote that Teller said Livermore had measured lithium-6 depletion within the first micron of the cathode surface.

The authors of the Livermore report wrote that the lithium-6 was not depleted but that it increased by several hundred percent. Either Teller misspoke, or Goodwin misheard him, or the report has a typographical error. The report, sent to John Huizenga, stated:

> SIMS showed the lithium to be greatly enriched in the amu 6 isotope; normal isotopic abundances are 7.52% (amu 6) and 92.5% (amu 7); the SIMS data showed the amu-6 peak to be 2-10 times that of the amu-7 peak. (LLNL Research on Cold Fusion, Sept. 14, 1989, 63).

NSF/EPRI Workshop in the News

Schneider was pleased with the workshop and wrote that it was extremely successful in promoting greater cross-disciplinary communication. He was only disappointed that so few of the formerly vocal critics attended. The dialogue at the workshop, he wrote, was "candid, open and occasionally intense, but the extreme adversarial nature of other meetings was avoided."

In a Nov. 4 letter, Werbos explained the background and development of the workshop. He also mentioned that Chu had, some months earlier, attempted his own "cold fusion" experiment but failed to get positive results. "Chu," Werbos wrote, "was no optimist on the subject."

Werbos, who had attended the Santa Fe workshop, was surprised at how many groups who had reported null results at Santa Fe now reported positive results at the NSF/EPRI workshop.

At 2 p.m., at the end of the workshop, Chu and Appleby, with the input of Schneider and Werbos, issued a simple three-paragraph press release at a news conference. About 20 members of the press attended. The essential information appears in the second paragraph of the press release. It was a direct assault on the ERAB panel:

> New, positive results in excess-heat production and nuclear product generation have been presented and reviewed in a logical, frank, open and orderly manner. Based on the information that we have, these effects cannot be explained as a result of artifacts, equipment or human errors. However, the predictability and reproducibility of the occurrence of these effects and possible correlations among the various effects, which are common for the accepted established scientific facts, are still lacking. Given the potential significance of the problem, further research is definitely desirable to improve the reproducibility of the effects and to unravel the mystery of the observations.

The organizers provided a separate statement from Teller, who also attended the press conference. He recommended that the "high-class work" be supported. Teller thought it was more likely that the experimental results were real, and he suggested the general concept of a neutron transfer process.

Paul Recer, writing for the Associated Press on Oct. 18, quoted Appleby and Chu, who said that there was no evidence that the process would ever be useful in energy devices, based on the limited knowledge to date, but they encouraged more research. "It is an anomalous phenomenon which cannot be black magic," Chu said, "but it cannot be explained by the normal rules of chemistry."

On Oct. 19, Warren E. Leary, the *New York Times* science correspondent in Washington, D.C., wrote a story that was certain to disturb the ERAB panel members. The headline of his story was "Recent Tests Said to Justify More Cold Fusion Research." His angle was the inverse of Bill Broad's July article "Panel Rejects Fusion Claim, Urging No Federal Spending."

Leary could not have made the contrast between the NSF/EPRI workshop and the ERAB panel any clearer. "Only three months after an Energy Department panel suggested playing down work in "cold fusion" experiments because evidence of such a process was 'not persuasive,'" Leary wrote, "experts who attended a workshop convened by the National Science Foundation and the Electric Power Research Institute said promising new work indicates that more research is desirable."

NSF Under Attack

The next day, Oct. 20, NSF administrators were under attack. Robert Park, the spokesman for the American Physical Society, in his weekly "What's New" newsletter, was the first to throw stones. He implied that NSF staff members had held a secret meeting by calling it a "closed meeting," rather than an invitation-only meeting. He didn't mention that opponents as well as proponents had been invited. He made an erroneous assumption about the purpose of the workshop, which had been planned long before the ERAB panel had issued its interim report.

"It was an apparent attempt," Park wrote, "to counteract the final report of the ERAB panel, due on Nov. 15, which is expected to recommend against any substantial new funding. A heavy majority of the 50 scientists invited to attend were drawn from the ranks of those who have claimed anomalous results of some sort. They were all sworn to secrecy about what transpired."

Erich Bloch, the director of NSF, on seeing Park's newsletter, wrote to several staff members within hours. "Please refer to the first item in 'What's New,'" Bloch wrote. "You have a major problem. NSF cannot be in a position where it holds secret meetings. Please see me Monday with a plan on how to counteract this perception — or maybe it is reality."

A few hours later, Carl Hall, with the NSF, issued a press release that made its supportive position clearer: "The conclusion of the workshop is that there is sufficient evidence of anomalous effects to warrant added research." The conclusion was nearly opposite of the conclusion that was about to be confirmed in 10 days by the ERAB panel.

Hall wrote that the meeting was not "secret," that it was widely attended by many people from the NSF, and that it had been reported in advance by Robert Pool in *Science* magazine and Ed Yeates on KSL-TV. More than 200 people had asked to attend the workshop, he said, but were turned away because the structure of the workshop required a focused group of experts. The press release did confirm that journalists had been discouraged from attending the workshop because their presence would have impeded the open scientific discussion.

Park, as well as Morrison, in his own newsletter, wrote that some people were upset that the organizers had issued a press release that included a position on the science, because the participants of the workshop did not come to an agreed-on conclusion. Specifically, certain attendees were upset that the NSF took an affirmative position. One of them, as he revealed in a Nov. 16, 1989, interview, was Lewis.

On Oct. 25, James A. Krumhansl, the president of the American Physical Society, drafted a letter to Mary L. Good, the chairperson of the National Science Board of the National Science Foundation. "I write to you on behalf of the American Physical Society," Krumhansl wrote, "to express our profound concern over the involvement of the National Science Foundation in a closed meeting, in apparent violation of its own guidelines. ... Moreover, the 50 scientists who were allowed to attend agreed not to disclose what they learned to the press."

Krumhansl was not a neutral party in this conflict. Back in May, he had presided over the American Physical Society's ugly Baltimore meeting. The NSF/EPRI workshop organizers and chairmen had learned from the mistakes of Krumhansl.

Krumhansl was correct that the attendees of the workshop had made a gentleman's agreement with each other not to discuss the details of the workshop to the press. He devoted a significant portion of his draft letter to complaining about and blaming Fleischmann and Pons for their shoddy science. According to Krumhansl, their work had been covered in an "ethical cloud" because it diverted hundreds of scientists "from productive lines of research" and they wasted "millions of dollars of public funds." He ended his letter with a call for an investigation:

> It is painful to us to see the NSF, which is the very heart of American science, involved in what appears to be a shoddy and improper activity. We ask, therefore, that the National Science Board examine the circumstances of this meeting. We believe that a complete transcript of what transpired at the meeting should be made public immediately. In addition, there should be a full accounting of the events that led up to the meeting. And finally, steps should be taken to ensure that there is no repetition. Nothing less than the integrity of the NSF is at stake.

Krumhansl sent a more-subdued final version to Good on Oct. 27. But now, he also said that the Council of the American Physical Society planned to discuss the matter at its forthcoming meeting. Krumhansl asked Good to examine the circumstances of the workshop and inform him of the board's findings in advance of the APS Council meeting.

On Nov. 16, John A. White, the assistant director for engineering at the NSF, responded, correcting the "apparent misconceptions" in Krumhansl's letter to Mary Good. Letters were exchanged between Robert Park and people at NSF, and Krumhansl and NSF, to little satisfaction of Park and Krumhansl.

Eventually, on Jan. 18, 1990, Krumhansl took the "nuclear" option: He sent a letter to Rob Ketchum, the chief counsel of the House of Representatives Space, Science and Technology Committee, apparently trying to provoke a congressional investigation. There is no evidence that the committee initiated an investigation.

CHAPTER 32

Panel Prepares Burial

Defending the Status Quo

Before completing its final report, the ERAB panel had to assess the implications of the NSF/EPRI workshop for the panel's own work. Two panelists attended the workshop and reported to the full panel.

Bard Report

Allen Bard distributed his two-page report, through Woodard, to the panel members on Oct. 19. He reported that researchers continued to see neutrons, tritium and excess heat, though his preconception was evident in his identifying them as "believers."

For many scientists, including Bard, the radical idea of nuclear reactions in low-energy experiments was inconceivable. When they could not find scientific ways to dismiss the data, they dismissed the observers as "believers."

Goodwin Report

Goodwin, the ERAB panel technical advisor, attended the first two days of the NSF/EPRI workshop. Woodard sent Goodwin's assessment to the ERAB panel on Oct. 19, as well. His report was a technical, nuts-and-bolts listing of the highlights. He paid particular attention to the reports of excess power and, where available, integrated excess energy, reported by six of the groups, including the government labs of Los Alamos and Oak Ridge. At the top of Goodwin's list, however, he briefly mentioned the dramatic increase of the apparent palladium-106 isotope reported by the Naval Research Laboratory. He also mentioned the lithium-6 depletion at Livermore (Chapter 31). Had it not been for Teller, the confirmatory results from this government lab may never have seen the light of day.

NRL researchers Rolison and O'Grady were apparently worried about reporting a confirmation and becoming part of the controversy. As both Bard and Goodwin wrote in their reports, the NRL researchers pleaded with workshop attendees to refrain from publicly releasing information about their experiments. Goodwin wrote in his report that the national labs should try to confirm positive results and that funding was required. His recommendations were on a collision course with the forthcoming recommendations from the panel members. His encouragement fell on deaf ears.

Unresolved Issues

Panel members Schiffer and Garwin had been discussing how to incorporate the various "unresolved issues" into the final ERAB report. Now they also had the NSF/EPRI workshop with which to contend.

"A more general issue," Schiffer wrote on Oct. 22, "will be how to react, if at all, to the NSF/EPRI workshop and the publicity they raised. I am told that they are planning to have a report for release about the same time as ours — the tone is apparent from the *NYT* article [by Leary] last week."

"As for NSF/EPRI," Garwin wrote on Oct. 22, "that seems to me to be a counter-offensive, come-up-with-anything activity. Nate Lewis was there, and I have heard fourth-hand from him. It is ridiculous to imagine doubling the 25% Pd-106 level. And I would rather believe isotope enrichment reducing Li-6 in the first micron than nuclear reactions. Mostly, I don't believe either."

"On the NSF workshop," Schiffer replied on Oct. 23, "I do not take any of the data seriously. At this point, it is more a matter of public relations. My inclination is to ignore it and proceed as we had planned."

Which they did.

Another "Unresolved Issue," From India

Woodard had told the sub-group coordinators that they needed to get their sections to him by Oct. 24. But the nuclear products sub-group didn't make the deadline. Schiffer wrote to Garwin on Oct. 22 and asked him for his neutron section. "I hesitate to badger you," Schiffer wrote, "but Woodard wants a disk sent tomorrow with the latest version. If I do not hear from you, I will send what I have and then incorporate changes after that."

Garwin was still trying to resolve the "unresolved issue" with Menlove's neutrons and the BARC neutrons and tritium. "They can't process all disks simultaneously," Garwin protested. "We should do our work right and submit the disk on Oct. 25." Schiffer relented and delayed, and Woodard's staff compiled all the sections on Oct. 24 without the nuclear products section.

On Oct. 25, Garwin received the package of materials that he had expected from Iyengar by Airborne Express. The package included a cover letter and a paper of yet another confirmation at BARC. The BARC researchers had performed a titanium-deuterium gas experiment but different from the one performed by the Frascati group. Rather than cycle the temperature of the cells between liquid nitrogen and room temperature, the BARC researchers went in the other direction. They ran the cells through three cycles of heating to 600° C and cooling to room temperature in an atmosphere of hydrogen, followed by another three cycles in an atmosphere of deuterium. Also, whereas the Frascati researchers looked primarily for neutrons in their Ti-D experiment, BARC searched primarily for tritium.

The BARC researchers measured tritium activity at 50-1,000 Becquerels, as compared to a background of less than 0.2 Becquerels. Additionally, after the experiments, they placed the titanium samples, which varied in geometry, onto films to produce autoradiographs. In what may be the 20th century equivalent of Roentgen's discovery of the X-ray, the circular outline of one of the samples appeared.

The authors, Rabindra Kumar Rout, Mahadeva Srinivasan and Anurag Shyam explained the images:

> The fogging observed in the autoradiographs is the combined effect of tritium and characteristic X-rays of the host material. The radiograph of the disc sample indicates evidence of tritium localized in the form of micro-structures. These spots are unevenly distributed on the face of the titanium; there are about 60 to 70 spots in all.

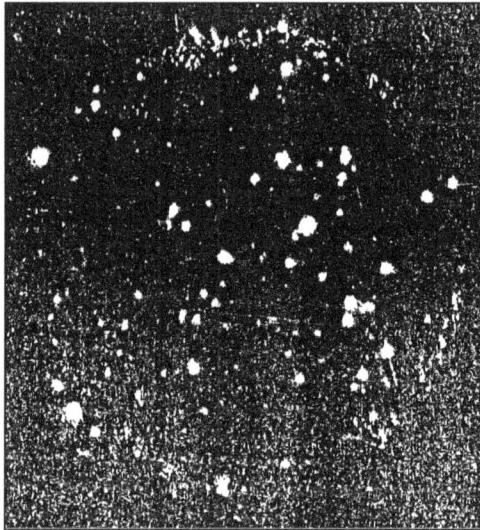

Autoradiograph of titanium disc that had been heated in vacuum up to 900° C and then loaded with deuterium gas

Iyengar was, as usual, enthusiastic and offered Garwin constructive ideas about the science. He finished with this provocative paragraph:

> It seems to me certain that there is no way of escaping the fact that deuterium-deuterium cold fusion does occur in metallic lattices. It is like discovering that neutron-induced fission takes place in uranium. It took a long time before appropriate conditions were invented to make it a source of energy. Perhaps it will take some time before we can obtain

appropriate conditions in the metal lattices, or perhaps using external agents by which energy production along with tritium will be feasible. There is definitely a hope. "Science Marches On." With best wishes, Yours sincerely, P.K. Iyengar.

Final Nail in the Coffin

Woodard sent the draft final report, without the nuclear products section, to the panel members on Oct. 25 by express mail. The panel was to meet in person at DOE headquarters on Oct. 30 and 31 to review and agree on the draft. Before the meeting, the nuclear products subgroup turned in its section.

Woodard notified the panel members of a special briefing they would receive. "One item which does not appear on the agenda," Woodard wrote, "is a briefing by the Naval Research Laboratory on their cold fusion research. They are not prepared to make their presentation during our open sessions, and therefore it is scheduled to take place during lunch on Monday."

Woodard attempted to avoid a repeat of the leak that occurred with the draft version of the interim report. He placed stern warning language on the cover of the Oct. 25 draft. The warning said that the draft had not been approved and did not constitute the panel's findings or the position of the U.S. Department of Energy.

Three days before the last ERAB panel meetings, the House Science Investigations and Oversight Subcommittee held a hearing on thermonuclear fusion research.

At issue was the Department of Energy's interest in promoting greater competition for funds between magnetic and laser fusion research. Fusion research scientists were opposed to such funding competition and lobbied for more funds, according to Barton Reppert, writing for the Associated Press. Reppert quoted Robert L. McCrory, the director of the Laboratory for Laser Energetics at the University of Rochester, Huizenga's university.

"A billion dollars a year to me," McCrory said, "does not seem like too much to have an energy future."

For that fiscal year, fusion researchers were getting only half a billion taxpayer dollars.

At 8:30 a.m. on Monday, Oct. 30, the ERAB panel began its final task. Garwin was unable to attend but sent a letter on Nov. 1 to Woodard and some of the panel members. He expressed his doubts about the excess-heat results presented at the NSF/EPRI workshop. Garwin was also concerned because, according to notes of the meeting he had read, Bard had said that "the calorimetric evidence was getting firmer." The notes also quoted Lewis, who said that the excess-heat results from Oriani (University of Minnesota) and Ernest Yeager (Case Western University) were the "most convincing to date."

Details of the two-day final meeting are scarce. The discussion in Huizenga's book appears weighted toward his point of view. Twenty-five years later, in November 2014, I spoke with Goodwin on the phone. He told me that the meeting was about as exciting as watching paint dry, apart from one crucial 15-minute period.

Ramsey's Prescient Preamble

Ramsey had been abroad for several months and rejoined the panel for its last meeting. On the second day of the meeting, Ramsey wrestled it to a standstill. He had brought with him two documents. The first was his resignation letter; the second was a preamble that he wanted the panel to include in its final report. He offered the panel a choice: accept one or the other.

Goodwin told me that, for a few minutes during the final meeting, sparks flew between Ramsey and Huizenga and the battle of words was fierce. The resignation of Ramsey, the only Nobel laureate on the panel, would have cast a cloud over the panel's final report. Huizenga, as he wrote in his book, thought he had no choice but to allow the inclusion of Ramsey's preamble. Huizenga resented its inclusion and thought that it watered down the impact of the panel's work. On the contrary, Ramsey's preamble was the most important part of the report and this history:

Ordinarily, new scientific discoveries are claimed to be consistent and reproducible; as a result, if the experiments are not complicated, the discovery can usually be confirmed or disproved in a few months. The claims of cold fusion, however, are unusual in that even the strongest proponents of cold fusion assert that the experiments, for unknown reasons, are not consistent and reproducible at the present time.

Norman Ramsey, courtesy Harvard University

However, even a single short but valid cold fusion period would be revolutionary. As a result, it is difficult to resolve all cold fusion claims convincingly since, for example, any good experiment that fails to find cold fusion can be discounted as merely not working for unknown reasons. Likewise, the failure of a theory to account for cold fusion can be discounted on the grounds that the correct explanation and theory have not been provided. Consequently, with the many contradictory existing claims, it is not possible at this time to state categorically that all the claims for cold fusion have been either proved or disproved convincingly. Nonetheless, on balance, the panel has reached the following conclusions and recommendations.

The panel, as it wrote in its top conclusion, was not convinced that "cold fusion" would ever lead to a source of energy:

> Based on the examination of published reports, reprints, numerous communications to the panel and several site visits, the panel concludes that the experimental results of excess heat from calorimetric cells reported to date do not present convincing evidence that useful sources of energy will result from the phenomena attributed to cold fusion.

The panel's top recommendation was that the government should not fund special "cold fusion" research programs — but it left the door open to give money to researchers who were already funded by DOE:

> The panel recommends against any special funding for the investigation of phenomena attributed to cold fusion. Hence, we recommend against the establishment of special programs or research centers to develop cold fusion. The panel was sympathetic toward modest support for carefully focused and cooperative experiments within the present funding system.

The panel members' level of incredulity at the experimental data given to it was reflected in this profound statement in the report:

> Nuclear fusion at room temperature, of the type discussed in this report, would be contrary to all understanding gained of nuclear reactions in the last half century; it would require the invention of an entirely new nuclear process.

There it was, articulated again, the inconceivable: an entirely new nuclear reaction process. The panel members knew they could not get away with saying that "cold fusion" was complete nonsense, but, clearly, that was the message they telegraphed.

The news media and the public interpreted the panel's report as an official government edict that "cold fusion" was a delusion and a mistake. In reality, the report was the opinion and position of the panel members, who simply reflected the prevailing scientific consensus.

Official Government Edict

On Nov. 2, Woodard sent the draft of the final report to the ERAB for approval. His cover letter again made it clear that it was still a work in progress: "This draft report has not been approved by the Energy Research Advisory Board. Likewise this draft does not constitute the position of the U.S. Department of Energy." Nevertheless, that language did not deter its premature release. Two days later, reporter Bill Broad, with the *New York Times,* interviewed Huizenga and included the news in his article titled "U.S. Panel Finds No Evidence of Cold Fusion."

On Nov. 8, Huizenga delivered the final 69-page report to John W. Landis, the chairman of the ERAB. The full board (not the panel) after making minor revisions, unanimously approved the report. With the final approved report in hand, the Associated Press ran its story of the conclusion of the ERAB cold fusion panel. On Nov. 26, the board sent the final report to Admiral Watkins.

Thus, the scientific controversy or, as Huizenga wrote, the presumed scientific fiasco of the century was closed: scientific advancement deterred by human nature.

Defending the Scientific Consensus

Nature Editor David Lindley was satisfied with the outcome and the relief it provided from the potentially disruptive uncertainty of recognizing the possibility of a new nuclear process. In his Nov. 16 article titled "Official Thumbs Down," Lindley reported Huizenga's blanket dismissal of calorimetry and then made two profound statements.

First, Lindley wrote that "discounting the calorimetry, the only anomalous claim still needing an explanation was that of Kevin Wolf of Texas A&M University, who repeatedly finds tritium in his electrochemical cells." Lindley knew nothing about the BARC tritium or about the NRL or Livermore isotopic shifts. There was so much that he didn't seem or want to understand. He certainly was aware of the Bockris and Storms tritium.

Cartoon by Pat Bagley in the style of the Spanish Inquisition. Caption reads "Brother Pons, just admit it's all the work of the devil, and we'll try to forget the whole thing." Depicted "inquisitors" represent MIT, Nature, *and the* New York Times.

Second, Lindley wrote that "new evidence from Wolf, discussed at the ERAB meeting, argues against a nuclear origin. ... His tritium, therefore, must have some kind of low-energy origin." This nonsensical statement revealed the extent to which Lindley had lost his objectivity. If the tritium was produced by some other non-nuclear mechanism, that in itself would have been a major discovery.

"Nevertheless," Lindley wrote, "Wolf has also searched minutely for sources of contamination and found nothing there, either." For the U.S., U.K. and European defenders of the scientific consensus, Wolf was the single outlier whom they could not easily dismiss. ERAB panel member Bigeleisen's Sept. 15 words became ever more meaningful: "It comes to the same conclusion: Kevin Wolf is in trouble." Wolf, therefore, had to know they were coming for him.

A few days later on Nov. 23, *Nature* published the paper "Upper Bounds on 'Cold Fusion' in Electrolytic Cells" from David Williams' group at Harwell. Williams' group didn't just report that they had failed to confirm any of the "cold fusion" claims in their own experiments; they also wrote that their work could authoritatively "establish clear bounds" for the phenomena reported in "cold fusion" experiments *universally*.

Defying the Scientific Consensus

By Nov. 8, 1989, the world had learned that the U.S. Department of Energy had reviewed "cold fusion" and determined that it was an illusion and that it provided no hope as a new source of clean nuclear energy. In spite of this decision, a few independent-thinking scientists refused to go along with the consensus. Among the more-independent thinkers in the U.S. were researchers at two Department of Energy national laboratories.

The first group, at Oak Ridge not only defied the now-official Department of Energy position but also had to contend with the hostile climate by its own lab's management. Mike Saltmarsh, the associate director of Oak Ridge's fusion energy division, had been emphatically denying reports of successful "cold fusion" results at Oak Ridge, according to an Associated Press news story by Lee Dye on June 14.

JoAnn Jacobsen-Wells, a science writer for the *Deseret News*, described the odd situation on Dec. 9 1989. "Scientists at a Department of Energy laboratory," Jacobsen-Wells wrote, "have announced confirmation of the University of Utah's cold nuclear-fusion experiments — even though the department's Energy Research Advisory Board says fusion is an illusion."

Jacobsen-Wells also reported, as she had on June 22, the tritium confirmation from Edmund Storms and Carol Talcott, researchers at the Los Alamos National Laboratory.

On Dec. 12, five scientists presented experimental and theoretical "cold fusion" papers at the American Society of Mechanical Engineers meeting in San Francisco. Gordon Michaels, a member of Charles Scott's team at Oak Ridge, presented the group's paper. An Associated Press writer covered the story on Dec. 13 and clarified the contrast between the Oak Ridge report and the ERAB panel. The AP story said, in part:

> Evidence showing "cold fusion" may be a real way to produce energy was presented Tuesday by Tennessee scientists. ...
>
> Advisory board co-chairman John R. Huizenga, a chemistry and physics professor at the University of Rochester, N.Y., said the panel remains unconvinced by the Oak Ridge experiments.

I never thought that Huizenga might have anything more to say about "cold fusion" than he had said in his book. The only contact between the two of us came through a third party. In January 2006, a few days after I published an online issue of *New Energy Times* magazine that reported new low-energy nuclear transmutation results, Camillo Franchini, an Italian physicist, sent Huizenga a link to my magazine. Franchini wanted to know whether Huizenga was convinced that the new results were real.

"I've paid my dues," Huizenga wrote, "in terms of time and energy spent on examining various outlandish claims made by cold-fusion proponents. If Mr. Krivit has some new and credible results, he can relay them to me through you."

Franchini extended the offer to me. I said no thanks. It was clear to me that Huizenga had made up his mind.

MIT Scientist Pays Penalty

One of the first scientists to pay a penalty for his support of "cold fusion" was Peter L. Hagelstein, an associate professor of electrical engineering at MIT who is often misidentified as a physicist. Hagelstein had been quick to suggest theoretical explanations for "cold fusion" — as fusion — as early as April 6, 1989, based solely on the limited data reported by Fleischmann and Pons.

Now, in December, Hagelstein was presenting his latest ideas at the American Society of Mechanical Engineers meeting. He had many theoretical ideas, but he did not have a comprehensive and complete model to fully explain the reported phenomena. Plenty of experimental evidence by this time showed that the results looked nothing like fusion. Hagelstein nevertheless proposed a mechanism for the heat transfer in the palladium lattice based on a "virtual fusion reaction" that produced helium as the dominant product: $^aH + {}^bH \mathrel{-\!\!-\!\!>} {}^{a+b}He.$

Anthony Flint, reporting for the *Boston Globe* on Dec. 6, wrote that, when Hagelstein's colleagues at MIT learned about the forthcoming paper, they began raising concerns. "Hagelstein is being opposed for tenure," Flint wrote, "by factions who believed that he embarrassed the institute by defending cold fusion in the spring. ... Critics say it was not so much Hagelstein's ideas as the way they were presented — through the news media ... rather than in the confines of peer review among scientists."

Flint quoted Ronald Parker, the director of the MIT Plasma Fusion Center, who said, "It just smacked of a lot of grandstanding." This is the same Parker who invited reporter Nick Tate to his office and told him that Fleischmann and Pons had committed fraud.

Hagelstein did nothing different on April 12 than Mark Wrighton did on April 6: Both had issued press releases, Hagelstein in support of "cold fusion" research, Wrighton in opposition. Both press releases included official statements from MIT's provost, John Deutch. Robert Pool explained the story on Dec. 15, 1989.

"Ironically," Pool wrote, "although it was publicity surrounding Hagelstein's ["cold fusion"] work that irritated many of the MIT faculty rather than the fact that he was doing it, Hagelstein did little to encourage the publicity. He never held a press conference and spoke only grudgingly to the press, giving few details of his work." As Pool suggested, and as the Wrighton press release shows, the real problem was not the way Hagelstein conducted himself but the subject matter.

Hagelstein's support — and therefore the association of MIT —with the now-officially discredited "cold fusion" made his more-conservative colleagues at MIT uncomfortable. Despite this friction, Hagelstein was granted tenure in 1990.

BARC-1500 Report

Sometime in December 1989, Padmanabha Krishnagopala Iyengar, the director of India's largest nuclear research center, Bhabha Atomic Research Centre, in Trombay, Bombay, India, and his friend and colleague, Mahadeva Srinivasan, the director of the Neutron Physics Division and assistant director of the Physics Group, published the BARC-1500 Report.

The 153-page book contains 11 papers on electrolytic cell experiments, four papers on gas-loading experiments, and five papers on theoretical ideas. As of 2015, the BARC-1500 report remains the largest and most significant set of reports of direct nuclear measurements in the field.

A Scientist Imagines Errors

In early winter 2014, I asked Richard Garwin why he, as a member of the ERAB panel, discounted the tritium measured in "cold fusion" experiments at BARC in India and at Texas A&M in the U.S. Here are the relevant excerpts from our conversation:

> RICHARD GARWIN: I took it seriously. I pointed out that I don't think it's real because of the fact that they have tritium experts there at the lab. That means they have tritium there. And it's very easy to get contamination depending on which technicians are doing what. It's really very easy to contaminate the samples, when they're being produced or transferred or measured. So you really have to go into detail as to how to measure it. I don't know that that was their problem, but this is such a hot field, such an important question, and it's really quite surprising that they haven't followed up.
>
> STEVEN B. KRIVIT: Did you have any sort of evidence that they

had contamination? Or is this just a wild speculation on your part?

GARWIN: It's not wild speculation. I have a lot of experience in these fields.

KRIVIT: On the point of contamination, is there anything specific that you know about this set of experiments that gave you some reason to think that they were the result of contamination?

GARWIN: If they are making measurements and they're finding tritium, then either it's real or it's — either it comes from what they're doing, or it comes from contamination. That's what I'm saying. Let's go on. I don't want to argue about it.

KRIVIT: What about the Bockris group's tritium at Texas A&M University?

GARWIN: I have great suspicions about the Bockris work. I think there was a problem with a technician there. Let me give you an example. I had conceived an experiment in 1951 at University of Chicago. I had the first opportunity work on it when I was at IBM in 1953. My objective was to make artificial diamonds.

In the next few minutes, Garwin told a long story of how the technician who worked on the experiment for him appeared to observe positive results, but later, the technician couldn't repeat the results.

GARWIN: My own feeling about that was that his technician really caught the fervor of making these things and may have put small diamonds into the experiment. So anyhow, I'm always suspicious about whoever is doing the experiment: not only the principal investigators but the others. And I was worried about the Bockris work for that reason.

Now the two BARC laboratories, I've never been to BARC; well, I was actually. I was in Bombay in the summer of 1960 for two weeks, but that was long before such activities and before there was much expertise there.

You say, in all cases, the tritium was measured by tritium

experts; yes, but there's always a possibility of contamination. Varying amounts of tritium were measured. The ratio of neutrons to tritium was different [than expected by theory]. It was hard to believe that this was the same phenomenon. There should be fast neutrons accompanying the tritium if the tritium is the result of a fusion reaction. All of which doesn't hang together; but I'm not doing the experiment, so it's up to them to reproduce the experiment and to find some way to have a convincing result.

KRIVIT: I'm trying to ascertain the significance of a few experimental findings. You seem to dismiss them.

GARWIN: Oh, yeah, there are all kinds of experimental findings that are wrong, and they're wrong for various reasons. So you wait until they can be reproduced, until they're a physical phenomenon, not when they're just some numerical word phenomenon.

KRIVIT: What was wrong with the Texas tritium?

GARWIN: I don't know. Where is it now?

KRIVIT: What was wrong with the BARC tritium?

GARWIN: I don't know. If it was real, that's fine. If it's not real, if they haven't produced it, then it's probably contamination. Sometimes, people just lie about their results, or the people who report to them lie. There are all kinds of such experiences in science. People paint their mice, and that was at Rockefeller University.

KRIVIT: So you're just imagining that there would have been some errors?

GARWIN: Yes.

KRIVIT: You know the saying that extraordinary claims require extraordinary evidence. When you approach an extraordinary claim from a critical perspective, do you also concur with the idea that critics have a responsibility to back up their assertions?

GARWIN: Certainly. They have to say everything they know and do the work required.

Denial of New Science

Garwin denied the evidence of new science based on imagined errors, imagined sloppiness, and imagined fraud. Garwin was an advisor to the federal government, a recognized authority in science, and a public figure often cited by the news media. His denial was not just a matter of personal opinion; it carried the weight necessary to influence public opinion, the media, and federal science research policy. Garwin did not act alone. He had the support of other physicists on the ERAB panel: John Huizenga, Jacob Bigeleisen and John Schiffer, all of whom had achieved stature and success as a result of their work during the Cold War era.

Clarke's First Law of Prediction

Sir Arthur C. Clarke, (1917-2008), a British science fiction writer, science writer and futurist, followed "cold fusion" research with interest. He was better able to connect the significance of the early research with the possible future of its benefit to society. He wrote the preface for my first book on "cold fusion." Two of his paragraphs are germane.

> As for the [critics], I can do no better than to quote my own First Law, which I first expressed more than 40 years ago: "When a distinguished but elderly scientist says something is possible, (s)he is almost certainly right. But when (s)he says something is impossible, (s)he is very probably wrong."

> Perhaps the most disappointing outcome would be if cold fusion turns out to be merely a laboratory curiosity, of some theoretical interest but of no practical importance. But this seems unlikely; anything so novel would indicate a major breakthrough. The energy produced by the first uranium fission experiments was trivial, but everyone with any imagination knew what it would lead to.

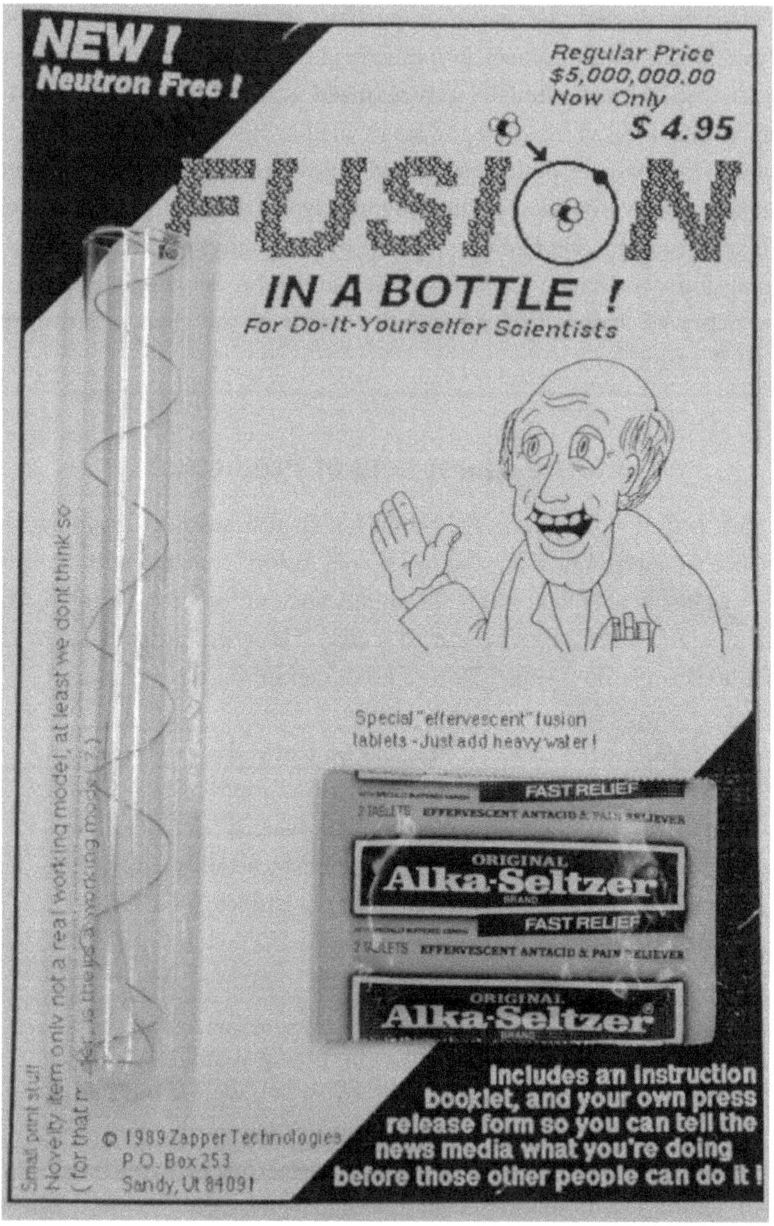

Satirical novelty item making fun of "cold fusion"

CHAPTER 33

Paradigm Paralysis

Inexplicable Data; Understandable Behavior

By March 1990, opinions about "cold fusion" had become deeply divided. On one side, people thought it was a valid nuclear phenomenon; on the other side, people thought it was all just a mistake, a delusion or fraud. This division was the result of paradigm paralysis.

John C. Harrison, the director of the National Stuttering Project, described this concept in a lecture at the First World Congress on Fluency Disorders in Munich, Germany, on August 1-5, 1994:

> A paradigm is a model or a pattern. It's a shared set of assumptions that have to do with how we perceive the world. Paradigms are very helpful because they allow us to develop expectations about what will probably occur based on these assumptions. But when data falls outside our paradigm, we find it hard to see and accept. This is called the paradigm effect. When the paradigm effect is so strong that we are prevented from actually seeing what is under our very noses, we are said to be suffering from paradigm paralysis.

Nuclear reactions, according to the dominant scientific paradigm, take place only under high-energy conditions, such as particle bombardment, or under extreme temperatures. However, emerging scientific evidence suggests that nuclear reactions may also take place under certain unusual and much milder, low-energy conditions, for example in electrolytic or gas-loading experiments with hydrogen or deuterium in the presence of certain metals.

My perspective on the "cold fusion" controversy has been influenced primarily by a quote from German physicist Max Planck (1858-1947):

> A new scientific truth does not triumph by convincing its opponents and making them see the light, but rather because its opponents eventually die, and a new generation grows up that is familiar with it.

People have condensed his quote to "Science advances one funeral at a time." The release of the ERAB panel's report, the final nail in the coffin for "cold fusion," reinforced the views of people who followed the dominant scientific paradigm. Now, they could simply point to the official report rather than argue why "cold fusion" was wrong.

Paradigm Threatened

The ERAB panel had collected extensive evidence about the new phenomena. Hundreds of pages of scientific information were available to the 22 members of the panel, but they failed to recognize the infrequent evidence of nuclear reactions buried among the failures. Until 2015, the only public document from the 1989 Department of Energy "cold fusion" review was the ERAB panel's final report.

Researcher Joe C. Farmer and colleagues at Lawrence Livermore National Laboratory filled out the panel's survey form around Aug. 15. The researchers wrote that with energy-dispersive X-ray spectroscopy (EDX) they only found palladium in the starting material. On their survey form, the LLNL researchers wrote simply that they had seen "no evidence of nuclear fusion." When asked about this activity, even 26 years later, Farmer declined to discuss the matter. In a more detailed report on Sept. 14, 1989, they told the Department of Energy that they had observed significant shifts in lithium-6 and lithium-7 isotopes.

CHARACTERIZATION EDX of starting material —
STRUCTURAL   only Pd detected.
CHEMICAL
BEFORE OR AFTER USE Auger, SIMS, &XPS of
METHODS   surface after use. Elements
RESULTS   detected included Si, S, Cl, C,
                     Ca, Pd, O, Fe, and possibly Cu & Mg by Auger.
                     SIMS detected H, Li, C, Na, Al, K, Ca, Fe, Pd,
                     and possibly Ti, Cr, & Cu. XPS showed Fe,
                     O, Ca, Pd, C, Si, and a trace of Al.

NOTABLE
OBSERVATIONS

              No evidence of nuclear fusion.

D / METAL RATIO ATTAINED

EXPERIMENT YIELDED HEAT   _____yes   __X__no
               NEUTRONS   _____yes   __X__no
               TRITIUM   _____yes   __X__no
               HELIUM   _____yes   __X__no

Portion of survey response by Joe C. Farmer and colleagues at Lawrence Livermore National Laboratory sent to ERAB panel reporting an array of transmutation data.

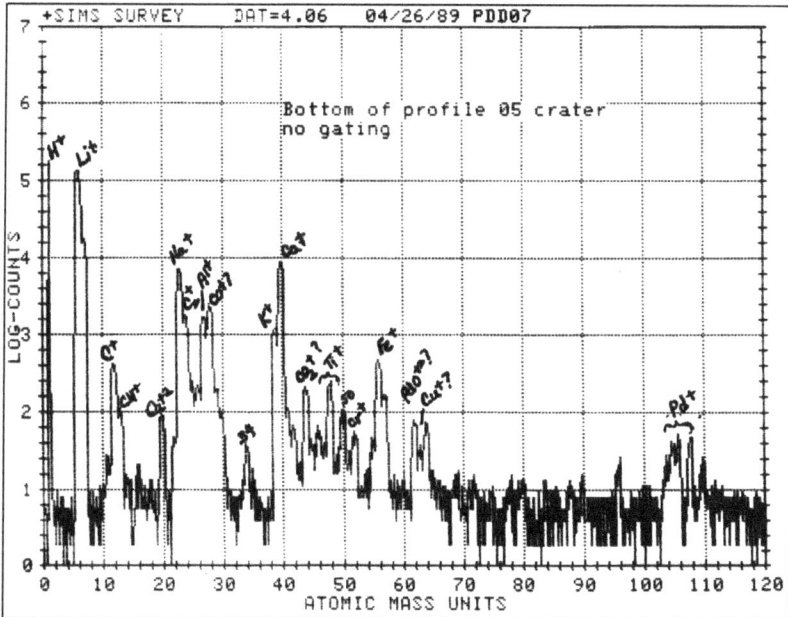

Spectrum of atomic abundances observed after a Lawrence Livermore National Laboratory "cold fusion" experiment.

BARC Pioneers in Rare Company

A dozen groups in India's Bhabha Atomic Research Centre (BARC) had observed evidence of nuclear reactions, but not everyone at the lab enthusiastically embraced the new science. In addition to terminating domestic "cold fusion" research, the U.S. panel's report brought the international research, particularly in Italy and India, nearly to a standstill.

Soon after Mahadeva Srinivasan and Padmanabha Krishnagopala Iyengar published their BARC-1500 Report, "cold fusion" lost a crucial ally in India. Iyengar's term as director expired on Jan. 31, 1990. He followed the traditional career path of all BARC directors and was appointed the chairman of the Atomic Energy Commission of India. Rajagopala Chidambaram was slated to be the next director, although he wasn't officially appointed until April 5, 1990.

Chidambaram had been the director of the Physics Group, of which Srinivasan had been the assistant director. Chidambaram had never been fond of "cold fusion." On June 5, 1989, he and another nuclear physicist, Vinod C. Sahni, published a paper in India's *Current Science* journal. In their one-page paper, they guessed that the formation of palladium-deuteride in electrolytic cells, a chemical process, produced heat that was of the same order of magnitude as the heat claimed by Fleischmann and Pons. Using the diplomatic language of science, they had said that Fleischmann and Pons' heat claims were bogus.

Chidambaram and Sahni did not indicate that they had analyzed the Fleischmann-Pons data or performed any of their own experiments. They, like Garwin, had dismissed "cold fusion" with prejudice within weeks of the initial announcement.

"If it is finally confirmed," Chidambaram and Sahni wrote, "this so-called 'cold fusion' would be physically very interesting; the possibility that it will lead to a significant new energy source appears doubtful at present."

Srinivasan told me what happened the day that Chidambaram took charge of BARC:

> He announced that there would be no further institutional support for cold fusion research. He pulled me into his office and lectured me for 2.5 hours that we must stop the work. This decision was a direct consequence of the DOE report.
>
> Nevertheless, Iyengar and I stuck to our guns. We were allowed to do what we liked using resources already available. Many of the other groups did not want to risk their careers, and so they brought their cold fusion work to an end. We came down from a level of 50 scientists actively engaged in cold fusion to about 15 who had the courage to continue quietly.

Many years later, in 2011, I had the chance to speak with Srinivasan about this. I had gone to India to report on the 16th International Conference on Cold Fusion, in Chennai. My flight, as is typical from the U.S. to India, got me there at 2 in the morning. A few hours and a short nap later, I was having tea with Srinivasan on the lawn of the Gandhinagar guest house. Srinivasan was excited about a news story on "cold fusion" that was coming out in the next day's *Times of India*. I told Srini, as people often call him, that I had seen an article in the *Deccan Chronicle* a few months earlier that had quoted Chidambaram speaking unfavorably about the research. In a flash, Srini's mood changed. I could see the anger come over his face and hear the irritation in his voice. Without knowing any of the history, I had inadvertently triggered a nerve.

Srini told me how Chidambaram shamed him many years earlier during a meeting in front of all the associate directors. Srinivasan had shown his newest results to Chidambaram in order to get official approval to present them at a science conference. He did give his approval, but not without spite. "Srinivasan," Chidambaram said, "you are a loner."

"Cold Fusion" Scientists Convene

In Utah, on March 29-31, 1990, researchers in support of "cold fusion" gathered for what would become the first of a long-running series of conferences that were later named the International Conference on Cold Fusion and denoted with the acronym ICCF.

Several of the outspoken opponents, despite having declared "cold fusion" just a huge mistake, attended the conference, according to an article in MIT's *Tech Talk*, written by Gene Mallove. John Huizenga, who led the ERAB "cold fusion" panel, was there. So were MIT professors Richard Petrasso and Ronald Ballinger. Douglas Morrison, the author of a series of newsletters about "cold fusion," was there, too.

The conference had been sponsored by the National Cold Fusion Institute. By this time, Pons had changed his mind and begun to work with some of the staff at the institute. The first director of the institute, Hugo Rossi, had resigned in the fall of 1989 after no progress had been made. Rossi was replaced by electrochemist Fritz Will in February 1990. Since 1960, Will had worked at General Electric's Development Center, in Schenectady, New York. Several presentations are worth noting.

Reproducibility

Although many people rejected "cold fusion" experiments based on the conflict with fusion theory, Will, was more open-minded. Even though he did not understand the underlying mechanisms, he still recognized the validity of the experimental evidence. The reproducibility problem didn't discourage him, either, as he explained in his opening address at the ICCF-1 conference:

> The history of science and technology has many examples where irreproducibility had been experienced for years. A prominent case is the metal-oxide semiconductor. It took years of effort with multimillion-dollar expenditures to achieve reproducible performance of such semiconductor devices. What ultimately led to reproducibility was the careful control of the level of impurities, most notably sodium.

Johnson Matthey Presentation

Johnson Matthey, the supplier of Fleischmann and Pons' precious metals, revealed interesting surface anomalies in the post-experiment cathodes used by Fleischmann and Pons.

The most significant anomaly they noticed was that one of the rods used for the cathode showed a "wrought microstructure which would ordinarily require a temperature of more than 200°C. Another rod showed recrystallisation, [which] would normally require a temperature of more than 300°C."

The company could not provide any conventional explanation for the morphological changes. As with the cathodes from the Naval Research Laboratory and the researchers at Livermore, one of Fleischmann and Pons' cathodes showed an anomaly with lithium-6 that was "consistently different from the expected values." The Johnson Matthey scientists offered no more detail about the lithium-6.

Defiant Hutchinson

A scientist from Oak Ridge had revealed to John Bockris at the May 1989 Santa Fe workshop that he had seen some positive results. That scientist was probably Donald P. Hutchinson, from the Oak Ridge Physics Division. On Sept. 14, 1989, he submitted a report to the ERAB panel about his positive excess-heat result.

Hutchinson presented his results on the first day of the ICCF conference, according to the program. The proceedings, however, state that his paper was not submitted. The paper, "Initial Calorimetry Experiments in the Physics Division-ORNL," was published as an internal report by Oak Ridge in May 1990 and is available on the U.S. Office of Science and Technical Information (OSTI) Web site.

Hagelstein Switches to Neutrons

On Dec. 12, Peter Hagelstein had presented his theoretical ideas about "cold fusion" in a paper titled "Coherent Fusion Theory" at the American Society of Mechanical Engineers meeting. Hagelstein had proposed a mechanism for the heat transfer in the palladium lattice based on a "virtual fusion reaction" that produced helium as the dominant product: $^{a}H + \,^{b}H \longrightarrow \,^{a+b}He.$

Now, three months later, at ICCF-1, in a paper with almost the same name, "Status of Coherent Fusion Theory," Hagelstein, the sole author on both papers, did an about-face, switching from the concept of "cold fusion" to the concept of neutron-catalyzed reactions.

"We," Hagelstein wrote, "are investigating two-step coherent reactions which begin through weak-interaction-mediated electron capture, which, in hydrogen isotopes, would produce off-shell virtual neutrons. No coulomb repulsion occurs for virtual neutrons."

Fusion was out. Weak interactions were in. Although he didn't say so, Hagelstein's ideas were a more-detailed explanation of the general concept presented by Edward Teller in October 1989 at the NSF/EPRI Workshop. Teller had suggested the concept of an as-yet-undiscovered neutral particle that might act as a catalyst to transfer neutrons among nearby nuclei.

Teller's idea was common knowledge by this time, as evidenced by the fact that physicist Robert T. Bush, at California State Polytechnic University-Pomona, presented his own neutron-transfer theory at this conference and credited Teller with the original idea. (Bush, 1990)

ICCF-1 — Fleischmann-Pons

In all the chaos of 1989, Fleischmann and Pons were not able to complete an adequate paper until the end of the year. They had submitted their manuscript to the *Journal of Electroanalytical Chemistry* on Dec. 21, 1989, and revised it on March 28, 1990. The 55-page paper was published in July 1990. (Fleischmann et al., 1990) They presented a version of it at ICCF-1.

One of the biggest deficiencies in their 1989 preliminary note had been the lack of clear evidence of integrated (cumulative) excess energy over the full duration of an experiment. Without such evidence, other people had no way of knowing whether the excess heat claimed by Fleischmann and Pons was real or was simply an illusion caused by a previously unrecognized energy storage-and-release mechanism.

Measurements of power (heat) and measurements of peak excess power (heat) were insufficient to rule out this possibility. The following data given in their 1990 paper resolve this uncertainty.

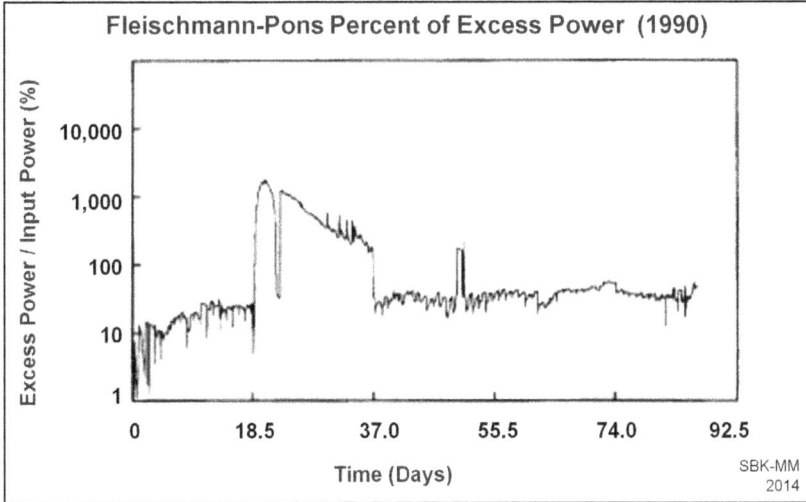

Measurements of rate of excess-heat generation as a function of time.
Calculation is (Power Out - Power In)/ Power In

The first graph shows that, throughout the duration of the run, there was no endothermic period; heat output was equal to or greater than electrical input. Here are the highlights of the run:

- From day 0 to day 5, excess power (heat) ranged between zero and 10%.
- From day 5 to day 18, excess power (heat) ranged from 10% to 20%.
- At day 18, a heat burst produced 1,000% excess power (heat) for four days.
- The heat burst remained above 100% excess power for 18 days.
- The cell continued to produce about 30% excess power (heat) for another 50 days.

There was no period of endothermic response, therefore no energy storage mechanism. As they stated in their 1990 paper, they had observed an unidentified heat-producing process that was 100 to 1,000 times larger than that possible by any known chemical processes.

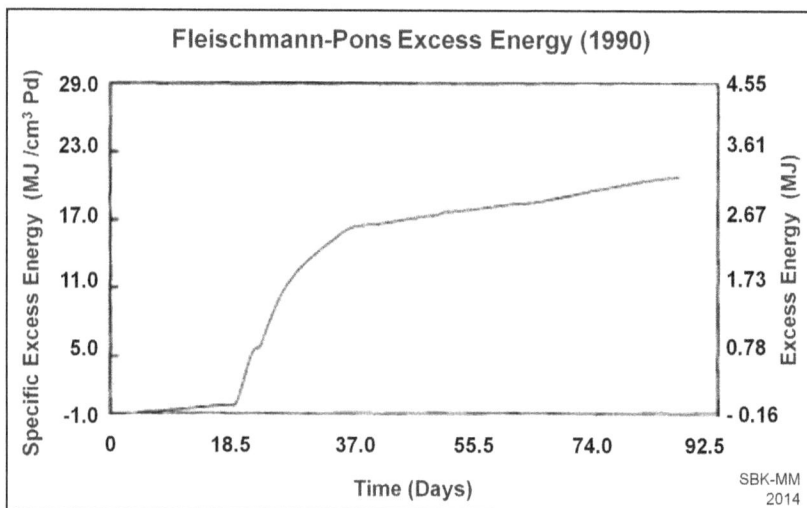

Measurements of total specific excess-energy output (left scale) and total excess energy (right scale) as a function of time. (Scale on right added by S. B. Krivit)

Fleischmann and Pons thought that the excess heat depended on cathode volume, rather than cathode surface. Thus, when they reported their heat values, they normalized excess energy to the volume of palladium. This represents the scale on the left in the excess-energy graph.

Their preference for equating energy release to the amount of presumed reactant material was a reasonable way to look at the energy production. It also made comparing heat production easy across dissimilar cathode geometries and sizes.

That way of representing data is confusing to nonspecialists. At my request, electrochemist Melvin Miles, a close colleague of Fleischmann's, calculated simple values for the excess energy. This represents the scale on the right.

However, the fuel appears to be hydrogen or deuterium, and for this reason, Fleischmann and Pons were probably wrong in equating energy release to the amount of palladium.

The measurements shown in the previous two graphs provide a quantitative analysis of the heat. A different way to look at Fleischmann and Pons' anomalous heating phenomenon is qualitatively. Fleischmann's 1991 paper provides an example. (Fleischmann, 1991)

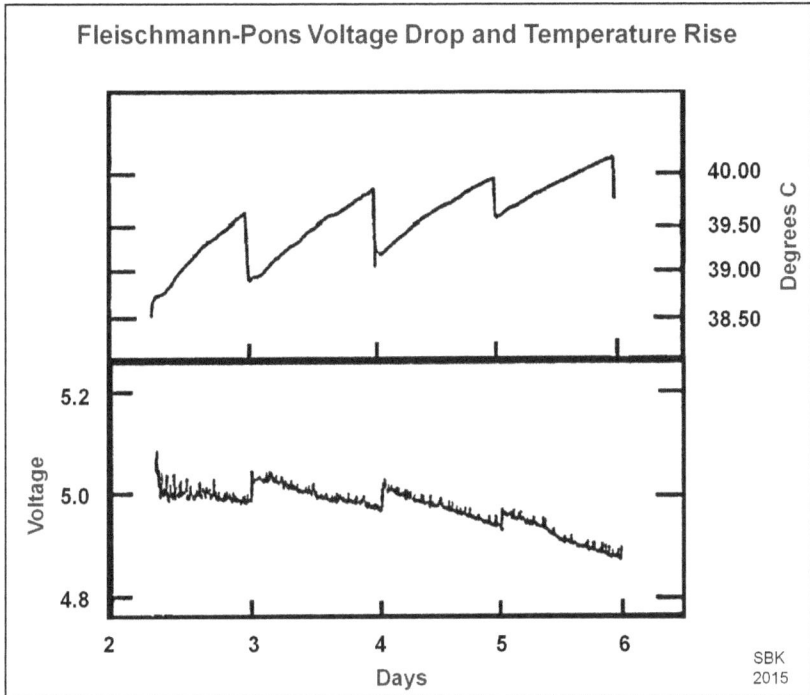

Measurements of anomalous heat phenomenon reported by Fleischmann in 1991. Cell voltage and cell temperature run in opposite directions. Cell current was 400 mA, water bath temperature was 30.00° C, room temperature was 21° C. The rate of excess heat generated at the end of days 3-6 was 0.045 W, 0.066 W, 0.086W, and 0.115 W. Integrated (cumulative) excess enthalpy was approximately 26 kJ.

In the experiment reported in the 1991 paper, simultaneous data for voltage and temperature are displayed. While heat output is increasing, cell voltage is decreasing.

Fleischmann and Pons used constant-current to supply electricity to their cells. Because current is fixed, and the cell is getting hotter — thereby reducing resistance — voltage and therefore input power must drop.

This simple graph is the clearest illustration of Fleischmann and Pons' anomalous heat phenomenon. According to all existing scientific understanding, cell output temperature should go down when electrical input power goes down.

Many of Fleischmann and Pons' critics had suggested that they had made errors in their calorimetry or mathematics. Unlike the quantitative analysis, this qualitative analysis does not rely on mathematics or calorimetry. The qualitative analysis makes such criticisms moot.

The best reference for a comprehensive analysis of Fleischmann and Pons' excess-heat phenomenon is Charles Beaudette's book *Excess Heat & Why Cold Fusion Research Prevailed*, 2nd ed., particularly pages 5-8 and 357-60. (Beaudette, 2002)

Lewis and Koonin's Parting Shot

On March 23, 1990, Nathan Lewis and Steven Koonin, of Caltech, used the *Los Angeles Times* newspaper to depict their version of the history.

"Manipulation of the popular media was instrumental in creating and sustaining the furor," Lewis and Koonin wrote. "As the facts became fully known, all of the highly touted claims were withdrawn or severely qualified. To its credit, the federal government sought better scientific advice. An expert group commissioned by the Department of Energy visited all of the laboratories claiming to have observed cold fusion."

Here are half a dozen things that were true about Lewis and Koonin's statement:

1. Fleischmann and Pons informally withdrew their neutron-gamma data, although the data were never a major part of their claims.
2. Fleischmann and Pons stopped publicly touting their excess-heat claims as extrapolations.
3. Fleischmann and Pons corrected a few calculation errors that appeared in their 1989 preliminary note.
4. Charles Martin, from Texas A&M University, developed uncertainty about his excess-heat claims.
5. Georgia Tech retracted its claim of neutrons.
6. Two graduate students at the University of Washington who thought they had seen tritium retracted their claim.

Here are half a dozen things that were not true about Lewis and Koonin's statement:

1. Neutrons were confirmed and detected by many groups, in much larger rates, and using much better detectors than by Fleischmann and Pons.
2. Tritium was confirmed and detected by many groups and in much larger amounts than by Fleischmann and Pons.
3. The Lawrence Livermore and Belgian fusion research laboratories had observed cathode heating events that defied easy explanation.
4. Livermore and the Naval Research Laboratory had seen isotopic shifts that were virtually impossible to explain by anything but nuclear reactions.
5. Excess heat had been confirmed by at least half a dozen laboratories.
6. Nearly everyone who looked for gamma-rays confirmed that they didn't exist or were negligible.

Lewis and Koonin explained the lessons of the "cold fusion" story:

Scientists would have been ecstatic if cold fusion had been real. But publicity alone didn't make it true. Confirmation by independent researchers was the only way to be sure. Regrettably, the present consensus is that Pons and Fleischmann were simply wrong. This follows not from prejudice but from many careful experiments performed and thoroughly documented by interdisciplinary teams using apparatus far more sophisticated than that available to the Utah researchers. The continuing claims of a few "believers" are plagued by irreproducibility, inconsistencies and the absence of peer-reviewed publication. Most telling, however, is that nobody else can make cold fusion work.

True progress withstands the test of time. Although cold fusion excited our imagination, in the end it was just another corrected mistake. Thus, the lessons it teaches are more important than the experiments themselves. We scientists, the media and the curious public would do well to remember them when trumpets herald the next unverified discovery.

This was the story as Lewis and Koonin wished to believe it. It wasn't so. The "careful experiments" at Caltech that measured no excess heat, no neutrons, no helium and no tritium corrected no mistake. Lewis and Koonin had pulled the wool over the public's eyes.

John Maddox: Parting Shots

Sir John Maddox, the editor of *Nature*, the most prominent science journal in the world, expressed his opinions about "cold fusion." Maddox had no advanced degree, let alone one in science. He earned the equivalent of a bachelor's degree in chemistry in 1947 from Christ Church College, in Oxford, England. Nevertheless, he earned respect from the scientific community for his broad knowledge of science and his eloquence. Here are his comments from the BBC-TV *Horizon* program titled *Cold Fusion*, which aired on March 26, 1990:

> I think it will turn out, after two or three more years' investigation, that this is just spurious and unconnected with anything you could call nuclear fusion, thermonuclear fusion. I think that, broadly speaking, it's dead, and it will remain dead for a long, long time.
>
> My own belief is that Pons and Fleischmann, for a very particular reason, had come to nurture a delusion about what was happening. ... They were convinced that this discovery was going to be of immense importance, as indeed it would have been if it were true, and was going to make a great fortune for somebody, themselves included. And so they kept it very secret. As a consequence, they were entirely isolated from the natural day-to-day skepticism of the scientific community, and they didn't get the help from the scientific community that they would normally have had.

In his March 29, 1990, editorial, "Farewell (Not Fond) to Cold Fusion," Maddox likened the claims to magic and unicorns and called scientists who had confirmed the results "believers." However, in 1989, by a very strong majority, confirming scientists said that their phenomena represented anomalous nuclear effects, not fusion. The term "believers" was unjustified at this time, although Fleischmann was still playing word games, as he told the *Deseret News* on March 30, 1990.

"We are convinced there are nuclear processes, but we are not convinced that it is fusion," Fleischmann said. "Our position is exactly what it was last spring. We have not backtracked."

They had, of course, backtracked from the fateful day on March 23, 1989, when they stood in front of a swarm of reporters and television cameras and declared that they had achieved a sustained fusion reaction.

In his editorial, Maddox also encouraged scientists to join with him and deny the possible social benefit of the research:

> What has irretrievably foundered is the notion that cold fusion has great economic potential. The time has come to acknowledge that. It would be a cruel deception of a largely amused public not to admit that simple truth. And it would be a serious perversion of the process of science to obfuscate the failures of the past year by reference to the difficulties of measurement in an admittedly difficult field.

BBC *Horizon* produced another show, "Too Close to the Sun," which aired on March 21, 1994. Maddox's final comments about Fleischmann revealed Maddox's deep commitment to defend the dominant science paradigm, so much so that he depicted Fleischmann as a cult leader:

> I think it's inevitable that if you have a persuasive, articulate, charismatic man who believes he's made a very important discovery that challenges scientific orthodoxy, his students, his colleagues, will have to take a view: Are we for him or against him? And in a way, it becomes quite quickly like the founding of a religion. You have a charismatic leader who persuades some people, who become his followers, that

they have seen the true light; and there's a large element of that, I fear, in what's happened.

David Williams' Parting Shots

David Williams, in 1989, was an electrochemist at the Harwell laboratory in the U.K. He had "only" seen occasional bursts of excess heat and neutrons, as discussed in Chapter 25, and denied that he had seen any positive results. Williams appeared on the 1990 and 1994 Horizon "cold fusion" shows. On the 1994 show, he dismissed the new science with ridicule:

> We weren't able to see it because the results are inherently irreproducible. This, of course, makes it hard for people to confirm or deny. And this then makes it very like all sorts of other phenomena, like, um, flying saucers! I mean, I could say, "Ha, I just saw a flying saucer just go past the window just there." And you could say, "You didn't see a flying saucer? Of course you didn't see it. You weren't looking!"

Williams wrote an obituary for Fleischmann that was published by the Royal Society of Chemistry. He had been a student of Fleischmann and, aside from "cold fusion," regarded Fleischmann fondly. But he included one sentence in the obituary that, like the statement by Lewis and Koonin, was history as he wished it to be.

"All of the results in the original [Fleischmann-Pons preliminary note,]" Williams wrote, "turned out to be of insufficient accuracy to support the claims that have been made."

The preliminary note was deficient in that it did not enable other scientists to replicate the experiment. However, that had nothing to do with Williams' point, which was that Fleischmann and Pons didn't have the data to support their claim of excess heat.

I wrote to Williams and asked whether he could defend that statement. I explained to him my understanding of the data reported by

Fleischmann and Pons in the preliminary note, which remained valid and which had been corrected. Williams did not respond.

In that single sentence, couched in the language of science, Williams had given the grossly misleading impression that all aspects of Fleischmann and Pons' original claims had been disproved. In writing the obituary, Williams had not only a historical responsibility but also a moral responsibility to honor Fleischmann's life and achievements.

Williams explained that, in his laboratory, using calorimeters that didn't have "the major error sources" present in the calorimeters of Fleischmann and Pons, "nothing unexpected" appeared. Williams was trying to write history the way he believed it had happened.

David Lindley's Parting Shot

In the same March 29, 1990, issue of *Nature*, Editor David Lindley wrote an obituary for "cold fusion" in an article titled "The Embarrassment of Cold Fusion":

> What was reprehensible a year ago has now become absurd. Still, there are whispers of a 100-page manuscript, replete with facts and figures, which the world will soon see. Most of the world, sadly for Pons and Fleischmann, is unlikely to care, except perhaps out of historical curiosity and a desire that the tale be neatly ended. ... Perhaps science has become too polite. ... Would a measure of unrestrained mockery, even a little unqualified vituperation, have speeded cold fusion's demise?

The behavior of Lewis, Koonin, Garwin, Happer, Huizenga, Lindley, Williams, Maddox and many others revealed the ugly side of science. For many of them, their words were vindictive. For all of them, their actions were counterproductive. But this doesn't mean that they are or were exceptionally mean-spirited people. Their behavior is understandable: Sometimes, when scientists are confronted with a threat to their current paradigm, self-preservation takes precedence over their altruistic intentions for scientific progress. As journalists, Lindley and Maddox had an additional duty of care; their behavior is less understandable.

Nature editor David Lindley

David Williams (Harwell laboratory)

CHAPTER 34

Not With a 10-Foot Pole

Disgraced and Discarded

"**K**evin Wolf is in trouble," ERAB panel member Jacob Bigeleisen wrote to his colleagues on Sept. 15, 1989. Wolf, a nuclear chemist in the Texas A&M Cyclotron Institute, had reported observing neutrons and tritium in his "cold fusion" cells in a presentation at the Department of Energy's (DOE) Santa Fe workshop.

By the time Wolf reported his results at the October 1989 NSF/EPRI workshop, he had observed tritium in three cells in his lab and had collaborated with researchers in the Chemistry Department who had observed tritium in a dozen of their cells.

Wolf, like Debra Rolison and William O'Grady at the Naval Research Laboratory, was doubtful about his own results. Nevertheless, at the Santa Fe workshop and at the NSF/EPRI workshop, Wolf reported the data as he observed it. His papers show that he tried everything he could think of to come up with an explanation other than a new kind of nuclear process.

In the fall, the ERAB panel members tried to find faults in the reports of positive "cold fusion" results. Wolf's claim was high on their list, but they couldn't find an obvious way to dismiss his results.

Wolf's story is intertwined with that of two other chemists at the university. On June 15, 1990, freelance investigative journalist Gary Taubes, 34, sold a story to the prominent U.S. magazine *Science*. In his article, Taubes depicted as frauds two chemists in the Chemistry Department who had measured tritium in their "cold fusion" cells.

Taubes focused his story on John Bockris, a world-famous electrochemist, and his graduate student, Nigel Packham. But Wolf was spared because, during the time Taubes was looking for dirt on Bockris and Packham, Wolf recanted. He spoke to a staff reporter at *Science*, Robert Pool, who reported Wolf's retraction of his tritium claim in a sidebar to Taubes' article. However, Wolf never published a journal article in which he reported scientific evidence to support his retraction.

Charles Martin, another electrochemist at Texas A&M University, had measured tritium in only one cell, but Martin had disavowed that result before 1990.

Taubes and the Texas Tritium

Taubes' story suggested that Packham had taken a tritium tracer solution and spiked the Chemistry Department's "cold fusion" cells. Had this been true, Packham would have had to have spiked 13 cells, some multiple times, over a period of weeks, without anyone noticing.

Taubes was convinced that the other reports of tritium around the world were not real. I discussed Taubes' 1990 article with him in June 2015 to understand his 1990 perspective. Here is what Taubes wrote when I asked him what evidence he had for his accusation of fraud:

> If you believe in the laws of nuclear physics, then the presence of the tritium in the cells, particularly so without neutrons being generated, as in Wolf's laboratory was, in effect, hard evidence of fraud.
>
> If the tritium were legitimate, this was a historic discovery virtually without parallel in science. So considering the controversy swirling around the Texas A&M experiments and what the DOE panel had come to believe, this was the most likely explanation.

Wolf presented a full summary (see table) of the Texas A&M tritium results to the October 1989 NSF/EPRI workshop members and it was published four years later, in August 1993.

Tritium Data Reported by Wolf at NSF/EPRI Workshop

Cell #	Notes	Activity *
C-D		4.9×10^6
C-B		3.7×10^6
C-C	After charging for 4 wks	64
C-C	After charging for 2 hrs	5290
C-C	After charging for 6 hrs	5.0×10^5
C-C	After charging for 12 hrs	7.6×10^5
C-D		1.2×10^5
C-E		3.8×10^4
C-F		6.3×10^4
C-G	After charging for 4 wks	120
C-G	After charging for 1 wk	250
C-G	+ 24 hrs with current increase	1.5×10^4
C-2		6.3×10^4
C-3		0
C-1		69
C-7		7.5×10^3
M-1		6.4×10^3
C-8	In electrolyte	5.0×10^5
C-8	In recombiner	1.5×10^8
C-9	In electrolyte	6.7×10^4
C-9	In recombiner	2.5×10^5
C-9	In electrolyte	1.9×10^5
C-9	In recombiner	2.4×10^5
D-6	In electrolyte	4.0×10^5
D-6	In recombiner	1.3×10^6

* Activity = d min^{-1} ml^{-1}, Cx-series possibly Chemistry Dept; M-Series possibly Martin's group; Cn- and Dn-series, Wolf's group.

Taubes' Process

Taubes' interviews at Texas A&M began months before he published his story in *Science*. He had visited the university, to Bockris' knowledge, three times and had told people there that he was writing a book on "cold fusion."

By his own admission, Taubes had "no smoking gun." His article had no direct scientific evidence, no written records, no confessions, and no witnesses to support research misconduct, let alone his more serious depiction of fraud.

When Taubes interviewed Packham in early March 1990, he accused Packham of fraud, sought a confession, didn't get one, looked for hard evidence, didn't find any, then wrote his article. In the article, he cited circumstantial evidence that Packham had committed fraud. He also quoted professor Charles Martin, at Texas A&M, who expressed his suspicions about fraud.

Taubes reviewed the paragraph above in June 2015 and did not dispute its accuracy. In a March 19, 1990, letter to Taubes, Bockris had discussed the possibility of contamination from pre-existing tritium. It seems to have been a follow-up discussion he had had with Taubes. Bockris agreed that contamination from within the palladium was possible but highly unlikely.

Bockris explained to Taubes some of the extraordinary circumstances that would have had to occur to embed tritium into the palladium cathodes. Even still, Bockris explained that the diffusion coefficient of tritium in palladium would result in a rapid decrease, quickly approaching zero, of any remaining tritium within the material.

He also explained to Taubes that, when palladium, in the form of a wire, is manufactured, the melting process also makes it very unlikely that any tritium would be retained through the heating process.

Bockris also told Taubes about the high levels of tritium the BARC researchers had found in their cathodes, not just in palladium but also in titanium cathodes. Bockris had made clear to Taubes the improbability of pre-existing tritium. Bockris, however, had no idea that Taubes was preparing a news article, rather than just a book:

In my view, you should write Packham an apology for voicing suspicions which are groundless. They are also rather stupid. How could Packham add tritium to solutions at Case-Western, Oak Ridge and Los Alamos, to say nothing of 4-5 foreign institutions including the massive reports from the tritium center at the Bhabha Atomic Research Institute?

Bockris' letter had no effect. Taubes paid no attention to the BARC, Los Alamos and Oak Ridge reports of tritium. However, by March 28, Bockris could see a change in Wolf's behavior. "Wolf's enthusiasm for cold fusion decreased inexplicably," Bockris wrote, "and he excused himself from a plenary lecture at the first Annual Conference on Cold Fusion in Salt Lake City."

Taubes told me that, in late March or early April, he submitted his story to the British journal *Nature*. But *Nature's* attorneys thought the article was potentially too damaging to Bockris' and Packham's reputations. British libel laws were tougher than U.S. laws then. After three revisions with the attorneys, he gave up and pitched his story to *Science*.

Meanwhile, Wolf was trying to figure out how to renounce his data. For a year, Wolf hadn't been able to find an alternative way to explain his tritium, despite his strenuous efforts. In the weeks preceding the Taubes article in *Science*, Wolf finally came up with one.

He had given his retraction to Robert Pool, who included it in a sidebar titled "Wolf: My Tritium Was an Impurity" to Taubes' forthcoming June 15 article. Sometime before June 7, a week before the *Science* article published, Wolf called reporter Jerry Bishop at the *Wall Street Journal*. Wolf told Bishop that he thought his tritium "might have" been the result of pre-existing contamination in his palladium cathodes.

Broad's Tritium Tale

The day after Wolf's recantation appeared in the *Wall Street Journal*, journalist Bill Broad at the *New York Times* wrote an article titled

"Contamination at 3 Labs Casts Doubt on Results Pointing to Cold Fusion."

Broad began his story with this erroneous and misleading statement: "Three top laboratories whose work supported low-temperature nuclear fusion said yesterday that their experiments had used palladium metal that was contaminated with tritium, a byproduct of fusion."

No laboratory had issued an official statement. Instead, Broad had spoken with two scientists. The first was Wolf, who, in 24 hours, had changed his story from "*there might have*" been pre-existing contamination to "*our results are consistent with contamination. ... We can offer no support for tritium being produced by cold fusion.*"

It was a striking assertion, considering that none of Wolf's H_2O cells and only three of his D_2O cells (C-8, C-9 and D-6) showed evidence of tritium — from any source.

Broad also asked Edmund Storms at the Los Alamos National Laboratory about the possibility of contamination. But Storms told Broad that, although one of his experiments was wrong, Storms had 11 other experiments that succeeded. Therefore, the second of Broad's labs that allegedly "cast doubt" was still strongly confirming the tritium.

The third lab was that of Bockris, whom Broad did not quote directly. Instead, Broad referred to statements by Bockris in Bishop's article. Bockris, too, was still strongly confirming the tritium. Thus, two-thirds of Broad's story was not true. Only one lab, not three, cast doubt on the results.

In 2015, I attempted to contact Broad by e-mail to better understand the circumstances of his article. He did not respond.

Wolf's Paper Trail

Wolf's explanation about pre-existing tritium contamination was full of holes. It was contradicted by these facts:

1. Tritium, a radioactive material with a 12-year half-life, doesn't exist as a natural contaminant on Earth, with the possible exception of the Oklo natural reactor.

2. Tritium in the U.S. in 1989 was made only in a nuclear reactor at the DOE's Savannah River Site.

3. Tritium is a restricted and highly controlled material.

4. Palladium is permeable to tritium; it tends not to retain tritium.

5. Wolf obtained his palladium from Hoover & Strong, a precious-metals company.

6. Hoover & Strong recycled its palladium in such a way that palladium could not retain tritium, if any had existed.

7. Wolf's dozens of null results showed no "contamination."

Moreover, Wolf had left a substantial paper trail that revealed the falsity of his retraction. His concern about pre-existing tritium as a contaminant in the palladium was not news. He had considered that possibility for much of the previous year. Four scientific papers, one of them a preprint, explain in great detail that Wolf had performed extensive checks for tritium contamination and found none. (Wolf, May 1989; Packham 1989; Wolf, October 1989; Wolf 1990)

When Wolf spoke with Bishop at the *Wall Street Journal* on June 6, 1990, he told him "only two of his cells had ever produced any tritium." He was now dissociating himself from the other cells that produced tritium in the Bockris lab, in which Wolf had been a collaborator.

Nearly the entire last page of his June 1990 paper explains the extensive efforts he made to test for pre-existing tritium. Test after test showed no evidence.

Wolf acknowledged that the dozens of experiments that failed to produce tritium were also evidence against pre-existing tritium:

> The negative result in itself constitutes a type of blank, and there are at least 20 more with no tritium. The program of blanks and testing continues, but at present the strongest argument against tritium contamination is the magnitude of the yields detected at locations where little or no tritium is used, and the unlikeliness of contamination in the refining and manufacture of materials.

Mahadeva Srinivasan, the former associate director of the physics group at BARC, and an expert on heavy water nuclear reactor technology, explained to me how palladium is used at reactor sites.

"Palladium membranes," Srinivasan wrote, "are used to separate tritium from its decay product helium-3 in nuclear weapons establishments. All hydrogenous isotopes easily pass through palladium membranes. This property of differential diffusion is used to enrich and remove tritium from the heavy-water moderator of CANDU reactors."

Srinivasan had examined Wolf's contamination claim and discussed it in a 1991 review of the field. According to Wolf, the tritium he measured was produced not in his experiment but in a nuclear reactor. Wolf's idea was that palladium used to separate the reaction product tritium from its decay product helium at the reactor site was recycled and wound up in his lab.

"Hoover and Strong Co.," Srinivasan wrote, "has since clarified that their manufacturing process is such that it is impossible for any tritium to remain in the palladium after the tortuous treatment to which it is subjected." (Srinivasan, 1991) Wolf's contamination scenario was fiction, but he was now no longer in hot water with his scientific peers.

Wolf Tries to Cover His Tracks

Wolf does not appear to have made a scientific examination of the palladium that he told the *Wall Street Journal* and the *New York Times* had pre-existing tritium. Instead, in an October 1990 paper he presented at the Anomalous Nuclear Effects in Deuterium/Solid Systems conference in Provo, Utah, he reported the results of a new series of experiments.

He had produced tritium again, but now he dismissed it because of statistics; only two of 50 D_2O cells produced tritium. He did more tests and came up with more convoluted ideas about pre-existing tritium in his cathodes. "The latent contamination of palladium," Wolf wrote, "is only one of the many ways that false results can be generated."

In his paper, Wolf berated other scientists who had confirmed evidence of anomalous nuclear reactions not just by tritium but also from a broad array of analytical approaches. He made a sweeping conclusion about all experimental reports of tritium in the field and dismissed them categorically.

Because Wolf provided no tabular data and stated no values for the tritium measurements in the text of his paper, his report lacks scientific credibility. His contamination claim wasn't taken seriously during the discussion following the presentation of his paper, according to Vladimir Aleksandrovich Tsarev, a scientist at Lebedev Physical Institute of the Academy of Sciences, USSR, and David Worledge, with the Electric Power Research Institute. Worledge was one of the conference chairmen. (Tsarev and Worledge, 1991)

A group of researchers at Utah's National Cold Fusion Institute performed the tests for hidden tritium that Wolf had failed to do. They reported the data at the October 1990 conference. The lead author on the paper was Krystyna Cedzynska. They checked 135 palladium samples with different metallurgical histories, from different manufacturers, including Wolf's source. They found no evidence of contamination and reported the specific values of tritium they measured as well as the sensitivity of their instruments. (Cedzynska, 1991)

"By the end of the Provo meeting of October 1990," Srinivasan wrote in his 1991 review, "the accusation that contamination is the cause of tritium observations in so many countries has been virtually rejected."

To round out the Wolf story, we go back to the very beginning. Thomas Maugh, reporting for the *Los Angeles Times,* was at the DOE's Santa Fe workshop on "cold fusion." He wrote on May 24, 1989, that Wolf "reported that his group had obtained evidence indicating the presence of large quantities of tritium and small quantities of neutrons in a cell that produced excess energy. The presence of tritium was confirmed, he said, by researchers from General Motors Corp. and the Los Alamos National Laboratory." Excess heat, neutrons and tritium: all in the same beautiful experiment.

Taubes' Tritium-Spiking Depiction

Taubes' fraud story published on June 15, 1990. He hadn't gotten the confession he had expected from Packham, but Wolf had caved in to the pressure. Despite the lack of credibility with his pre-existing tritium contamination scenario, Wolf recanted. Wolf's retraction was essential for Taubes. Without it, Taubes' story could not have worked. Ironically, Wolf's *Journal of Fusion Energy* paper published the same month as Taubes' *Science* article.

In his paper, Wolf reported that "tritium has been identified in nine Pd-Ni electrolytic cells at ... levels that are 10^2-10^6 above background."

In contrast, Taubes' wrote that, "after a year of ambiguous or simply negative experiments, Bockris' tritium data remain not only the single most extraordinary "cold fusion" effect but also the only compelling evidence in support of the original cold fusion claims."

Taubes knew about the tritium at Los Alamos reported by Storms and Talcott and by another Los Alamos researcher. He knew about the tritium reported from university labs in Florida, Utah, and Ohio.

Taubes also knew about the tritium reported at Oak Ridge National Laboratory. I asked him whether he had a copy of the Oak Ridge report that cited an "increase of tritium in the electrolyte by at least a factor of 25," but he could not remember. He knew about the BARC tritium, but he could not remember whether he had a copy of the full BARC-1500 report. Bockris had explained all of this to Taubes in his March 19 letter, nearly three months before the article published. Bockris' letter made no difference to Taubes and the story he wanted to write.

Chemist Charles Martin, one of Bockris' colleagues at Texas A&M, was the person who supplied Taubes with the concerns of fraud:

> Martin went to Mike Hall, head of the chemistry department, and voiced his suspicions. "I warned Hall that I thought there was a very good chance the experimental results were the result of fraud," Martin recalls. Hall then checked with Fackler about A&M's policy toward fraud.
>
> At the time, the A&M administration was revising its fraud policy. The current version seemed to have no provision for an investigation without a faculty member willing to press the case.
>
> "I had to publicly act as an accuser," Martin says. Although Martin was seriously concerned about possible

fraud, he says, [he] felt that all the evidence was circumstantial. "I can't go before a committee and accuse anyone of scientific fraud when all I have is circumstantial evidence."

The first part of Martin's quote in the last paragraph likely is wrong for three reasons. First, there is no evidence that Martin made a formal accusation. A half-dozen news stories in the *Dallas Morning News* make it clear that the university initiated an inquiry only after and in response to Taubes' article.

Second, none of the Dallas articles about the university inquiry mentioned Martin as an accuser. Third, Martin's quote says that he didn't have sufficient evidence to accuse anyone of fraud.

Either Taubes misquoted Martin to make it look like Martin was the public accuser or Martin did not communicate clearly to Taubes. A statement from Martin more consistent with the other facts would have been, "I would have had to publicly act as an accuser."

Accuser was an ugly word, redolent of witch-hunts. *Whistle-blower* would have been marginally less loaded, but not by much. Over the next ten days, Martin briefed both [Kenneth Hall, the deputy director of the Texas Engineering Experiment Station,] and John Fackler, the dean of the College of Science, on most of the suspicious details of the tritium work. ...

In the midst of the uproar, [ERAB panel member Allen] Bard tried to calm Martin's anxieties. "It's all going to fall apart like a house of cards on these people," he assured Martin. "It's only a matter of time. Why are you driving yourself to distraction over this?" Martin replied that he felt like Dr. Frankenstein: "I helped create the monster, I should help kill the monster." ...

Fackler and Hall decided that ultimate responsibility for scientific honesty lay with the principal investigator, in this case Bockris. Without hard evidence, any investigative act would constitute a threat to academic freedom. They needed

Martin to accuse publicly or, as Hall put it, "to put his own reputation on the line." Barring such an accusation, they needed hard evidence.

The university needed an accuser or hard evidence; it had neither. But Martin's suspicions were good enough for Taubes and *Science*.

I tried to discuss "cold fusion" history with Martin in 2010, but he refused. When I interviewed Packham, he provided some insight into how Taubes approached his sources:

> I was a naïve 28-year-old graduate student. I think that the tactics that Gary employed, he knew would work, and they did, and he got to me, no doubt. But the other thing I would say is that I think, if there would have been any hint of truth in what Gary was suggesting, I think he would have found it.
>
> He would have found a way to get me to tell him something that he wanted me to tell him. He's able to get stories out of people that I think other people can't get. But I think the very fact that I didn't tell him something that he wanted to hear is testament that there was no story to tell.
>
> Bockris is a character. A lot of people don't understand him; a lot of people certainly don't like him. That's not his fault. Bockris does have some radical ideas, but sometimes radical ideas are the ones that move this crazy world we have along.
>
> As far as me, if in hindsight I knew what I was going to go through, would I do it again? Knowing the character it developed in me, yes, I probably would. I think I grew a lot. It was traumatic, but again, if A&M was the only group out there that could produce excess heat and radiological effects, that would have been very different. The fact was there were people from, literally, all around the world — India, Japan — that were seeing something. Could they explain it? No. Did they necessarily need to at that point? In my opinion, and I think Bockris would echo this, probably not. There are

many things in this world that have unexplained reasoning behind them, and some of them are curious enough to investigate, and some will create a lot of problems.

Taubes certainly did not act alone. He resolved for the ERAB panel members the seemingly final "unresolved issues" that they were not able to dismiss. I do not know whether scientists outside of Texas A&M helped Taubes in his attack against Bockris and Packham in 1990. But I do know that, in his book, Taubes collaborated with the ERAB panel members Allen Bard, Richard Garwin, William Happer, John Huizenga, and Steven Koonin.

Taubes also collaborated with Nathan Lewis, Charles Martin, and Del Lawson (a graduate student in Martin's lab). All of them, as Taubes wrote in his book, read drafts of the book and suggested changes and corrections to him.

In an odd way, Taubes himself may have been a victim of the ERAB panel, for they most likely did not share with him everything they knew and the documents they had been given by the national laboratories.

Not With a 10-Foot Pole

I wasn't there 25 years ago. Taubes was. I made my best attempt to see things from his perspective. Taubes has been kind enough to help me in the last year, to help answer a few questions I've had about this history. We've had a cordial, though occasionally bristly rapport, particularly when I have suggested to him that he was, effectively, Bockris' accuser. In response to many of my questions, he said he would talk only off the record. "I still get nervous that any response will be cast in the most negative possible light," he explained.

After Taubes' article published in *Science*, Bockris sent a letter to the editor. So did Edmund Storms. Neither was published. Apparently, the editors of *Science* didn't want to touch the subject.

In response to Taubes' article in *Science*, and the ensuing national media attention, Texas A&M University was forced to convene an inquiry. The three-member committee began its work, according to the

Dallas Morning News, in July and completed its work in November. On Nov. 19, 1990, the paper reported that the university found no evidence of fraud.

Panel member Joseph Natowitz, a nuclear chemist and former head of the university's Chemistry Department, said he had no reason to believe that anybody spiked the samples.

John Bockris

On July 10, 2004, I traveled to Texas and met with Bockris and his wife, Lillian Bockris, to record his story. Bockris had skin as thick as a rhinoceros and a strong ego, to boot. The treatment he had received after he performed his tritium research and other daring experiments didn't faze him. He knew the work he had done and the contributions he had made to science. The events took a toll on Lillian, though, and I have one of her letters on the *New Energy Times* Web site. Lillian Bockris died at 82 in 2005. John O'Mara Bockris died at 90 in 2013.

Taubes' article in *Science* put a chill in the field of "cold fusion," as it was known then. It was a difficult science problem, hard to reproduce and hard to understand. But its reality was obvious to almost all of the researchers who had seen the evidence with their own eyes.

By March 1990, according to the *Wall Street Journal*, government funding was no longer available. The Electric Power Research Institute remained the only U.S. source of research grants for "cold fusion." Publication in top-tier journals was impossible. Access to laboratory resources became scarce. Graduate students avoided the research.

In January 1991, Frank Close's book *Too Hot to Handle: The Race for Cold Fusion* published. Close nearly accused Fleischmann and Pons of fraud. On March 17, Bill Broad published a front-page story in the *New York Times* on Close's book and depicted Fleischmann and Pons as unethical and possibly frauds. The headline was "Cold Fusion Claim Is Faulted on Ethics as Well as Science," Broad's article also appeared in the *International Herald Tribune*, with the headline "Cold Fusion Claims Are Branded as Bogus."

With the specter of groundless accusations of fraud by news organizations like *Science* and the *New York Times*, only the most stubborn scientists had the courage to continue with the new science. The field became politically radioactive. No one else would touch "cold fusion," even with a 10-foot pole.

Science Marches On

After "cold fusion" was declared dead in March 1990, some people in Utah became hostile to Pons and his family. (More on that in a moment.) Pons prepared for major changes. Toyota had been courting Fleischmann and Pons since the summer of 1989, when the company offered to pay their travel expenses to a conference in Japan, in exchange for a private meeting.

At some point, a former student of Fleischmann's, electrochemist Keiji Kunimatsu, introduced Fleischmann and Pons to Minoru Toyoda, a founding member of the Toyota automobile conglomerate.

By the fall of 1990, Pons' discussions with the Japanese were well under way. Then, in early October, Pons disappeared. On Oct. 16, JoAnn Jacobsen-Wells, at the *Deseret News,* reported what she was told by University of Utah officials: Pons was overseas, and nobody knew where he was or when he was expected to return. But they knew he was visiting foreign laboratories and he had made arrangements for another professor to take over his classes. Jacobsen-Wells reported that, for several weeks, rumors had been flying around campus that Pons was preparing to accept a job with a Japanese research institute. By this time, Pons was handling all of his communication with people in Utah through his attorney, Gary Triggs. Fleischmann was back home in England.

On Oct. 24, the *Deseret News* reported that, according to Triggs, Pons would not be at an important Utah State Fusion/Energy Advisory Council meeting on Oct. 25. Joe Tesch, the chief deputy attorney general for the state, who was responsible for overseeing the state's portion of the "fusion" patent claims, reassured the newspaper that Pons had "no intention of leaving the University of Utah or the state of Utah as his home base and residence."

Hours after the news of Pons' disappearance went public, Triggs sent a fax to the university and requested a one-year sabbatical on Pons' behalf. Jacobsen-Wells reported that news on Oct. 25, as well as this: "Pons' house has been put up for sale. His telephones were disconnected this week, and family, neighbors and friends said the Pons family has moved to France. Several U. officials have said they have heard that Pons is in the Mediterranean city of Nice." That news got widespread attention.

If Utah legislators and members of the public hadn't yet been galvanized against Fleischmann and Pons, they were now. Randy Moon, the state's science advisor and member of the fusion council, spoke with the Associated Press on Oct. 25 about the $5 million that had been granted and partially spent by the National Cold Fusion Institute. Fleischmann and Pons were now being blamed by Utah officials for the lack of progress in the institute, which they had advised the University of Utah not to form and which they initially refused to work in.

"Stan owes it to a lot of people to be able to report on his results," Moon said. "A lot of people at the university have been counting on Pons to show the research is viable."

On Oct. 26, Bill Broad, at the *New York Times,* reported that the state of Utah formulated a "plan to track down the enigmatic researcher and hold him accountable for his work." The threats of being hunted down worked. Pons met the fusion council on Nov. 8 and asked to continue his research abroad as well as at the university. According to the *Deseret News* on Nov. 9, the council and Pons came to terms, but by all appearances, Pons was done with Utah and, for that matter, the U.S. He and his family restarted their lives in France. Pons is still in hiding.

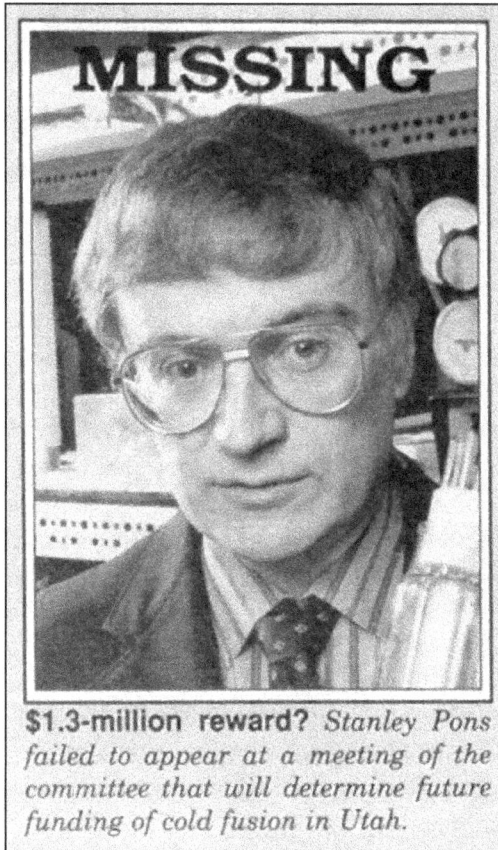

$1.3-million reward? *Stanley Pons failed to appear at a meeting of the committee that will determine future funding of cold fusion in Utah.*

Nov. 9, 1990, image from Science *depicting Pons as a wanted criminal.*

In July 1985, Technova, part of the Toyota Motor Corp. conglomerate, established a research and development laboratory in France, called the Institut Minoru de Recherche Avancées, named in honor of Minoru Toyoda. It was located in Sophia Antipolis, near Nice, and called IMRA-Europe. The lab opened its doors in June 1988. On receiving a research proposal from Fleischmann and Pons, Toyoda invited the chemists to continue their work there. (White, 1992) The two chemists had the funds to hire staff and set up advanced experiments. Fleischmann spoke about this in 1994 to the BBC:

> It was exciting to make a move. It was a relief to get away from that terrible atmosphere in the United States and to have the opportunity to work — totally without public attention for a period of time.

Fleischmann and Pons worked at the IMRA for several years. I do not know when or why they stopped. I have a strong hunch that a division occurred between them. There also may have been a disagreement over performance objectives and the expectations of the Japanese sponsor. I met Pons and his wife once, in Marseille, France, in 2004. Even when he spoke briefly about the ugly period in 1989, his expression was calm and peaceful. But when I mentioned Fleischmann's name to Pons, he got livid. I saw the same reaction, years later, when I was with Fleischmann and asked him about Pons. Neither would tell me more about the breakup or their professional partnership.

Despite the way they were treated, Fleischmann and Pons, to my knowledge, never lashed out at their critics publicly. On at least two occasions, Pons' attorney contacted scientists who had been publicly critical of the two electrochemists' work.

Perhaps the most illuminating summation of their experience was expressed by Fleischmann on Jan. 5, 1991, when he was interviewed by Eugene Mallove. At the time, Mallove was the chief science writer with the MIT News office:

> If Stan and I write anything up about this history, which I'm not sure we will, we would write a summary which would be a popular book. We would publish some of the documentation, letters and so on, in a second volume because I think that's of interest to people. We won't publish everything. I'm sure that some of this correspondence is so hurtful to people that it will have to be archived for 50 years. They have to be dead and gone, including ourselves.

On March 21, 1994, the BBC-TV show "Horizon" aired a program on "cold fusion." Fleischmann, quoted in the show, had no regrets:

> If it had been anything else, we would have said, "People don't want us to do it. Forget it. Let's just leave it alone." But this is not in that category. This is interesting science, new science, with a hint of a possibility of a very useful technology. Therefore, if you've got any integrity, you don't give up. You only give up if you find you are wrong. But as long as you believe that you are right, you have to continue. And you have to take the consequences."

Perhaps Pons' most insightful comment about the controversy is a statement he made to JoAnn Jacobsen-Wells, at the *Deseret News,* on May 28, 1989:

> It appears that the people who would benefit most by this work being discredited have taken the initiative to cause us great difficulty. They might cause us difficulty, but they will not stop the science.

He was right: The science continued.

Glossary of Scientific Terms

Absorber: *See* neutron absorber.

Accelerator: *See* particle accelerator.

Activation: A process in which a non-radioactive material is subjected to nuclear radiation and becomes radioactive.

Activity: A measure of the level of radioactivity of a material. Measured by the number of spontaneous nuclear disintegrations in a specific amount of material during a specific interval.

Alchemy: Primarily a reference to ancient methods and practices intended to effect elemental or personal transformation.

Alpha (particle, emission): A Greek letter used to describe one of the first types of radioactive emissions. It is emitted during a nuclear reaction and was later identified as a helium-4 nucleus. (*See also*: Appendix D)

Alpha decay: Radioactive decay in which an alpha particle is emitted. Each emitted alpha lowers the atomic number of the nucleus by two and its atomic mass by four.

AMU: Atomic Mass Unit (See atomic mass)

Anode: In electrolysis, the metal contact point of the electrical circuit that attracts the flow of electrons.

Atom (atomic): Basic building block of all matter. Atoms comprise three elementary particles: protons, neutrons and electrons. Each atom has one nucleus in its center containing the protons and neutrons. The nucleus is surrounded by electrons, normally equal in number to the number of protons in the nucleus of a neutral atom.

Atomic energy: *See* nuclear energy.

Atomic number: Measured by the total number of protons in an atom's nucleus; determines the type of chemical element. *See also* atomic mass.

Atomic mass: Effectively measured by the total number of protons and neutrons in an atom's nucleus; determines the type of isotope within a specific range of possible isotopes. *See also* atomic number.

Atomic transformation: *See* transmutation.

Beta (particle, emission): A Greek letter used to describe one of the first types of radioactive emissions. It is emitted during a nuclear reaction and was later identified as an electron. (*See also*: Appendix D)

Beta decay: A weak interaction in which a neutron inside a nucleus decays into a proton, an energetic electron and a neutrino. The energetic electron released in a beta decay exits the nucleus as a beta particle.

Binding energy: For a nucleus, the energy required to pull the neutrons and protons apart.

BF_3 detector (counter): Boron tetrafluoride detector; used to measure neutron emissions. Detector consists of a cylindrical tube filled with boron trifluoride gas, which is used to detect low energy "thermal" neutrons. With the addition of a neutron moderator surrounding the detector (to bring down neutron energy), the detector can also be used to detect higher-energy "fast neutrons."

Branching ratio (D+D fusion): According to the well-understood theory of deuterium-deuterium nuclear fusion, the reaction paths occur through one of three possible branches. The first branch produces a neutron. The second branch produces tritium. The third branch produces helium-4. In D+D fusion reactions, on average, a neutron is produced almost 50% of the time, tritium is produced almost 50% of the time, and helium-4 is produced less than 1% of the time. Since the discovery of D+D fusion, these ratios have always been consistent.

Capture: *See* nuclear capture.

Cathode: In electrolysis, the metal contact point of the electrical circuit that emits the flow of electrons and attracts positive ions or protons.

Charged particle: A fundamental particle such as an electron, proton or positron, or a compound particle that carries a net positive or negative electrical charge.

Chemistry: An area of science primarily involved with interactions between atoms and electrons, their structures and properties. There are two historical exceptions. The first was in the early 20th century, when chemists were as involved in nuclear research as physicists were. The second period began in 1989, with the introduction of what was later called LENRs.

Cold fusion (idea): The proposed concept that nuclear fusion reactions occur at or near room temperature. An unproven idea that deuterium nuclei overcome the Coulomb barrier at room temperature at high reaction rates and undergo deuterium-deuterium nuclear fusion. Fusion relies primarily on strong-force interactions and normally requires temperatures in the millions of degrees.

Cold fusion (history): Historical events that took place primarily in 1989 in the aftermath of the announcement of the claim by electrochemists Martin Fleischmann and Stanley Pons at the University of Utah of a "sustained nuclear fusion reaction" at room temperature.

Collective effects: Describes the interaction of many-body groups of essentially identical items, such as elementary particles. When the items interact as a group, they create different effects than they would produce either alone or with a few others. The concept can apply to many-body physics, such as electrons oscillating together, or to a flock of birds flying in formation and thus creating lift efficiencies that none of the birds could create individually.

Coulomb barrier: An electrostatic barrier surrounding positively charged nuclei that, under normal temperatures and pressures, prevents nuclei from interacting with each other.

Cross-section: A measure of the probability of a specified interaction between an incident photon or particle radiation and a target particle or system of particles. It is the reaction rate per target particle for a specified process divided by the flux density of the incident radiation.*

Decay: *See* radioactive decay.

Deuterium: A stable isotope of hydrogen that has one proton and one neutron in its nucleus. Also known as heavy hydrogen.

Deuteron: The nucleus of a deuterium atom, comprising a proton and a neutron.

Disintegration: A process in which constituent parts of the nucleus of an atom separate from the nucleus and fly off, leaving a smaller atom in place.

Electrochemist: Person who works in the field of electrochemistry.

Electrochemistry: The study of electricity and how it relates to chemical reactions. In electrochemistry, electricity can be generated by movements of electrons from one element to another in a reaction known as redox reaction, or oxidation-reduction reaction. (U.C. Davis)

Electrode: In electrochemistry, the metal contact point of the electrical circuit that conducts the flow of electrons.

Electrolysis: Chemical decomposition by an electric current of a liquid, or solution containing ions, into constituent elements.

Electrolytic fusion: The idea of creating nuclear fusion by electrolysis.

Electromagnetic force: One of the four fundamental physics forces. Repels protons from one another and keeps atomic nuclei separate from one another. *See also* strong force.

Electron: A stable elementary particle that is a component of an atom. It possesses a negative electrical charge and exists outside and orbits around the nucleus of an atom.

Electron-volt (MeV, Mega-electron-volt): A unit of energy equal to the change in energy of one electron in passing through a voltage difference of 1 volt.*

Electroweak interaction: A term used in particle physics that describes the unified behavior of electromagnetism and weak interactions.

Element: Designates a form of matter that is distinguished by a unique number of protons in its nucleus and unique chemical properties.

Emissions: *See* alpha, beta and gamma rays.

Energy: Power during a given period.

Exploding electrical conductors (Exploding wire phenomenon): A phenomenon in which an electrical conductor such as wire, foil, or film is deliberately subjected to a very high current with near instantaneous rise time (less than 2 microseconds), causing it to explode loudly and violently, momentarily forming a plasma, and leaving behind a cloud of metal vapor.

Fission: *See* nuclear fission.

Fusion: *See* nuclear fusion.

Gamma rays (gamma radiation, gamma emission): Gamma rays are highly penetrating forms of electromagnetic radiation emitted from nuclear transitions. Gamma rays are a class of photons (a larger group of massless entities) that, according to quantum mechanics, behave both as waves and as particles. On Earth, they are encountered from radioactive material decays and a few rare terrestrial events. Gamma rays are identified by their energy from the so-called photo-peak. A range of various-energy gamma ray interaction-related peaks and continua are depicted in a typical gamma spectrum ranging from the photo-peak at the upper end of the energy scale down to zero. (*See also*: Appendix D)

Gamow factor: The probability that two nuclear particles will overcome the Coulomb barrier and undergo nuclear fusion reactions.

Gas-loading: An experimental method in which molecules of a gas, typically deuterium or hydrogen, dissociate, ionize and then move into hydride-forming sites in a host metal or metal-oxide structure.

Half-life: The time required for half of a given quantity of a radioactive material to decay.

Heavy hydrogen: *See* deuterium.

Heavy water (D_2O): Composed of deuterium and oxygen. Deuterium atoms each contain one proton, but deuterium also has a neutron, making it twice as heavy, and the resulting heavy-water molecule is slightly heavier than normal water. *See also* light water.

Hydride (Deuteride): Compounds that hydrogen (or deuterium) form with other chemical elements, typically within metals or alloys. Some metals can absorb hundreds of times their own volume of hydrogen or deuterium.

Hydrogen: A chemical element with a single proton in its nucleus. Its normal isotope, known as protium, is stable. Its second isotope, known as deuterium, is also stable. Its third isotope, known as tritium is unstable.

Ion: An atom that has either an excess or shortage of an electron or electrons. Ordinarily, neutral atoms have an equal number of electrons to their protons.

Ionization: A process by which an atom gains or loses an electron or electrons.

Isotope: A variation of an element that contains the same number of protons but a different number of neutrons from the most abundant version of that element. Isotopes have the same atomic number but a different atomic mass.

Isotope, stable: An isotope that is not undergoing radioactive decay or emitting gamma radiation.

Isotope, unstable: An isotope that is undergoing radioactive decay involving emission of particles and/or gamma radiation.

Isotopic abundance: The relative number of atoms of a specific isotope among all the isotopes of a given element, expressed as a fraction of all the isotopes of that element.

Isotopic shift: A change in the ratios among isotopes of one species of elements away from the isotopic abundance of the same species that exists in nature.

LENR, LENRs: *See* low-energy nuclear reactions.

Light water (H_2O) (normal water): Water composed of the normal hydrogen isotope, which contains one proton and no neutron. The term "normal water" is used sometimes synonymously; however, one of every 6,000 molecules of normal water is a molecule of heavy water.

Loading: The process of placing atoms of deuterium (or hydrogen) interstitially into vacant spaces within the crystalline lattice of metallic elements.

Loading ratio: The ratio between the number of atoms of deuterium (or hydrogen) and the number of atoms of the host metal into which they have been loaded.

Low-energy nuclear reactions (LENRs): A class of nuclear reactions that occur at or near room temperature that are based on non-fusion reactions, for example, neutron-based reactions. LENRs are based on Standard Model physics and can occur in condensed matter under moderate (room-temperature) conditions. Key steps in LENR processes, unlike nuclear fusion or fission, are based primarily on electroweak interactions rather than strong-force interactions. Unlike fission reactions, low-energy nuclear reactions do not produce nuclear chain reactions. (*See also* Appendix A)

Metal hydrides or deuterides: Metals that have absorbed hydrogen or deuterium in their atomic structure, or lattice. *See also* loading.

Moderator: *See* neutron moderator.

Neutrino: An elementary particle having virtually no mass. Like a neutron, it has no electrical charge; it barely interacts with ordinary matter.

Neutron: An unstable (when outside a nucleus) elementary particle that is a component of an atom and exists inside the nucleus of an atom. It has no electrical charge. A free neutron outside of a nucleus has a half-life of approximately 10.3 minutes before it decays into a proton, an electron, and an electron antineutrino.

Neutron absorber: A material or object with which neutrons interact, resulting in their disappearance as free particles without production of other neutrons.

Neutron capture: A nuclear reaction in which an atomic nucleus and one or more neutrons collide and merge to form a heavier nucleus.

Neutron, cold: Neutrons of kinetic energy on the order of 1 milli-electron-volt or less (0.001 eV).

Neutron, fast: Neutrons having kinetic energy between 1 MeV and 20 MeV.

Neutron moderator: Material used to reduce the speed of neutrons, without absorbing them into the moderator material.

Neutron, prompt: Neutrons emitted from a nuclear process, at the time of the reaction, without measurable delay.

Neutron, slow: Neutrons having kinetic energy between 1 eV and 10 eV.

Neutron, thermal: A free neutron that has been slowed down by a moderator, is in equilibrium with its surroundings, has an energy between 0.025 eV and 0.2 eV.

Neutron, ultra-low-momentum (ULMN): A neutron with kinetic energies that are effectively zero, on the order of 10^{-12} eV or less — that is, .000000000001 eV. The kinetic energy of ULMNs is an estimated value because it has never been measured. ULMNs are extremely slow neutrons with extremely low kinetic energies and commensurately large DeBroglie wavelengths because they are created through a many-body collective process (as opposed to being produced by a two-body nuclear reaction that occurs inside a star). ULMNs are thus orders of magnitude slower than so-called "ultra-cold" neutrons, which are typically produced for experiments aiming to better measure the lifetime of free neutrons located outside nuclei. (Courtesy: Lewis Larsen)

Nuclear: Activity or properties having to do with characteristics of or changes in an atomic nucleus.

Nuclear capture: A nuclear process by which an atom acquires an additional particle.

Nuclear chemistry: Chemistry-related aspects of nuclear and atomic research.

Nuclear energy: Energy that is released during a nuclear reaction, such as nuclear fission, nuclear fusion, radioactive decay or a variety of nuclear processes that capture nuclear particles.

Nuclear fission: The process in which a larger nucleus is split into two (or, rarely, more) parts. The process is usually accompanied by the emission of neutrons, gamma radiation and, rarely, small charged nuclear fragments.

Nuclear fission, spontaneous: Nuclear fission that occurs spontaneously, without the addition of particles or energy to the nucleus.*

Nuclear fusion: The process in which two light nuclei overcome electrostatic repulsion and form one newer, heavier atom.

Nuclear physics: Approaches to and studies of nuclear science and technology based on principles, processes and devices common to physics.

Nuclear process: A mechanism, such as fission, fusion, or radioactive decay, which changes the energy, form, or structure of the nucleus.

Nuclear reaction: An event, occurring from a nuclear process, in which the energy, form, or structure of the nucleus is changed.

Nuclear science: The study of nuclear processes and reactions.

Nuclear transformation: *See* nuclear transmutation.

Nuclear transmutation: A nuclear process in which an element changes into another element by the increase or decrease in the number of protons in its nucleus.

Nuclei: *See* nucleus.

Nucleosynthesis: The formation of new nuclides by any number of nuclear processes, including nuclear decay.

Nucleus (nuclei, pl.): Center part of an atom that contains protons and neutrons. Comprises nearly the entire mass of the atom but only a tiny part of its total volume.

Nuclide: A distinct species of an atom identified by the number of protons and neutrons in its nucleus and its nuclear energy state.

Particles: *See* charged particle and neutron.

Particle accelerator: A device for imparting kinetic energy to charged particles.*

Photon: A massless elementary particle that is a unit of light and other forms of electromagnetic radiation. It can have properties of both waves and particles.

Physics: A field of science and technology that measures, studies and influences matter, motion and energy.

Power: The rate of doing work; equivalent to the amount of energy consumed per unit of time.

Protium: *See* hydrogen

Proton: Stable elementary particle that is a component of atoms. The proton has a positive electrical charge and exists in the nucleus of an atom.

Quantum mechanics: The branch of physics that deals with the mathematical description of the motion and interaction of subatomic particles, incorporating the concepts of quantization of energy, wave-particle duality, the uncertainty principle, and the correspondence principle. (Source: Oxford Dictionary)

Radiation, nuclear (Radioactive Emission): Emission of charged or uncharged particles or electromagnetic rays, including alphas, betas, neutrons, and gamma-rays.

Radiation, prompt: Prompt radiation is produced and emitted from its source immediately. When the reaction stops, so does the prompt radiation. *See also* radioactive decay.

Radioactive: The property of an unstable material that spontaneously emits particles or gamma rays.

Radioactive decay: A form of nuclear radiation that emits alpha and/or beta particles from radioactive materials. The emissions may take place during nuclear reactions as well as after the reactions stop. The decay causes the radioactive interior to lose some of its constituent material. An element that undergoes radioactive decay will change into a new element or a new isotope.

Radioactive half-life: For a single radioactive decay process, the time required for the activity to decrease to half its value by that process.*

Radioactivity: A naturally occurring process in which unstable elements spontaneously emit particles or gamma rays. In addition to naturally occurring radiation, man-made nuclear processes can cause some non-radioactive elements to become radioactive.

Radiochemist: Person who works in the field of radiochemistry.

Radiochemistry: The part of chemistry that deals with radioactive materials.*

Radioisotope: *see* Isotope, unstable

Radionuclide: A radioactive nuclide.

Radium: An unstable radioactive chemical element with a half-life of about 1,600 years, and 88 protons in its nucleus.

Radon: An unstable chemical element and radioactive gas with a half-life of 3.8 days, and 86 protons in its nucleus. It is produced from the decay of uranium or thorium.

Scattering: A process in which a change in direction or energy of an incident particle or incident radiation is caused by a collision with a particle or a system of particles.*

Scattering, elastic: Scattering in which the total kinetic energy is unchanged.*

Sonic implantation: An experimental method that uses acoustic cavitation to stimulate activity on metal surfaces and induce low-energy nuclear reactions.

Spectral lines: Bright and dark lines — seen in spectra of photon-emitting items, such as candle flames, glowing gas, or stars — that are characteristic of a given atom or molecule.

Standard Model: The Standard Model of particle physics is a theory that explains the physics of the world and what holds it together. It encompasses the behavior of fundamental particles across three of the four fundamental forces in physics: electromagnetic, weak interactions, and strong interactions (not gravity).

Strong force: One of the four fundamental physics forces; works only at very short distances within nuclei. Keeps protons and neutrons bound together inside atomic nuclei. *See also* electromagnetic force.

Surface plasmon electrons: A collective many-body effect; coherent oscillations of entangled electrons that take place at the surface of metals and at other interfaces.

Transformation: *See* transmutation.

Transmutation, biological: A process that shows the increase of elements in human and other living matter that cannot be explained environmentally or by conventional biology.

Transmutation, natural: Spontaneous nuclear transmutation that occurs by the natural activity of a radioactive element.

Transmutation, nuclear: The changing of one element to another by a change in the number of protons in its nucleus.

Transmutation, man-made: Human-triggered nuclear transmutation; traditionally occurs by exposure to radioactive sources. Can also occur by nontraditional processes, specifically LENRs.

Tritium: An unstable isotope of hydrogen. A chemical element with a single proton in its nucleus and two neutrons. It is radioactive with a half-life of 12.3 years.

Weak force: A fundamental force of physics that produces weak interactions.

Weak interaction: An elementary particle interaction that is involved in many forms of nuclear decay (radioactivity), for example a beta decay process. In all such interactions, neutrinos are emitted or absorbed. Weak interactions are distinct from strong-force interactions because, at low average particle energies, weak-interaction cross sections are vastly lower than strong-force interactions. Weak interactions are not necessarily weak energetically, and some can involve very large releases of energy. For example, beta decays of some extremely neutron-rich nitrogen isotopes can release more than 20 MeV of nuclear binding energy. For comparison, the strong-force deuterium-tritium fusion reaction releases 17.6 MeV.

* Source: Glossary of Terms in Nuclear Science and Technology, American Nuclear Society, ISBN 0894485539

Cartoon depicting John Huizenga, Gary Taubes, Douglas Morrison and Steven Jones, courtesy Carol Talcott

Appendix A — Definition* of Low-Energy Nuclear Reactions

Low-energy nuclear reactions (LENRs) are a class of nuclear reactions — based on Standard Model physics — that can occur in condensed matter under mild macrophysical conditions. Key steps in LENR processes, unlike nuclear fusion or fission, are based primarily on electroweak interactions rather than strong-force interactions. Unlike fission reactions, low-energy nuclear reactions do not produce nuclear chain reactions.

LENRs involve a broad set of nuclear phenomena spanning many length-scales that have two characteristics in common: a) production of neutrons from electroweak reactions; and b) many-body collective effects between oppositely charged particles. (In condensed-matter systems, these particles are typically quantum mechanically entangled).

LENRs take place in three realms: a) electrically dominated reactions, in which nuclear-strength local electric fields on micron scales in condensed matter enable electroweak neutron production; b) magnetically dominated reactions, in which many-body collective magnetic-field effects directly accelerate charged particles in plasmas; and c) mixed reactions, in which components of dusty plasmas behave in ways characteristic of the electrically dominated reactions and the magnetically dominated reactions.

The word "low" in "low-energy nuclear reactions" refers to the magnitude of input energies that are required to trigger LENR reactions; the magnitude of output energies released after triggering may be either low or high. Researchers chose this term to distinguish it from the field of high-energy particle physics, which uses very high temperatures or particle accelerators to trigger nuclear reactions.

The two most unusual characteristics of LENRs are that neutron-catalyzed transmutation reactions, which typically occur only in stars, fission reactors, or high-energy particle accelerators, can be initiated in tabletop condensed-matter experimental systems without releasing biologically dangerous amounts of energetic neutron or gamma radiation.

Electrically Dominated LENR Reactions

Electrically dominated reactions take place in condensed matter. These

LENRs take place under relatively mild conditions — that is, without the requirement of using large nuclear fission reactors, extremely high temperatures, or high-energy particle accelerators.

Given proper types and amounts of input energy, these LENRs take place when specific conditions are present on the surfaces of metals or at metal-oxide interfaces, in the presence of hydrogen or ones of its isotopes, deuterium or tritium. No radioactive seed elements are required. Neutrons produced in an electroweak reaction at micron-scale LENR-active sites on surfaces or at interfaces are subsequently captured by nearby atoms; these energy-releasing captures induce nuclear transmutations. Neutrons produced in LENRs have ultra-low energy, so almost all of them are captured locally; externally detectible emissions of deadly energetic neutrons are thus also avoided.

LENR experiments typically produce a variety of nuclear transmutation products and various types of effects and may produce macroscopically measurable excess heat. A variety of elements may be synthesized from one another, and isotopic shifts may occur; these transmutation products are generally stable elements produced by beta decays of short-lived, neutron-rich unstable isotopes created by previous neutron captures.

According to the Widom-Larsen theory, LENRs in the electrically dominated realm have two unique characteristics: a) produced neutrons have ultra-low-momentum and b) unreacted heavy electrons present in LENR-active sites suppress dangerous energetic gamma emission by locally converting incident gamma radiation from any source directly into infrared radiation (heat). (Widom and Larsen, 2006; Srivastava et al. 2010)

Magnetically Dominated LENR Reactions

Magnetically dominated reactions take place in plasmas. These LENRs can, for example, occur in magnetic flux tubes of solar flares; these processes may produce GeV neutrons, other elementary particles, and energetic gamma rays that are not suppressed.

Mixed LENR Reactions

Mixed reactions take place in dusty plasmas. These LENRs can occur in organized magnetic fields present in a plasma as well as on solid surfaces of micron- to nanometer-sized dust particles of condensed matter, which are embedded in such plasmas. Examples include exploding wire experiments and in natural lightning. * *(See also concise LENRs definition in Glossary)*

Appendix B — Timeline of Related 20th Century Events

1920

Charles Galton Darwin describes collective many-body excitations of electrons. His ideas lay the foundation for the understanding of the collective behavior of electrons. (Darwin, 1920)

1923 (November)

Robert Millikan speculates that transmutations going from lighter to heavier elements might occur in the stars. The idea of nuclear disintegration going from heavy to light elements is known at this time. (Millikan, 1923)

1929

Fritz Houtermans and Robert Atkinson propose that thermal kinetic energies inside stars are high enough to allow nuclei of light elements to overcome the Coulomb barrier and form heavier elements. The concept is later identified as thermonuclear fusion. In 1933, Oliphant experimentally confirms nuclear fusion. (Atkinson and Houtermans, 1929)

1932

James Chadwick experimentally confirms the existence of the neutron. In 1910, Rutherford had theorized its existence. (Chadwick, 1932)

1932

John Cockcroft and Ernest Walton, at the Cavendish laboratory, build the first apparatus for accelerating atomic particles to high energies. They report the first man-made transmutation by artificially accelerated particles: the disintegration of lithium by fast hydrogen protons. (Cockcroft and Walton, 1932)

1933

Leó Szilárd conceives the idea of the nuclear chain reaction. (Rhodes, 1986)

1933

Mark Oliphant experimentally confirms fusion of deuterons into various targets to create helium-3 and tritium, using a particle accelerator at Cavendish. Concurrently, he observes the liberation of excess nuclear binding energy, which prompts him to speculate that fusion is the process that powers the sun. (Oliphant, Harteck and Rutherford, 1934)

1934 (January)

Frédéric and Irène Joliot-Curie create artificial radioactivity in previously stable elements. (Joliot and Joliot-Curie, 1934)

1934 (March 3)

Gian-Carlo Wick proposes the concept of electron capture. (Wick, 1934)

1935

Harold John Taylor reports the first neutron-capture-based transmutations. He experimentally demonstrates that boron-10 nuclei capture thermal neutrons and fission into helium-4 and lithium-7. (Taylor, 1935)

1938 (Nov. 18)

Otto Hahn and Fritz Strassmann bombard uranium with neutrons and transmute uranium into smaller atoms. In 1939, Lise Meitner and Otto Frisch identify the effect as fission. (Hahn and Strassmann, 1938)

1939 (Feb. 11)

Lise Meitner and Otto Frisch propose the concept of neutron-induced uranium fission. (Meitner and Frisch, 1939)

1939 (March 1)

Hans Bethe proposes that 1) hydrogen-hydrogen fusion is the process that powers the stars, 2) no elements heavier than helium-4 could have been formed in stars, and 3) the production of neutrons in stars is negligible. His first proposal was correct; the second and third were not. (Bethe, 1939)

1946 (March 14)

Fred Hoyle makes an early contribution to the theory of supernovae-exploding stars. He proposes that neutron creation in the hot cores of collapsing stars can be explained by the reaction of an electron with a proton ($e + p \rightarrow n + \nu$). (Hoyle, 1946)

1951

Ernest Sternglass observes neutron production in keV-energy (low-energy) electric discharge experiments in a hydrogen-filled X-ray tube directed at targets of silver and indium. *See* Darwin 1920. Albert Einstein suggests that collective effects may explain the results.(Sternglass, 1997; Trost, 2013)

1957

Geoffrey and Margaret Burbidge, William Fowler and Fred Hoyle propose what is later regarded as the modern concept of nucleosynthesis of elements in stars. They theorize that fusion reactions create elements up to the atomic mass of iron and that neutron capture processes and decays create heavier elements beyond iron. Prior to their work fusion-based concepts alone were unable to fully explain the production of heavier elements The group later propose that nucleosynthesis also occurred outside the cores of stars. (Burbidge, Burbidge, Fowler and Hoyle, 1957)

1960

Corentin Louis Kervran (1901-1983) begins publicly discussing his research in biological transmutation. (*See also:* 1971)

1968

Sheldon Glashow, Abdus Salam, and Steven Weinberg develop modern electroweak theory. (Nobelprize.org)

1971

Kervran publishes a biological transmutation book. (Kervran, 1971)

1973

The Gargamelle collaboration at the European Organization for Nuclear Research performs the first stage of experimental confirmation of electroweak interactions. (CERN)

1983

The "UA1" and the "UA2" collaborations of the European Organization for Nuclear Research perform the second stage of experimental confirmation of electroweak interactions. (CERN)

Appendix C — Timeline of Early "Cold Fusion" History

Late 1960s-Early 1970s

Martin Fleischmann and his colleagues conduct research on the separation of hydrogen and deuterium isotopes (*Infinite Energy*, **6**, 113)

Late 1970s

Stanley Pons examines isotopic separation in palladium electrodes and is puzzled by some of his results. (*Infinite Energy*, **6**, 113)

Mid-1980s

Fleischmann studies the 1929 experiments of German physics professor Alfred Coehn who ran currents across palladium wires in the presence of hydrogen gas. Fleischmann studies the 1930s work of Percy Bridgman; cold explosions resulting from the compression and shear of metal lattices.

1983

Martin Fleischmann begins working as a research professor at the University of Utah.

Early 1985

Fleischmann and Stanley Pons observe the partial vaporization of a 1 cm palladium cube. (either January or February) (Beaudette, 2002, 35-36)

May 1988

Fleischmann and Pons submit a research proposal on electrolytic fusion to the U.S. Office of Naval Research.

August 1988

Ryszard Gajewski receives the Fleischmann-Pons proposal.

1988-1989

A significant set of events takes place between Fleischmann-Pons and physicist Steven Earl Jones at Brigham Young University. These are displayed on a separate timeline at the *New Energy Times* Web site under the "1989 Archives" section.

March 2, 1989

DOE approves a Fleischmann-Pons proposal for a $322,000 grant.

March 11, 1989

Fleischmann and Pons submit the first version of their preliminary note to the *Journal of Electroanalytical Chemistry* (JEAC). The journal receives it on March 13.

March 13, 1989

The University of Utah files its first U.S. patent application on the Fleischmann-Pons claim.

March 16, 1989

University of Utah administrators meet and decide to schedule a press conference.

March 20, 1989
Fleischmann and Pons submit the second version of their preliminary note to JEAC. The journal receives it on March 22. The editors accept the manuscript sometime before March 23.

March 21, 1989
The University of Utah files its second U.S. patent application on the Fleischmann-Pons claim.

March 23, 1989, 1p.m.
Fleischmann and Pons announce a "fusion" claim at a University of Utah press conference.

March 24, 1989
Edward Teller calls Pons and gets a preprint of the JEAC preliminary note. Teller appears on television and supports the idea.

March 24, 1989
Fleischmann and Pons submit a scientific manuscript to *Nature*.

March 26, 1989
Pons tells the *Wall Street Journal* that other reactions in addition to fusion may be taking place.

March 27, 1989
James F. Ziegler, at IBM Research in Yorktown Heights, N.Y., informs other IBM scientists that his attempt to duplicate the Fleischmann-Pons experiment has failed.

March 28, 1989
Fleischmann lectures at the Harwell laboratory (U.K. Atomic Energy Research Establishment).

March 30, 1989
Fleischmann and Pons change the peak on their gamma-ray graph from 2.5 MeV to 2.2 MeV and fax new graph to *Nature*.

March 31, 1989
Fleischmann lectures at the European Organization for Nuclear Research (CERN) and speaks afterward at a press conference.

March 31, 1989
After hearing Fleischmann's lecture, Douglas Morrison proposes in his newsletter that the dominant reaction in "cold fusion" is $d + d \rightarrow Helium\text{-}4$.

March 31, 1989
Pons lectures at the University of Utah. Cameras and recording devices are banned.

March 31, 1989
Pons releases five confidential preprint copies of the revised JEAC manuscript.

March 31, 1989
Jones lectures at Columbia University and speaks afterward at a press conference.

March 31, 1989
Physicists Gyula Csikai and Tibor Sztaricskai, at Lajos Kossuth University, part of the University of Debrecen, Hungary, confirm neutrons in a Fleischmann-Pons experiment.

April 4, 1989

Pons lectures at Indiana University and speaks afterward at a press conference. (He had been scheduled to give a lecture on other topics long before the fusion announcement.)

April 4, 1989

Fleischmann submits corrections to the proof of his and Pons' JEAC preliminary note.

April 5, 1989

Peter Hagelstein, at MIT, submits the first version of a manuscript for "A Simple Model for Coherent DD Fusion in the Presence of a Lattice" to *Physical Review Letters*. By April 12, he had submitted four manuscripts.

April 6, 1989

MIT issues a press release and announces that Mark Wrighton failed to repeat the Fleischmann-Pons experiment. MIT Provost John Deutch says he does not think the claims of "cold fusion" are true. MIT team submits manuscript to journal in July 1989, and the paper is published in June 1990.

April 6, 1989

A congressional subcommittee authorizes the redirection of $5 million from the Department of Energy's magnetic fusion program for "cold fusion."

April 6, 1989

Elsevier Publishing Co. releases the Fleischmann-Pons JEAC preprint.

April 7, 1989

Steven Koonin and Michael Nauenberg, at Caltech, submit a manuscript to *Nature*, proposing a theory that attempts to explain "cold fusion."

April 7, 1989

The Utah state Legislature approves a $5 million appropriation for "cold fusion" research.

April 10, 1989

Elsevier's JEAC publishes the Fleischmann-Pons preliminary note.

April 10, 1989

Charles Martin, Kenneth Marsh and Bruce Gammon, at Texas A&M University, announce confirmation of Fleischmann-Pons' excess heat at a press conference. They do not claim that it is fusion. They are the first U.S. researchers to announce a confirmation and the first worldwide to announce a confirmation of excess heat in a Fleischmann-Pons experiment. This group at Texas A&M later questions the validity of its results. (Two other groups at Texas A&M worked on replication attempts.)

April 10, 1989

James Mahaffey's group at Georgia Institute of Technology announces at a press conference the confirmation of neutrons in a Fleischmann-Pons experiment. His group reports an error on April 14 and issues a formal retraction in a press release and press conference on April 25.

April 10, 1989

The University of Utah files its third U.S. patent application on the Fleischmann-Pons claim.

April 10, 1989

Peter Hagelstein, in a journal submission, proposes the model $d + d \longrightarrow$ *Helium-4 + 23.8 MeV* to explain "cold fusion." No measured data exists to support the model.

April 12, 1989

Eugene Mallove, chief science writer for the MIT News Office, issues a press release to announce that Peter Hagelstein has proposed ideas to explain "cold fusion" and that MIT has filed patents based on them. MIT Provost John Deutch, quoted in the press release, is pleased to see Hagelstein's work. (Some news outlets erroneously report this as a press conference.)

April 12, 1989

Fleischmann and Jones speak at the Erice Fusion Forum workshop in Erice, Italy. A press conference takes place afterward.

April 12, 1989

Physicist Runar Kuzmin, at the University of Moscow, reports confirmation of the Fleischmann-Pons experiment. (News reports do not specify which aspect.)

April 12, 1989

The American Chemical Society holds a special session on "cold fusion," followed by a press conference. Pons speaks at both.

April 12, 1989

James Mahaffey, at the Georgia Institute of Technology, retracts his April 10 claim at a press conference.

April 12, 1989

An explosion occurs at the Department of Energy's Lawrence Livermore National Laboratory during a "cold fusion" experiment.

April 13, 1989

Prominent nuclear chemist Glenn Seaborg, of the Department of Energy's Lawrence Berkeley National Laboratory, is asked to come to the White House.

April 13, 1989

Van L. Eden and Wei Liu, graduate students at the University of Washington, report at a press conference the confirmation of tritium in a Fleischmann-Pons experiment. (Retracted on April 25.)

April 14, 1989

The *Journal of Physical Chemistry* receives Cheves Walling and Jack Simons' manuscript with the $d + d \longrightarrow Helium\text{-}4 +23.8 \, MeV$ "cold fusion" idea.

April 14, 1989

Peter Hagelstein gives a lecture at MIT about his $d + d \longrightarrow Helium\text{-}4 +23.8$ *MeV* "cold fusion" explanation. Hagelstein acknowledges that his idea is highly speculative and that it is not supported by measured data. (Mallove, April 21, 1989, Cornell Cold Fusion Archive)

April 14, 1989

Seaborg meets with Admiral James Watkins, secretary of Energy, and Bob Hunter, director of the Office of Energy Research of the Department of Energy. They prepare a briefing on "cold fusion" and discuss it with John H. Sununu, White House chief of staff, and his aides. They agree to form a panel to look into "cold fusion." After the meeting, Seaborg meets socially with President George H.W. Bush.

April 14, 1989

The University of Utah files its fourth U.S. patent application on the Fleischmann-Pons claim.

April 16, 1989

Cherian K. Matthews and two associates at the Indira Gandhi Center for Atomic Research confirm excess heat in a Fleischmann-Pons experiment using differential calorimetry. They also confirm evidence of neutrons. They report their excess-heat confirmation to the news media on April 25.

April 17, 1989

Pons speaks at the first of weekly follow-up news conferences at the University of Utah. Cheves Walling and John Simons report to the *Los Angeles Times* that they have detected helium-4 using one of the Fleischmann-Pons experiments and suggest that helium-4 is the dominant nuclear product. Walling and Simons are the first to claim that the experimentally measured amount of helium-4 corresponds to the amount of heat that should be present if the heat was from deuterium-deuterium fusion.

April 17, 1989

Pavol Povinc and a team of physicists at Comenius University in Bratislava, Czechoslovakia, report confirmation of the Fleischmann-Pons experiment. (News reports do not specify which aspect.)

April 18, 1989

IBM scientists submit a manuscript to *Physical Review Letters* reporting their failure to confirm the Fleischmann-Pons experiment. They also distribute preprints of their preliminary note.

April 18, 1989

Francesco Scaramuzzi, at ENEA-Frascati, announces confirmation of "cold fusion" neutrons from a cell containing titanium shavings loaded with deuterium. ENEA management organizes a press conference and files a patent.

April 18, 1989

The University of Utah files its fifth U.S. patent application on the Fleischmann-Pons claim.

April 18, 1989

Humberto Arriola and Jesus Soberon, of National Autonomous University of Mexico, announce at a press conference their confirmation of the Fleischmann-Pons experiment.

April 18, 1989

Robert Huggins, at Stanford University, reports confirmation of excess heat in a Fleischmann-Pons experiment at a press conference. Huggins shows the difference between D_2O and H_2O experiments using differential rather than absolute calorimetry. He does not claim fusion. Stanford issues a press release two days later.

April 18, 1989
Pons lectures on his and Fleischmann's "cold fusion" work at the Los Alamos National Laboratory.

April 19, 1989
Spero Penha Morato, at the Physics Institute of the University of São Paulo (Brazil), announces at a press conference his confirmation of neutrons in a Fleischmann-Pons experiment.

April 19, 1989
C. V. Sundaram, the director of the Indira Gandhi Center for Atomic Research, tells the news media that the labs' researchers had confirmed neutrons in a variation of the Fleischmann-Pons experiment.

April 19, 1989
Admiral James Watkins, the secretary of the U.S. Department of Energy, meets with directors of 10 DOE national labs to Washington, D.C., for an emergency meeting to discuss "cold fusion" research. (Rosenblatt, 1989)

April 20, 1989
Glen J. Schoessow and John A. Wethington Jr., at the University of Florida, report confirmation of tritium in a Fleischmann-Pons experiment.

April 20, 1989
Dieter Seeliger, at Technical University in Dresden, reports confirmation of neutrons in a Fleischmann-Pons experiment.

April 20, 1989
The editors of *Nature* announce that Fleischmann and Pons have decided to not to publish their manuscript submitted to the journal.

April 20, 1989
Richard Garwin, in a report on the Erice Fusion Forum workshop published in *Nature*, is the first prominent scientist to bet against confirmation of the Fleischmann-Pons claim.

April 21, 1989
Nathan Lewis lectures at Caltech on "cold fusion" and announces that he failed to confirm the Fleischmann-Pons experiment.

April 21, 1989
Researchers in the Neutron Physics Division (under the direction of Assistant Director Mahadeva Srinivasan) at Bhabha Atomic Research Center observe first neutron burst in a palladium-deuterium electrolytic experiment.

April 21, 1989
German scientists Gerhard Kreysa, Günter Marx, and Waldfried Plieth, at the Institute for Organic Chemistry at the University of Berlin, issue a news release and report that they failed to confirm the Fleischmann-Pons experiment. They also say they can explain the Fleischmann-Pons results by ordinary non-nuclear processes.

April 21, 1989
U.S. Secretary of Energy Admiral James Watkins issues a press release announcing three initiatives to pursue and encourage research into the phenomena claimed by Fleischmann and Pons.

April 21, 1989
Robert A. Roe (D-NJ) publishes a press release announcing a congressional

hearing on "Recent Developments in Fusion Energy Research."

April 21, 1989

Gerson Otto Ludwig, at the Institute of Space Research in São Jose dos Campos (São Paulo, Brazil), reports at a press conference his confirmation of helium-3 from a Fleischmann-Pons experiment.

April 22, 1989

Two independent teams led by Yoon Kyong-Sok and Lee Kyu-Ho, at the Korea Advanced Institute of Science and Technology, report confirmation of excess heat, helium and tritium in Fleischmann-Pons experiments. The Korean minister of science and technology announces an initial funding program of $1.5 million.

April 23-24, 1989

Physicists convene at various U.S. Department of Energy laboratories to discuss a five-year plan for nuclear and high-energy physics.

April 24, 1989

John Dash, at Portland State University, observes excess heat in a Fleischmann-Pons experiment.

April 24, 1989

U.S. Secretary of Energy Admiral James Watkins issues a memorandum to 10 DOE lab directors to intensify their research efforts and report their progress to him weekly.

April 24, 1989

U.S. Secretary of Energy Admiral James Watkins directs John H. Schoettler, chairman of the DOE ERAB, to form a panel to investigate "cold fusion."

April 25, 1989

The *Deseret News* reports that Utah Governor Norm Bangerter, in a trip to the Far East, met with officials from companies in Japan, Korea and Taiwan and encouraged them to make investments in Utah.

April 25, 1989

Anthony John Appleby, at Texas A&M University, sends a letter to Congress reporting confirmation of excess heat in a Fleischmann-Pons experiment.

April 25, 1989

John O'Mara Bockris observes tritium in a Fleischmann-Pons experiment.

April 26, 1989

Robert A. Roe (D-NJ), Chairman of the House Committee on Science, Space and Technology, convenes a congressional hearing on "cold fusion."

April 26, 1989

Huggins discusses Stanford experiments at a Materials Research Society meeting in San Diego, California

April 27, 1989

Chemist Uziel Landau, at Case Western Reserve University, reports confirmation of excess heat in a Fleischmann-Pons experiment.

April 27, 1989

K.S.V. Santhanam, at Tata Institute of Fundamental Research in Bombay, reports confirmation of excess heat in a Fleischmann-Pons experiment.

April 28, 1989
Moshe Gai lectures at Yale University. His talk is titled "Does 'Cold Fusion' Exist?"

April 28, 1989
Caltech sends out three press releases saying that, despite extensive efforts, its researchers could not repeat the Fleischmann-Pons claims.

April 28, 1989
MIT professors Ronald R. Parker and Ronald Ballinger are interviewed by Nick Tate, of the *Boston Herald*; they accuse Fleischmann and Pons of fraud. [1].

April 29, 1989
Kelvin Lynn at Brookhaven National Laboratory tells news media that Brookhaven researchers failed to confirm the Fleischmann-Pons claims.

April 29, 1989
Moshe Gai, on behalf of a Yale and Brookhaven team, issues a press release or announces in a press conference (news accounts do not specify which) that the team failed to confirm the Fleischmann-Pons experiment.

April 29, 1989
Gerson Otto Ludwig, at the Institute of Space Research in São Jose dos Campos (São Paulo, Brazil), reports confirmation of neutron bursts in a Fleischmann-Pons experiment.

April 29, 1989
Researchers at Central Electrochemical Research Institute, in India, report to their management the confirmation of excess heat in a Fleischmann-Pons experiment.

April 29, 1989
The Associated Press reports that Princeton University has failed to confirm the Fleischmann-Pons experiment.

May 1, 1989
MIT Professor Ronald Parker issues a press release and holds a press conference. Parker denies that he made accusations of fraud to the *Boston Herald*.

May 1, 1989, 4 p.m.
The American Physical Society conducts the first of three press conferences on "cold fusion." The featured speaker is Robert Perry.

May 1, 1989, 5 p.m.
The American Physical Society conducts the second of three press conferences on "cold fusion." The featured speaker is Nathan Lewis.

May 1, 1989, 7:30 p.m.
The American Physical Society holds the first of two special sessions on "cold fusion."

May 2, 1989, 10 a.m.
The American Physical Society conducts the third press conference on "cold fusion." The featured speakers are Moshe Gai, Johann Rafelski, Steven Jones, Steven Koonin, Richard N. Boyd, Douglas Morrison, and Walter Meyerhof.

May 2, 1989, 7:30 p.m.
The American Physical Society holds the second of two special sessions on "cold fusion."

May 2, 1989
The University of Utah files its sixth U.S. patent application on the Fleischmann-Pons claim.

May 3, 1989
Texas A&M issues a press release announcing Anthony John Appleby's confirmation of excess heat in a Fleischmann-Pons experiment.

May 4, 1989
Fleischmann and Pons were scheduled to meet with John Sununu, White House chief of staff for President George H.W. Bush Sununu cancels an hour before because of a "scheduling conflict."

May 8, 1989
The Princeton Plasma Physics Laboratory issues a statement to the press that it has failed to confirm the Fleischmann-Pons experiment.

May 8, 1989
The Electrochemical Society holds a special session on "cold fusion," followed by a press conference. Several scientists speak at the press conference.

May 16, 1989
The University of Utah files its seventh U.S. patent application on the Fleischmann-Pons claim.

May 23, 1989
Physicists Birger Emmoth, Magnus Jandel, Irena Gudowska and Waclaw Gudowski, at the Manne Siegbahn Institute for Physics in Sweden, issue a press release and report confirmation of neutrons in a Fleischmann-Pons experiment. They said they had submitted a manuscript to the journal *Physica Scripta*.

May 23-25, 1989
The Department of Energy holds a "Workshop on 'Cold Fusion' Phenomena" in Santa Fe, New Mexico, organized by the Los Alamos National Laboratory, held at the Sweeney Convention Center.

May 24, 1989
The first Department of Energy ERAB "cold fusion" panel meeting takes place in Santa Fe.

June 6, 1989
Six members of the ERAB "cold fusion" panel perform a site visit at the University of Utah.

June 9, 1989
The Los Alamos National Laboratory validates the tritium samples sent from John Bockris at Texas A&M. (Burnett-KUER)

June 12, 1989
The Los Alamos National Laboratory terminates negotiations with the University of Utah for collaboration, because of the university's "continued inaction."

June 15, 1989

The Harwell laboratory holds a press conference to announce the termination of its "cold fusion" research.

June 15, 1989

Congress proposes to cut $68 million from the fiscal year 1990 budget of $349 million for magnetic fusion research.

June 19, 1989

Eleven members of the ERAB "cold fusion" panel perform a site visit at Texas A&M University.

June 22, 1989

The second Department of Energy ERAB "cold fusion" panel meeting takes place in Washington, D.C.

June 23, 1989

Edmund Storms tells the *Deseret News* that he and Carol Talcott (representing Los Alamos National Laboratory) confirmed tritium in a Fleischmann-Pons experiment. In the same article, Los Alamos spokesman Jeff Schwartz denies the credibility of the confirmation.

June 29, 1989

General Electric signs a letter of understanding with the University of Utah to develop fusion research.

July 3-6, 1989

Padmanabha Krishnagopala Iyengar presents the first BARC results at the Fifth International Conference on Emerging Nuclear Energy Systems, in Karlsruhe, Germany.

July 11-12, 1989

The third Department of Energy ERAB "cold fusion" panel meeting takes place in Washington, D.C. Panel members prepare a draft of the interim report.

July 13-20, 1989

ERAB panel sends out a survey to laboratories and requests research summaries. Panel includes the draft interim report with the survey request.

July 13, 1989

Bill Broad, of the *New York Times,* interviews panel co-chairman John Huizenga. A news story runs with the headline "Panel Rejects Fusion Claim, Urging No Federal Spending."

July 20, 1989

After giving the full panel a week to review the draft interim report, Huizenga sends the report to John Schoettler, the vice chairman of the ERAB board.

August 7, 1989

General Electric signs a financial agreement with the University of Utah to develop fusion research.

August 7, 1989

The Utah Fusion/Energy Advisory Council approves a motion to give $4.5 million in state funds for the National Cold Fusion Institute. The institute officially opens, and staff begins moving in.

August 16, 1989

The Electric Power Research Institute holds a meeting on "cold fusion" at the University of Utah.

August 17, 1989

Schoettler, after getting full approval by the ERAB board members, sends the four-page interim report to Admiral Watkins, the secretary of Energy.

August 23, 1989

The ERAB panel learns about the first set of BARC experiments.

September 26, 1989

National Cold Fusion Institute Director Hugo Rossi tells the Associated Press that experiments at the NCFI have shown no sign of producing fusion.

October 13, 1989

The fourth Department of Energy ERAB "cold fusion" panel meeting takes place in Dulles Airport, Chicago.

October 15-20, 1989

The Electrochemical Society meets in Hollywood, Florida, and hosts a special session on "cold fusion." Positive results (such as excess heat, neutrons, and tritium) are reported from around the world.

October 16-18, 1989

NSF/EPRI workshop on "cold fusion" takes place in Washington, D.C. Researchers present a variety of positive results (such as excess heat, neutrons, tritium and isotopic shifts).

October 17-19, 1989

Researchers present "cold fusion" papers at the fall meeting of the Atomic Energy Society of Japan.

October 30-31, 1989

The fifth Department of Energy ERAB "cold fusion" panel meeting takes place in Washington, D.C.

November 3, 1989

National Cold Fusion Institute Director Hugo Rossi issues a press release saying the institute has found no evidence of "cold fusion."

November 4, 1989

The *New York Times* runs Bill Broad's story "U.S. Panel Finds No Evidence of Cold Fusion."

November 8, 1989

Huizenga delivers the final ERAB report to John W. Landis, chairman of the ERAB board, for final approval.

November 8, 1989

The Associated Press runs a story that announces the conclusion of the ERAB "cold fusion" panel.

November 20, 1989

Rossi resigns from the National Cold Fusion Institute.

November 26, 1989

John W. Landis, chairman of the ERAB board, sends the final ERAB report to Admiral Watkins.

December 1989

The Bhabha Atomic Research Center publishes its BARC-1500 report.

December 9, 1989

The *Deseret News* reports that scientists at Oak Ridge National Laboratory are scheduled to present confirmatory results at next week's American Society of Mechanical Engineers meeting, in contradiction to statements from lab officials that none of its researchers has confirmed the results of Fleishmann and Pons.

December 12, 1989

Oak Ridge and other researchers present "cold fusion" papers at the American Society of Mechanical Engineers meeting, in San Francisco.

December 19, 1989

Fleischmann and Pons submit their first complete (55-page) manuscript to the *Journal of Electroanalytical Chemistry*. (Publishes July, 1990)

December 21, 1989

Fritz G. Will, a scientist at General Electric Corp.'s New York development center, is appointed the new director of the "National Cold Fusion Institute" by the University of Utah.

March 1990

In early March, journalist Gary Taubes interviews Nigel Packham at Texas A&M University and accuses him of fraudulently spiking his "cold fusion" cells with tritium.

March 29-31, 1990

The First Annual Conference on "cold fusion" (ICCF-1) takes place in Salt Lake City, Utah. (It was later named the International Conference on Cold Fusion.)

March 29, 1990

Fleischmann and Pons speak at an invitation-only press briefing at ICCF-1.

March 29, 1990

Nature publishes a paper by University of Utah physicist Michael H. Salamon that states he found no neutrons, using his detector, on Fleischmann and Pons' experiments. Salamon had performed the measurements in May-June 1989.

May 1990

Kevin Wolf at Texas A&M tells journalist Robert Pool, at *Science*, that he retracts his tritium claim. He tells Pool that the tritium was in the palladium cathode before the start of his experiments. Wolf's retraction publishes June 15.

May 1990

The status of "cold fusion" research in China is reviewed at a national-level meeting held at Beijing. (Source: Srinivasan, 1991)

June 1, 1990

Tim Fitzpatrick, of the *Salt Lake Tribune,* reports that an anonymous donation of $500,000 to the National Cold Fusion Institute actually came from the University of Utah.

June 6, 1990

Wolf tells journalist Jerry Bishop at the *Wall Street Journal* that he retracts his tritium claim. He tells the same thing to Bill Broad at the *New York Times* the next day.

June 7, 1990

The Utah state Fusion/Energy Advisory Council initiates financial and scientific reviews of the National Cold Fusion Institute. The council determines that Chase Peterson hid the transfer of $500,000 to the National Cold Fusion Institute. The university Academic Senate questions Peterson's ability to lead the university and passes a resolution asking the state Board of Regents and the University Institutional Council to consider firing Peterson.

June 11, 1990

Peterson announces he will retire at the end of the 1990-91 academic year.

June 12, 1990

Science magazine issues a press release about a forthcoming news story by Gary Taubes suggesting that researchers at Texas A&M committed science fraud by spiking their cells with tritium.

June 15, 1990

Gary Taubes publishes an article in *Science* suggesting that John Bockris and Nigel Packham at Texas A&M committed science fraud by spiking their cells with tritium.

July 1990

Fleischmann and Pons' first complete paper publishes in the *Journal of Electroanalytical Chemistry*.

July 1990

Texas A&M administrators, responding to an article written by Gary Taubes in *Science*, appoint a faculty panel to review the university's "cold fusion" research, specifically the work of John Bockris and Ph.D. candidate Nigel Packham.

August 16, 1990

Packham defends his dissertation in front of his graduate committee, the news media and 50 onlookers from the university.

October 16, 1990

The *Deseret News* reports that Pons is somewhere overseas and nobody knows when he is expected to return.

October 22-24, 1990

The Anomalous Nuclear Effects in Deuterium/Solid Systems Conference takes place at Brigham Young University.

October 24, 1990

Gary Triggs, the personal attorney for Pons, faxes Pons' request to the university for a sabbatical.

October 25, 1990

The *Deseret News* reports that Pons is missing, his telephone is disconnected, and his home is up for sale. Neighbors say he has moved to southern France.

October 26, 1990

The state of Utah prepares to search for Pons and hold him accountable for his research and for the money spent by the state on "cold fusion."

November 8, 1990

Pons meets with the Utah fusion council.

November 1990

The three-member Texas A&M committee finds no evidence of fraud in its inquiry of "cold fusion" research at the university.

November 19, 1990

Texas A&M University releases the conclusions of its investigation, completed on Oct. 15, into allegations that researchers in John Bockris' lab spiked their experiment with tritium to fake "cold fusion." The university concludes that the allegations are unfounded.

January 8, 1991

The University of Utah announces the resignation of Pons from his tenured faculty position, effective January 1. The university appoints him a research professor.

January 19, 1991

Eugene Mallove discovers the July 1989 baseline shift of MIT excess-heat data.

June 30, 1991

The National Cold Fusion Institute, located in the University of Utah Research Park, closes.

1991

Several "cold fusion" researchers form Future Energy Applied Technology, a private company, located in the University of Utah Research Park. The company is later renamed ENECO.

December 1, 1993

ENECO acquires exclusive world-wide license rights to the University of Utah "cold fusion" patent applications. Federal DOD scientist Michael Melich is a shareholder. Federal DOD scientist Yan Kucherov is director of research. ENECO returned the rights to the university in 1997, at which time the university abandoned the patent applications. ENECO filed for bankruptcy in 2008.

Notes

1. See also Eugene Mallove's "Partial Chronology of Events Relating to MIT's Handling of Cold Fusion," in "MIT Special Report," Infinite Energy, 24, 1999.
2. See also Bruce Lewenstein's chronology in the Cornell Cold Fusion Archive Finding Aid for additional events between 1991 and 1994.
3. Primary sources used for this chronology include original press releases, manuscripts, published journal articles, news reports from the Associated Press, Los Angeles Times, Nature, New York Times, Wall Street Journal, and the Deseret News, original documents in the Cornell Cold Fusion Archive, and documents from the New Energy Times Garwin Archive.
5. This chronology does not reflect extensive international research efforts and news events. This chronology is limited by the author's access to resources in the United States and English-language news sources.
6. Future historians who want to follow or expand on this chronology likely will find additional events in the archives of the Salt Lake Tribune, Boston Herald, and Boston Globe.

Appendix D — Basic Types of Radioactive Emissions

Type	Nature of Radiation	Penetrating Power[1]	Ionizing Power [2]
Alpha α $_2^4He$	A helium nucleus of 2 protons and 2 neutrons. Mass = 4 Charge = +2	**Low** Particles are stopped by a few cm of air or a thin sheet of paper.	**Very High** The biggest mass and charge of the three. Packs the biggest punch.
Beta β $_{-1}^0e$	High kinetic energy electrons. Mass = 1/1850 Charge = -1	**Moderate** Most particles are stopped by a few mm of metals like aluminum.	**Moderate** Less than the alpha particle.
Gamma γ $_0^0\gamma$	Very high frequency electromagnetic radiation. Mass = 0 Charge = 0	**Very High** Most, but not all, gamma rays are stopped by a thick layer of steel or concrete, or a few cm of dense lead.	**Lowest** Carries no electric charge and has no mass, so it has very little punch when it collides with an atom.

1. When penetrating denser material, more radiation is absorbed and stopped than when penetrating less-dense material. However, as mass or charge decreases, the penetrating power increases.
2. Ionizing power is the ability to remove electrons from atoms and form positive ions. Ionizing radiation is harmful to living cells. Courtesy Georgia State University, adapted by *New Energy Times*.

Appendix E — Helium Permeation in Metals Analysis

Gas Behavior in Hydride-Forming Metals at or Near Standard Temperature and Pressure	Hydrogen	Helium
Readily permeates hydride-forming metals	Yes	No
Diffuses through defects, cracks or grain boundaries in metals	Yes	Yes
Soluble (dissolves) in hydride-forming metals	Yes	No

Bowman Jr., Robert C. (Feb. 7, 2007) "NMR Studies of 3He Retention and Release in Metal Tritides — A Review," Hydrogen & Helium Isotopes in Materials Conference, Albuquerque, N.M. [Helium does not outgas from metals easily or quickly.]

Chien, Chun-Ching, Hodko, Dalibor, Minevski, Zoran and Bockris, John O'M. (April 1992) "On an Electrode Producing Massive Quantities of Tritium and Helium," Journal of Electroanalytical Chemistry, 338, 189-212 [Helium on near-surface areas on cathode can be retained if quickly immersed in liquid nitrogen.]

Gozzi, D., Cellucci, F., Cignini, P.L., Gigli, G., Tomellini, M., Cisbani, E., Frullani, S., Urciuoli, G.M. (1998) "X-Ray, Heat Excess and 4He in the D:Pd System," Journal of Electroanalytical Chemistry, 452, 253, and Erratum, 452, 251-71 [Helium does not show up in the bulk if the cathode is vaporized.]

McKubre, Michael, et al., (June 1998) "Development of Energy Production Systems from Heat Produced in Deuterated Metals, Volume 1," Electric Power Research Institute, TR-107843 [Researchers hypothesized, but did not test, that helium was retained (occluded) in metal during experiment.]

Ramsay, W., and Travers, M.W. (January 1897) "An Attempt to Cause Helium or Argon to Pass Through Red-Hot Palladium, Platinum, or Iron." Proceedings of the Royal Society of London (1854-1905), 61(-1), 266-7 [Helium won't dissolve in metal even at high temperature.]

Schultheis, D. (2007) "Permeation Barrier for Lightweight Liquid Hydrogen Tanks," Ph.D. dissertation, University of Augsburg [Defect-free metal will not allow helium to pass through.]

Xia, Ji-xing, Hu, Wang-yu, Yang, Jian-yu, and Ao, Bing-yun (2006) "Diffusion Behaviors of Helium Atoms at Two Pd Grain Boundaries," Transactions of Nonferrous Metals Society of China, 16, S804-7 [Helium has low solubility in metals, grain boundaries support permeation.]

1989 cartoon by Gary Brookins depicting a "lemon" (failed reaction)

Appendix F — ERAB Panel Members and Meetings

Panel Members

Allen J. Bard, Professor of Chemistry, University of Texas
Jacob Bigeleisen, Distinguished Professor of Chemistry, SUNY, Stony Brook
Howard K. Birnbaum, Professor of Materials Science, University of Illinois
Michel Boudart, Professor of Chemistry, Stanford University
Clayton F. Callis, President, American Chemical Society
Mildred Dresselhaus, Institute Professor, MIT*
Larry R. Faulkner, Head, Chemistry Department, University of Illinois
T. Kenneth Fowler, Professor of Nuclear Engineering, University of California
Richard L. Garwin, Science Advisor to the Director of Research, IBM Corporation
Joseph Gavin, Jr., Senior Management Consultant, Grumman Corporation*
David Goodwin, Department of Energy, Panel Technical Advisor
William Happer Jr., Professor of Physics, Princeton University
Darleane C. Hoffman, Professor of Chemistry, Lawrence Berkeley Laboratory
John Huizenga, Co-Chairman, Professor of Chemistry and Physics, University of
    Rochester*
Steven E. Koonin, Professor of Theoretical Physics, Caltech
Peter Lipman, U.S. Geological Survey
Barry Miller, Supervisor, Analytical Chemical Research Department, AT&T Bell
    Laboratories
David Nelson, Professor of Physics, Harvard University
Norman Ramsey, Co-Chairman, Professor of Physics, Harvard University*
John P. Schiffer, Associate Director, Physics Division, Argonne National
    Laboratory
John Schoettler, Independent Petroleum Geologist* (Also vice chairman of ERAB)
Dale Stein, President, Michigan Technology University*
William Woodard, Department of Energy, Secretary, Cold Fusion Panel
Mark Wrighton, Head, Department of Chemistry, MIT
* = ERAB Members

Meetings and Site Visits

(Sometimes attended by ERAB Executive Director Thomas G. Finn, as well)

May 24: First Panel Meeting, Santa Fe (2 hours)
    Bard, Bigeleisen, Birnbaum, Finn, Fowler, Garwin, Goodwin, Happer,
    Hoffman, Huizenga, Koonin, Lipman, Ramsey, Schiffer, Schoettler, Stein,
    Wrighton
June 2: University of Utah Laboratory Visit (3 hours)
    Bard, Faulkner, Finn, Goodwin, Happer, Miller, Ramsey

June 9: ENEA Frascati Laboratory Visit
Garwin
June 13: Brigham Young University Laboratory Visit
Schiffer
June 16: ORNL (Scott) Laboratory Visit
Bigeleisen
June 19: Texas A&M University Laboratory Visit
Bard, Bigeleisen, Callis, Finn, Goodwin, Happer, Huizenga, Miller, Ramsey, Schiffer, Wrighton
June 20: Caltech Laboratory Visit
No information on attendees or reason for visit at a lab that had no results
June 22: Second Panel Meeting, Washington, D.C.
Most of the panel members
July 6: SRI International Laboratory Visit
No information on attendees
July 6: Stanford University Laboratory Visit (4 hours)
Faulkner, Happer
July 7: LANL Laboratory Visit
Fowler
July 11-12: Third Panel Meeting, Washington, D.C.
Most of the panel members
Oct. 13: Fourth Panel Meeting, Chicago
Nuclear products sub-group, possibly others
Oct. 30-31: Fifth Panel Meeting, Washington, D.C.
Most of the panel members

Subgroups for the Final Report

Preamble
Norman Ramsey
Introduction
T. Kenneth Fowler (Coordinator), Joseph Gavin Jr., John Huizenga and Steven E. Koonin
Calorimetry
Allen Bard (Coordinator), Larry R. Faulkner, William Happer Jr., Barry Miller and Mark Wrighton
Materials
Howard K. Birnbaum (Coordinator), Michel Boudart, Mildred Dresselhaus, David Nelson and Dale Stein
Fusion Products
John P. Schiffer (Coordinator), Jacob Bigeleisen, Richard L. Garwin, Darleane C. Hoffman, Steven E. Koonin and Peter Lipman
Conclusions and Recommendations
John Huizenga (Coordinator,) Allen J. Bard, Howard K. Birnbaum, Clayton F. Callis, Richard L. Garwin and John P. Schiffer.

Bibliography

Albagli, David, Ballinger, Ron, Cammarata, Vince, Chen, X., Crooks, Richard M., Fiore, Catherine, Gaudreau, Marcel P. J., Hwang, I., Li, C. K., Linsay, Paul, Luckhardt, Stanley C., Parker, Ronald R., Petrasso, Richard, D., Schloh, Martin O., Wenzel, Kevin W., and Wrighton, Mark S. (June 1990, July 1989 pre-print) "Measurement and Analysis of Neutron and Gamma-Ray Emission Rates, Other Fusion Products, and Power in Electrochemical Cells Having Pd Cathodes," *Journal of Fusion Energy*, **9**(2), p. 133-48

Atkinson, Robert, and Houtermans, Fritz (1929) "Aufbaumöglichkeit in Sternen," *Z. für Physik*, **54**, p. 656-65

Bailey, Ronald (December 1, 1993) "Political Science," *Reason*, **25**(7), p. 61-3

Beaudette, Charles G. (2002) *Excess Heat & Why Cold Fusion Research Prevailed*, 2nd ed., p. 188, 357-60, (South Bristol, Maine: Oak Grove Press)

Bethe, Hans (March 1, 1939) "Energy Production in Stars," *Physical Review*, **55**, p. 434

Biberian, Jean-Paul (Feb. 2, 2015) "Biological Transmutations," *Current Science*, **108**(4), p. 633-5

Bowman Jr., Robert C. (Feb. 7, 2007) "NMR Studies of ^{3}He Retention and Release in Metal Tritides — A Review," Hydrogen & Helium Isotopes in Materials Conference, Albuquerque, N.M.

Bridgman, Percy Williams (1947) "The Physics of High Pressure," International Textbooks of Exact Science, London

Burbidge, E. Margaret, Burbidge, Geoffrey R., Fowler, William A., and Hoyle, Fred (1957) "Synthesis of the Elements in the Stars," *Reviews of Modern Physics*, **29**(4), 547-650

Bush, Ben F., Lagowski, Joseph J., Miles, Melvin M., and Ostrom, Greg S. (1991) "Helium Production During the Electrolysis of D_2O in Cold Fusion Experiments," *Journal of Electroanalytical Chemistry*, **304**, p. 271-8

Bush, Robert T. (1990) "Isotopic Mass Shifts in Cathodically Driven Palladium Via Neutron Transfer Suggested by a Transmission Resonance Model to Explicate Enhanced Fusion Phenomena (Hot and Cold) within a Deuterated Matrix," *Proceedings of the First Annual Conference on Cold Fusion*, March 29-31, 1990, Salt Lake City, Utah, National Cold Fusion Institute: Salt Lake City

Caldwell, Roy, and Lindberg, David, (pub. online 2010) "Cold Fusion: A Case Study for Scientific Behavior," http://undsci.berkeley.edu/article/cold_fusion_01

Cedzynska, Krystyna, Barrowes, S.C., Bergeson, H.E., Knight, L.C., and Will, Fritz G. (1991) "Tritium Analysis in Palladium With an Open-System Analytical Procedure," *Proceedings of Anomalous Nuclear Effects in Deuterium/Solid Systems, Provo, Utah,* October 22-24, 1990, p. 463-6

CERN (retrieved Dec. 1, 2014)
http://home.web.cern.ch/about/experiments/gargamelle

Chadwick, James (1932) "The Existence of a Neutron," *Proceedings of the Royal Society of London*, **A136**, p. 692
Chadwick, James (1932) "Possible Existence of a Neutron," *Nature*, **129**, p. 312
Chien, Chun-Ching, Hodko, Dalibor, Mineuski, Zoran, and Bockris, John O'M. (April 1992) "On an Electrode Producing Massive Quantities of Tritium and Helium," *Journal of Electroanalytical Chemistry*, **338**, p. 189-212
Cockcroft, John, and Walton, Ernest (1932) "The Disintegration of Elements by High-Velocity Protons," *Proceedings of the Royal Society of London*, **137**, p. 229-42; Cockcroft, John, and Walton, Ernest (1932) "Possible Existence of a Neutron," *Nature*, **129**, p. 312
Coehn, Alfred (1929) "Nachweis Von Protonen In Metallen," ["Proof of Protons in Metals,"] Zeitschrift für Elektrochemie, 35, **676**

Darwin, Charles Galton (1920) "Motion of Charged Particles," *Philosophical Magazine*, Series 6, **39**, p. 537-51
De Ninno, Antonella, Frattolillo, Antonio, Lollobattista, Giuseppe, Martinis, Lorenzo, Martone, Marcello, Mori, Luciano, Podda, Salvatore, and Scaramuzzi, Francesco (1989) "Emissione Di Neutroni da un Sistema Deuterio-Titanio," *Energia Nucleare*, **6**(1), p. 9-11
De Ninno, Antonella, Frattolillo, Antonio, Lollobattista, Giuseppe, Martinis, Lorenzo, Martone, Marcello, Mori, Luciano, Podda, Salvatore, and Scaramuzzi, Francesco (May 1989) "Emission of Neutrons as a Consequence of Titanium Deuterium Interaction," *Il Nuovo Cimento*, **101a**(5), p. 841-4
De Ninno, Antonella, Frattolillo, Antonio, Lollobattista, Giuseppe, Martinis, Lorenzo, Martone, Marcello, Mori, Luciano, Podda, Salvatore, and Scaramuzzi, Francesco (June 1, 1989) "Evidence of Emission of Neutrons From a Titanium Deuterium System," *Europhysics Letters*, **9**(3), p. 221-4
Dickman, Steven (April 27, 1989) "1920s Discovery, Retraction," *Nature*, **338**, p. 692

Fleischmann, Martin (1991) "The Present Status of Research in Cold Fusion," *The Science of Cold Fusion: Proceedings of the Second Annual Conference on Cold Fusion*, June 29 - July 4, 1991, "Volta" Centre for Scientific Culture, Villa Olmo, Como, Italy. Eds. Tullio Bressani, Emilio Del Giudice, Giuliano Preparata, (Bologna, Italy: Societa Italiana di Fisica)
Fleischmann, Martin (2000) "Reflections on the Sociology of Science and Social Responsibility in Science, in Relationship to Cold Fusion," *Accountability in Research*, **8**, p. 19
Fleischmann, Martin, and Pons, Stanley (1993) "Calorimetry of the Pd-D$_2$O System; From Simplicity via Complications to Simplicity," *Physics Letters A*, **176**, p. 118
Fleischmann, Martin, and Pons, Stanley (pub. August 1993) "Calorimetry of the Palladium-D-D$_2$O System," *Proceedings of NSF/EPRI Workshop on Anomalous*

Effects in Deuterided Metals, October 16-18, 1989, Washington, D.C.

Fleischmann, Martin, Pons, Stanley, Anderson, Mark W., Li, Lian Jun, and Hawkins, Marvin (July 1990) "Calorimetry of the Palladium-Deuterium-Heavy Water System," *Journal of Electroanalytical Chemistry*, **287**, p. 293-348

Garwin, Richard L., Lederman, Leon M., and Weinrich, M. (Feb. 15, 1957) "Observations of the Failure of Conservation of Parity and Charge Conjugation in Meson Decays: The Magnetic Moment of the Free Muon," *Physical Review*, **105**(4), p. 1415-17

George, Russ (June 1994) "An Interview With Mahadeva Srinivasan," *Cold Fusion Magazine,* **1**(2), p. 18-25

Gozzi, D., Cellucci, F., Cignini, P.L., Gigli, G., Tomellini, M., Cisbani, E., Frullani, S., Urciuoli, G.M. (1998) "X-Ray, Heat Excess and 4He in the D:Pd System," *Journal of Electroanalytical Chemistry*, **452**, p. 253, and Erratum, **452**, 251-71

Hahn, Otto, and Strassmann, Fritz (Nov. 18, 1938) "Über die Entstehung von Radiumisotopen aus Uran durch Bestrahlen mit Schnellen und Verlangsamten Neutronen," *Naturwissenschaften*, **26**(46), p. 755-6

Happer, William, as quoted by Taubes, Gary (June 15, 1993) *Bad Science: The Short Life and Weird Times of Cold Fusion*, p. 305, (New York: Random House)

Hoyle, Fred (1946) "The Synthesis of the Elements from Hydrogen," The Monthly Notices of the Royal Astronomical Society, 106(5), p. 343-383

Iyengar, Padmanabha Krishnagopala (1989) "Cold Fusion Results in BARC Experiments," *Proceedings of the Fifth International Conference on Emerging Nuclear Energy Systems*, Karlsruhe, Germany, July 3-6, 1989, Ulrich von Mollendorff, Balbir Goel, eds., p. 291-5, (World Scientific)

Iyengar, Padmanabha Krishnagopala, and Srinivasan, Mahadeva, (1990) "Overview of BARC Studies in Cold Fusion," *Proceedings of the First Annual Conference on Cold Fusion*, Salt Lake City, Utah, March 29-31, 1990, p. 62-81, (National Cold Fusion Institute: Salt Lake City)

Joliot, Frédéric, and Joliot-Curie, Irène (1934) "Artificial Production of a New Kind of Radioelement," *Nature*, **133**, p. 201

Jones, Steven Earl, Palmer, Paul, Czirr, J. Bart, Decker, Daniel L., Jensen, G.L., Thorne, J.M., Taylor, S.F., and Rafelski, Johann (1989) "Observation of Cold Nuclear Fusion in Condensed Matter," *Nature*, **338**, p. 737

Kervran, Corentin L. (1971) *Biological Transmutations*, (Swan House Pub.)

Kluev, V.A., Lipson, A.G., Topornov, Y.P., Derjaguin, B., Lushikov, V.I., Strelkov, A.V., and Shabalin, E.P. (1986) "High-Energy Processes Accompanying the Fracture of Solids," *Soviet Technical Physics Letters,* **12**, p. 1333

Langmuir, Irving (October 1989) "Pathological Science," 1953 lecture transcr. and ed. by Robert N. Hall, in *Physics Today*, **42**, p. 36-48

Lewis, N.S., Barnes, C.A., Heben, M.J., Kumar, A., Lunt, S.R., McManis, G.E., Miskelly, G.M., Penner, R.M., Sailor, M.J., Santangelo, P.G., Shreve, G.A., Tufts, B.J., Youngquist, M.G., Kavanagh, R.W., Kellogg, S.E., Vogelaar, R.B., Wang, T.R., Kondrat, R., and New, R. (Aug. 17, 1989) "Searches for Low-Temperature Nuclear Fusion of Deuterium in Palladium," *Nature,* **340**, p. 525-30

McKubre, Michael, Rocha-Filho, Romeu C., Smedley, Stuart, Tanzella, Francis, Chao, Jason, Chexal, Bindi, Passell, Tom, and Santucci, Joseph (1990) "Calorimetry and Electrochemistry in the D/Pd System," *Proceedings of the First Annual Conference on Cold Fusion*, Salt Lake City, UT, March 29-31, 1990, (Salt Lake City: National Cold Fusion Institute)
Meitner, Lise, and Frisch, Otto (Feb. 11, 1939) "Disintegration of Uranium by Neutrons: A New Type of Nuclear Reaction," *Nature,* **143**, p. 239-40
Melich, Michael E., and Hansen, Wilford N. (1993) "Some Lessons From Three Years of Electrochemical Calorimetry," *Frontiers of Cold Fusion: Proceedings of the Third International Conference on Cold Fusion*, Nagoya, Japan, October 21-25, 1992; Ikegami, H., Ed., (Tokyo: Universal Academy Press)
Mellor, J.W., Ed. (1922) "*Comprehensive Treatise on Inorganic and Theoretical Chemistry, Vol. 1: H, O,*" Longmans, Green and Co.
Menlove, Harold O., Paciotti, M.A., Claytor, Thomas N., and Tuggle, Dale G. (1991) "Low Background Measurements of Neutron Emission From Ti Metal in Pressurized Deuterium Gas," *The Science of Cold Fusion: Proceedings of the Second Annual Conference on Cold Fusion*, June 29 - July 4, 1991, "Volta" Centre for Scientific Culture, Villa Olmo, Como, Italy. Eds. Tullio Bressani, Emilio Del Giudice, Giuliano Preparata, (Bologna, Italy: Societa Italiana di Fisica)
Miles, Melvin (2015) "Excerpts From Martin Fleischmann Letters," International Conference on Condensed Matter Nuclear Science, Padua, Italy
Millikan, Robert A. (Nov. 1923) "Gulliver's Travels in Science," *Scribner's Magazine,* **74**(5)

NNSA (National Nuclear Security Administration) (May 2016) "2015 Review of the Inertial Confinement Fusion and High Energy Density Science Portfolio: Volume I," DOE/NA-0040
Nobelprize.org (Retrieved Dec. 1, 2014) http://www.nobelprize.org/nobel_prizes/physics/laureates/1979

Oliphant, Mark, Harteck, P., and Rutherford, Ernest (1934) "Transmutation Effects Observed With Heavy Hydrogen," *Proceedings of the Royal Society of London A,* **144**, p. 692-703
Oliphant, Mark, and Rutherford, Ernest (1933) "Transmutation of Elements by Protons," *Proceedings of the Royal Society of London A,* **141**, p. 259

Packham, Nigel J.C., Wolf, Kevin L., Wass, Jeff C., Kainthla, Ramesh C., and Bockris, John O'Mara (Oct. 10, 1989) "Production of Tritium From D_2O Electrolysis at a Palladium Cathode," *Journal of Electroanalytical Chemistry,*

270, p. 451-8

Paneth, Fritz (May 14, 1927) "The Transmutation of Hydrogen Into Helium," *Nature*, **119**(3002), p. 706-7

Passell, Thomas O. (June 1998) "Development of Energy Production Systems From Heat Produced in Deuterated Metals, Volume 1," TR-107843-V1, Electric Power Research Institute, TR-107843-V1, p. 357

Pinch, Trevor J. (1995) "Rhetoric and the Cold Fusion Controversy," in *Science, Reason and Rhetoric*, (University of Pittsburgh Press)

Petrasso, R.D., Chen, X., Wenzel, K.W., Parker, R.R., Li, C.K., and Flore, C. (May 18, 1989) "Problems With the Gamma-Ray Spectrum in the Fleischmann et al. Experiments," *Nature*, **339**(6221), p. 183-5

Ramsay, W., and Travers, M.W. (January 1897) "An Attempt to Cause Helium or Argon to Pass Through Red-Hot Palladium, Platinum, or Iron." *Proceedings of the Royal Society of London* (1854-1905), **61**(-1), p. 266-7

Rhodes, Richard (1986) *The Making of the Atomic Bomb*, (New York: Simon and Schuster)

Rolison, Debra R., and O'Grady, William E. (pub. August 1993) "Mass/Charge Anomalies in Pd After Electrochemical Loading With Deuterium," *Proceedings of NSF/EPRI Workshop on Anomalous Effects in Deuterided Metals*, October 16-18, 1989, Washington, D.C.

Rosenblatt, Gerd M. (April 20, 1989) letter to division directors "Cold Fusion Research," Library of Congress, Seaborg Papers, Box 838, Cold Fusion Folders

Schmid, Peter, and Kellner, Gottfried (retrieved Nov. 2, 2014) Morrison obituary, CERN Web site

Schultheis, D. (2007) "Permeation Barrier for Lightweight Liquid Hydrogen Tanks," Ph.D. dissertation, University of Augsburg

Soddy, Frederick (1909) *The Interpretation of Radium*, (London: Putman and Sons)

Srinivasan, Mahadeva (1991) "Nuclear Fusion in an Atomic Lattice: An Update on the International Status of Cold Fusion Research," *Current Science*, **60**, p. 417-39

Sternglass, Ernest (1997) *Before the Big Bang — The Origin of the Universe*, (New York: Four Walls Eight Windows)

Sublette, Carey (retrieved Jan. 20, 2008) "India's First Bomb," http://nuclearweaponarchive.org/India/IndiaFirstBomb.html

Taubes, Gary (June 1993) *Bad Science: The Short Life and Weird Times of Cold Fusion*, Random House

Taylor, Harold John (1935) "The Disintegration of Boron by Neutrons," *Proceedings of the Physical Society*, 47(5), p. 873-76

Teller, Edward (pub. August 1993) "Anomalous Effects on Deuterided Metal," *Proceedings of NSF/EPRI Workshop on Anomalous Effects in Deuterided Metals*, October 16-18, 1989, Washington, D.C.

Teller, Edward, as quoted by Garwin (October. 23, 1986) "Interview With Richard

Garwin," by Finn Aaserud

Trost, Hans Jochen (Nov. 28, 2013), translation of Aug. 3, 1951 letter in "Einstein's Lost Hypothesis," *Nautilus* magazine, Winter 2014, 21-29

Tsarev, Vladimir A., and Worledge, David (December 1991) "New Results on Cold Nuclear Fusion: A Review of the Conference on Anomalous Nuclear Effects in Deuterium/Solid Systems, Provo, Utah, October 22-24, 1990," *Fusion Technology*, **20**, p. 484-508

Urutskoev, Leonid, Liksonov, V.I., Tsinoev, V.G. (2002) "Observation of Transformation of Chemical Elements," *Annales Fondation Louis de Broglie*, **27**(4), p. 701-726

Walling, Cheves, and Simons, John (June 15, 1989) "Two Innocent Chemists Look at Cold Fusion," *Journal of Physical Chemistry*, **93**(12), p. 4693-6

Watkins, James. (April 24, 1989) letter to DOE lab directors "Research on 'Cold' Fusion," Library of Congress, Seaborg Papers. Box 838, Cold Fusion Folders

White, Carol (Dec. 11, 1992) "Japan Cold Fusion Conference Sets New Direction for Science," *Executive Intelligence Review*, **19**(49), p. 20-24

Wick, Gian-Carlo (March 3, 1934) "Sugli Elementi Radioattivi di F. Joliot e I. Curie," Rendiconti Accademia, Lincei, Italy, 19, p. 319-24

Williams, David, E. (2013) "Obituary of Professor Martin Fleischmann," *Electrochemistry Newsletter*, Royal Society of Chemistry, **1**, p. 4

Williams, David E., Findlay, D.J.S., Craston, D.H., Sené, M.R., Bailey, M., Croft, S., Hooton, B.W., Jones, C.P., Kucernak, A.R.J., Mason, J.A., and Taylor, R.I. (Nov. 23, 1989) "Upper Bounds on 'Cold Fusion' in Electrolytic Cells," *Nature* **342**, p. 375-84

Wolf, Kevin L., Packham, Nigel J.C., Lawson, Del, Shoemaker, John, Cheng, Frank, and Wass, Jeff C. (presented May 1989) "Neutron Emission and the Tritium Content Associated With Deuterium-Loaded Palladium and Titanium Metals," *Journal of Fusion Energy*, **9**(2), p. 105-13, (pub. June, 1990)

Wolf, Kevin L., Lawson, D.R., Packham, Nigel J.C., and Wass, J.C. (presented October 1989) "A Search for Neutrons and Gamma Rays Associated With Tritium Production in Deuterided Metals," *Proceedings of NSF/EPRI Workshop on Anomalous Effects in Deuterided Metals,* October 16-18, 1989, Washington, D.C., p. 8-1-8-20, (pub. August, 1993)

Xia, Ji-xing, Hu, Wang-yu, Yang, Jian-yu, and Ao, Bing-yun (2006) "Diffusion Behaviors of Helium Atoms at Two Pd Grain Boundaries," *Transactions of Nonferrous Metals Society of China*, **16**, p. S804-7

Index

About the Author

Steven B. Krivit lives in San Rafael, California, and is an investigative science journalist and international speaker. He studied industrial design at the University of Bridgeport (Connecticut) and completed his bachelor's degree in business administration and information technology at National University (Los Angeles). He was a computer network systems engineer until 2000, when he became curious about low-energy nuclear reaction (LENR) research. He founded the *New Energy Times* Web site and online news service to share what he learned. By 2016, he had spoken with nearly all the scientists who were involved in the field. He has lectured nationally and international to scientific as well as lay audiences. He has advised the U.S. intelligence community, the U.S. Library of Congress, members of the Indian Atomic Energy Commission and the interim executive director of the American Nuclear Society. He is the leading author of review articles and chapters about LENRs, including invited papers for the Royal Society of Chemistry (2009), Elsevier (2009 and 2013) and John Wiley & Sons (2011). He was an editor for the American Chemical Society 2008 and 2009 technical reference books on LENRs and editor-in-chief for the 2011 Wiley *Nuclear Energy Encyclopedia.*

Krivit was the first science journalist to publicly identify and teach the distinctions between the unproven theory of "cold fusion" and the experimentally confirmed neutron-catalyzed LENRs. He did so in 2008 at the 236th national meeting of the American Chemical Society. His chapters in the *Elsevier Encyclopedia of Electrochemical Power Sources* were the first chapters on LENRs in a print encyclopedia.

Other Volumes in This Series

Hacking the Atom: Explorations in Nuclear Research, Vol. 1

This book shows, for the first time, why low-energy nuclear reaction phenomena are not the result of fusion, why they are the result of nuclear processes, and why they can now be explained by a feasible theory. The theory does not conflict with existing physics but expands scientific knowledge and reveals a new field of nuclear science.

Lost History: Explorations in Nuclear Research, Vol. 3

This book explores the story of forgotten chemical transmutation research during the 1910s and 1920s, a precursor to modern low-energy nuclear reactions research. This work has been obscured and absent in the dialogue of the scientific community for a century.

For More Information
www.stevenbkrivit.com

www.ingramcontent.com/pod-product-compliance
Lightning Source LLC
Chambersburg PA
CBHW020855210326

41598CB00018B/1668